B. Sahlins

11/89

-7. frontiers + boundaries

37. mountains on maps

42. mountains as passages, not barriers

44ff- claims of Fr. vs Sp.

64ff- maps.

85. territory, jurisdiction.

106. gauchos.

113. state structure.

127- resistance to state-building in the borderland.

137ff smuggling.

155ff. village identity

158. state of affair - rocks with holes.

160. "neutral road" from Spain to Llivia 177,203. 210,243

165. consequences of peasant use of national identities

147 though anti-new, French Cerdana émigrés seized v. French

229. Identification of self as well as other in national terms -
       border incidents. 243 (Chundo incidents became
                              border incidents?.)

286ff- were there real (= behavioral) cultural differences on
       the 2 side of the border, the 2 borders? not much -

*274-6 SUMMARY

292 "colligends" (jokingly?) threaten to join France - can we tell how

nation emerge on frontier/periphery before in
                    center. rural bias

Boundaries

Boundaries

The Making of France and Spain in the Pyrenees

Peter Sahlins

UNIVERSITY OF CALIFORNIA PRESS
Berkeley · Los Angeles · Oxford

The Publisher wishes to thank the Frederick W. Hilles Publication Fund of Yale University for their contribution toward the publication of this book.

The following illustrations have been reproduced with the kind permission of the archives and libraries: (1) G. Bouttats, "Disegno dell'Isola della Conferenza nel Fume Bidasoa" (1678), Bibliothèque Nationale, Cartes et Plans, Ge D 14207; (2) [Pierre du Val,] "Les 33 villages de Cerdagne cédés à la France par le Traité de Llívia," Bibliothèque Nationale, Cartes et Plans, Ge DD 1221–31, pl. 182; (3) "Carte de partie de Catalogne et de Roussillon," Bibliotheque de l'Arsènal, MS 6443 (246); (4) Du Cheylat, "Carte d'une partie de la Cerdagne avec les limites de France et d'Espagne" (1703), Bibliothèque de l'Arsenal, MS 6443 (254); (6) "Plano o descripción del Termino de la Villa de Llivía . . . " (1732), Archives Départementales des Pyrénées-Orientales, C 2046; (7) J. Rousseau, "Carte militaire des Pyrénées-Orientales (1795), Servicio Geográfico del Ejercito, Cartoteca Histórica, no. 315; (8) "Reconaissance militaire d'un portion de la frontière espagnole" (1840), Ministère de la Guerre, Archives de l'Armée de la Terre, Mémoires et Reconaissances 1223, no. 50; (9) "Copia exacta reducida del croquis remitido por el pueblo de Guils" (n.d., ca. 1860), Archivo del Ministerio de Relaciones Exteriores, Tratados y Negociaciones 221–222.

University of California Press
Berkeley and Los Angeles, California

University of California Press, Ltd.
Oxford, England

Library of Congress Cataloging-in-Publication Data
Sahlins, Peter.
  Boundaries: the making of France and Spain
in the Pyrenees / Peter Sahlins.
     p.   cm.
  Bibliography: p.
  Includes index.
  ISBN 0–520–06538-7 (alk. paper)
  1. Pyrenees (France and Spain)—History. 2. Cerdaña (Spain and France)—
History. 3. Catalonia (Spain)—History. 4. Roussillon (France: Province)—
History. 5. France—Boundaries—Spain. 6. Spain—Boundaries—France.
  7. Self-determination. National—Cerdaña (Spain and France)—History. I. Title.
DC611.P985S24 1989   946'.52—dc19   89–4711
                                                                            CIP

Printed in the United States of America

1 2 3 4 5 6 7 8 9

The paper used in this publication meets the minimum requirements of American National Standard for Information Sciences—Permanence of Paper for Printed Library Materials, ANSI Z39.48–1984 ∞

To Leslie

Contents

Figures, Tables, Maps, and Illustrations

Maps

Illustrations

Photographs

Abbreviations

Full references are given at the first citation of all archival series and published works. A complete listing of all archival series, abbreviations for archival subseries, and printed sources cited may be found in the bibliography at the end of the book.

ARCHIVES

ACA	Archivo de la Corona de Aragón (Archive of the Crown of Aragon, Barcelona)
ACSU	Arxiu Capitular de la Seu d'Urgell (Chapter Archive of the Seu d'Urgell)
ADPO	Archives Départementales des Pyrénées-Orientales (Archive of the Department of the Pyrénées-Orientales, Perpignan)
ADSU	Arxiu Diocesà de la Seu d'Urgell (Diocesan Archive of the Seu d'Urgell)
AGS	Archivo General de Simancas (National Archive, Simancas)
AHMB	Arxiu Històric Municipal de Barcelona (Municipal Archive, Barcelona)
AHP	Arxiu Històric de Puigcerdà (Puigcerdà Historical Archive)
AMLL	Arxiu Municipal de Llívia (Municipal Archive of Llívia)

FAIG	France, Archives de l'Inspection du Génie (Archive of the Superintendency of Engineers, Château de Vincennes)
FAMRE	France, Archives du Ministère des Relations Extérieures (Archive of the Ministry of Foreign Relations, Paris)
FAN	France, Archives Nationales (National Archive, Paris)
FMG AAT	France, Ministère de la Guerre, Archives de l'Armée de la Terre (War Ministry, Land Army Archive, Château de Vincennes)
SAHN	Spain, Archivo Histórico Nacional (National Archive, Madrid)
SAMRE	Spain, Archivo del Ministerio de Relaciones Exteriores (Archive of the Ministry of Foreign Relations, Madrid)

LIBRARIES

BC	Biblioteca de Catalunya (Catalan Library, Barcelona)
BRP	Biblioteca del Real Palacio (Library of the Royal Palace, Madrid)
FBA	France, Bibliothèque de l'Arsenal (Arsenal Library, Paris)
FBN	France, Bibliothèque National (National Library, Paris)
SBN	Spain, Biblioteca Nacional (National Library, Madrid)

COLLECTIONS OF LETTERS

| Lettres Pyrénées | *Lettres du Cardinal Mazarin où l'on voit le secret de la négotiation de la Paix des Pyrénées* (Amsterdam, 1690) |
| Lettres Mazarin | *Lettres du Cardinal Mazarin,* ed. M. A. Chéruel, 9 vols. (Paris, 1879–1906) |

JOURNALS

BSASLPO	*Bulletin de la société agricole, scientifique, et littéraire des Pyrénées-Orientales* (Perpignan)
CERCA	*Centre d'études et de recherches Catalans* (Perpignan)
RGPSO	*Revue géographique des Pyrénées et du Sud-Ouest* (Toulouse)
RHAR	*Revue historique et archéologique du Roussillon* (Perpignan)

Note

All place names in the Cerdanya, Catalonia, and Roussillon have been standardized in their modern Catalan spelling, except where these appear quoted in the footnotes, and except those already familiar to the English-speaking world, whether in their Castilian (Spanish) form (for example, Ebro, Lérida) or in French (Roussillon, Perpignan). When I or the sources refer to the entire valley or "County of Cerdanya," I have used the Catalan spelling; otherwise, I have written "French Cerdagne" and "Spanish Cerdaña." Concerning proper names other than the names of kings, I have opted to follow the language of the document cited while standardizing usage. All translations are my own unless otherwise noted.

I have tried to keep foreign language terms to a minimum and, when using them, have attempted to make their meaning clear. The problem of weights, measures, and coinage presents more difficulties. The unit of land measure used in the Cerdanya was the *jornal* (Catalan), equal to approximately .35 hectares or .9 acres, said to be the amount of land a person could sow in a day. Each *jornal* was divided into 16 *ares*. The *carga* (Catalan) was a measure of capacity, said to be the weight carried by a mule, equivalent to roughly 120 liters of liquid, 120 kilograms of wood, or 180 kilograms of rye. (In the eighteenth century, 1 jornal customarily yielded 1 carga of rye to the peasant proprietor, with deductions made for taxes, tithes, and reseeding.)

The monetary system in use in the Cerdanya was of a complexity that baffled provincial administrators during the Old Regime and which continues to baffle today. Catalonia, as part of the Crown of Aragon, used the pound system; as in France, financial accounts were kept in "pounds" (lliures in Catalan, livres in French), made up of 20 "shillings" (sous in both languages), or 144 "pence" (diners, deniers). The Catalan lliura in the later seventeenth and eighteenth century, however, was roughly three times the value of the French livre tournois, which itself was worth twice that of the livre perpignan in use in the Counties of Roussillon and Cerdanya until the French Revolution. In Catalonia, prices were often quoted in Catalan reals (1 real equaled 2 sous, thus 1 lliura equaled 10 reals). Whenever possible, I have given all monetary sums as French livres tournois.

For more detailed information on foreign language terms, weights, measures, and coinage in the early modern period, readers should consult the glossary and appendices in J. H. Elliott's *The Revolt of the Catalans: A Study in the Decline of Spain, 1598–1640* (London, Eng., 1963), 553–578. Additional material for the Cerdanya in the period covered by this book is provided in P. Sahlins, "Between France and Spain: Boundaries of Territory and Identity in a Pyrenean Valley, 1659–1868" (Ph.D., Princeton University, 1986), 903–907, where readers may also find further documentation and source material on the themes that are the subject of this book.

Preface

"The history of the world is best observed from the frontier."

Pierre Vilar

"Only by moving grandly on the macroscopic level can we satisfy our intellectual curiosities. But only by moving minutely on the molecular level can our observations and explanations be adequately connected. So, if we would have our cake and eat it too, we must shuttle between the macroscopic and the molecular levels in instituting the problem *and* in explaining it."

C. Wright Mills

This book is an account of two dimensions of state and nation building in France and Spain since the seventeenth century—the invention of a national boundary line and the making of Frenchmen and Spaniards. It is also a history of Catalan rural society in the Cerdanya, a valley in the eastern Pyrenees divided between Spain and France in 1659. This study shuttles between two levels, between the center and the periphery. It connects the "macroscopic" political and diplomatic history of France and Spain, from the Old Regime monarchies to the national territorial states of the later nineteenth century; and the "molecular" history—the historical ethnography—of Catalan village communities, rural nobles, and peasants in the borderland. On the frontier, these two histories come together, and they can be told as one.

By the Treaty of the Pyrenees and its addenda in 1659–1660, France annexed the northern Catalan County of Roussillon and thirty-three villages of the adjoining County of Cerdanya. The treaty divided the valley between France and Spain through the center of the plain and left the Spanish enclave of Llívia completely surrounded by French villages, as it remains today (see maps 1–3). But the 1660 treaty failed to define the exact territorial location of the Spanish–French boundary. Only the Treaties of Bayonne in 1866–1868 formally delimited the political boundary, as France and Spain placed border stones along an imaginary line demarcating their respective national territories. The intervening two centuries saw the formation and consolidation of two nations and two national territorial states in the Cerdanya, despite the persistence of a common Catalan culture and social relations of exchange across the boundary. Rather than proceeding with this story from the center to the periphery, as much of the literature has done, this book considers the history of the French and Spanish nation-states from the perspective of the borderland, at the intersection of state and society, of the political and civil orders. The protagonists in this drama are the statesmen and peasants, ministers and mayors, customs officials and smugglers, and generals and deserters who together participated in the making of France and Spain in the Cerdanya.

The two villages of Palau and Aja stand in plain sight of each other, separated by no more than a few hundred meters and, since the Treaty of the Pyrenees, by an invisible boundary dividing France from Spain. Like the two Cerdanyas, these two villages have had since 1659 two separate histories—in the limited sense of that which can be known about their past experiences. When beginning this inquiry, I intended to write a strictly comparative study of these experiences; but it quickly became apparent that different quantities and kinds of information were available about the French and Spanish Cerdanyas, and the focus of this study gradually came to rest on the French side of the boundary. In part this was a function of the survival of historical records, itself a reflection of the different historical experiences of the two Cerdanyas. Specifically, the turbulent political history of Spain since the late eighteenth century means that fewer sources have survived for the Spanish Cerdaña, as opposed to the French Cerdagne.

Yet historians have in the archives of the Spanish towns of Puigcerdà and Llívia a better historical record than for any single village of the French Cerdagne. Because these settlements were towns, they enjoyed more developed municipal administrations, replete with notaries and scribes who recorded efficiently the details of local life. Their archives have survived, although in a most disorganized manner. But Puigcerdà and Llívia represent a social difference of both type and scale (towns versus villages) inserted within a national distinction (Spanish versus French), which makes them less than representative of the largely rural Spanish Cerdaña.

The difference in the historical record of the two Cerdanyas has a further dimension: it is a function of the differences of political organization of the Spanish and French states, of the differences in the intentions and historical experiences of state formation and nation building in the borderland. Spanning the boundary, the Cerdanya has an unusually rich documentation. A few sheep straying across the boundary in the eighteenth or nineteenth century produced dozens of letters among officials and administrators at all levels of the French and Spanish governments. Because it is a borderland, we can learn much about local culture—religious practices or pastoral usages—from the archives of the ministries of foreign affairs. Yet in truth, the French archives provide greater and better material than those of Spain. This difference in quantity and quality suggests something of the nature of the problem addressed in this book. The French state "annexed," "integrated," and eventually "assimilated" part of the valley that it acquired in 1659; the Spanish state did so too, but starting from a very different point and ending in a very different form. A principle concern of this study is to understand what such an assimilation meant, particularly in France, from both a national and local perspective.

One Cerdanya or two Cerdanyas? However distinct their documented histories, the two Cerdanyas shared a similar historical experience as a borderland of Spain and France. The boundary eventually divided the Cerdanya into two, but it also functioned to bring the two sides of the valley closer together than they had been before. It is the tension between the unity of the Cerdanya as a borderland and the division of the valley between France and Spain that informs this study.

Acknowledgments

The research for this book in archives from Paris to Madrid was funded by grants from the Social Science Research Council and the American Council of Learned Societies; the Alliance Française de New York; the French–American Foundation; and the Graduate School of Princeton University. I gratefully acknowledge the generous support and encouragement of these institutions. I am indebted to Professor John H. Elliott and the Institute for Advanced Study in Princeton for the assistantship (1986–1987) that provided the opportunity to recast an unwieldy dissertation into book form, and to the Society of Fellows in the Humanities at Columbia University, New York, for the Mellon postdoctoral fellowship (1987–1988) that allowed me to complete the task, and to begin others.

I wish to thank the directors, archivists, and personnel whose patience and assistance greatly facilitated archival research in France and Spain. In particular, I am indebted to Michel Bouille of the Archives Nationales in Paris; Jules Lagarde, now retired from the staff of the Archives Départementales des Pyrénées-Orientales; Pascal Evens of the Archives du Ministère des Relations Extérieures; Josep Vinyet i Estebanell, Town Secretary of Llívia; and Salvador Galceran i Vigue of the Arxiu Històric de Puigcerdà. Thanks also to the families in the Cerdanya who kindly allowed me to consult the family papers in their possession: Mathias of Palau (for the Delcor papers); Naudi of Gorguja (Carbonell), Vilanova of La Tor de Carol (Garreta); and Montellà of

Santa Llocaya (Montellà). A special thanks to Sebastià Bosom i Isern for his invaluable and delightful assistance in exploring the archives of Puigcerdà and Llívia, and for his generous gifts of friendship and hospitality.

I am grateful to the editors of the *Journal of Modern History* for permission to reprint excerpts from a previously published article; and to the John F. Enders and A. Whitney Griswold funds of Yale University for assistance toward the preparation of maps and illustrations.

It gives me great pleasure to acknowledge the teachers, colleagues, and friends whose criticisms and suggestions at different stages have significantly improved this work. James Amelang, Robert Darnton, Natalie Zemon Davis, John Elliott, James Fernandez, Richard Kagan, L. A. Kauffman, Marshall Sahlins, and Paula Sanders all read chapters of the book or the dissertation on which it is based, and I regret their consistently useful comments were not always heeded. During my research and travels in France and Spain, I gained much from discussions with scholars and other knowledgeable people whose assistance and suggestions have helped to make this book a better one. I would like to thank Joan Becat, Andreu Balent, Antoni Cayrol, Mathias Delcor, Daniel Fabre, Josep María Fradera, Xavier Gil, Josep Llobera, Alice Marcet, Michel Martzluff, Lou Stein, and Peirs Willson for their support and encouragement. I am especially indebted to Louis Assier-Andrieu, Núria Sales, and Pierre Vilar, who in our discussions and correspondence helped me to clarify many points of Catalan culture and political history. In thanking them, I retain responsibility for all errors of fact and interpretation.

Without Pierre Etienne Manuellan's generosity, warmth, and friendship, my sojourns in Paris and France would have been less productive and pleasurable; Sooni Taraporevala contributed in important ways to my thinking about boundaries and identities, for which I am grateful. Andrew Apter, Shanti Assefa, James Boyden, David Eddy, Paul Freedman, Carla Hesse, Lawrence Pervin, Forest Reinhardt, and Louis Rose provided at various stages moral and intellectual support and friendship, as have my many colleagues, too numerous to mention, who have shared with me their ideas and criticisms over the years.

In Princeton, Barcelona, and the Cerdanya, Xavier Gil offered his enthusiastic support, extensive bibliographic skills, and warm friendship, without which the "center of the world" would be a less well-known and hospitable place. James Amelang gave me my first introduction to Catalonia and to the streets of Barcelona; I have since relied heavily on his knowledge of things Catalan and Spanish, and on his unflagging interest, support, editorial skills, and friendship. Natalie Zemon Davis has generously given of her time and passion as teacher, colleague, and friend in support of the dissertation and book. The support of my family has helped in more ways than they know, and my father's encouragement and example have inspired me from beginning to end. L. A. Kauffman has my deepest gratitude; as editor, as friend, and as partner, she has done more for this book than anyone else.

Introduction

BOUNDARIES AND TERRITORY

The Pyrenean frontier of France and Spain is one of the oldest and most stable political boundaries in western Europe: it has not shifted location since France annexed the province of Roussillon and part of the Cerdanya valley in 1659–1660. Twentieth-century theorists consider the French–Spanish boundary a "fossilized," "cold," or "dead" boundary, since it has rarely presented cause for major international contention.[1] Today, the official boundary of France and Spain has none of the political significance of the many contested borders throughout the Third World, or even the United States–Mexican boundary, to mention only the most newsworthy. Yet the reports of its "death," in the sense of the French–Spanish boundary's permanent lack of controversy, have been slightly exaggerated. The rights of fishermen near Hendaye, the protests of Roussillon wine-growers opposed to the entry of Spanish wines, and disputes over territorial competence in the repression of Basque terrorism are among the issues that continue to occupy the press and the foreign offices of Spain and France. For some, Spain's 1986 entry into the European Common Market may have re-

1. For example, J. Brunhes and C. Vallaux, *La géographie de l'histoire* (Paris, 1921), 353; M. Foucher, *L'invention des frontières* (Paris, 1987), 128; for a typology of classes of tension in border areas, see R. Gross, "Registering and Ranking of Tension Areas," in *Confini e Regioni: Il potenziale di sviluppo e di pace delle periferie* (Trieste, 1973), 317–328.

1

vived echos of Louis XIV's claim that "the Pyrenees are no more"; for others, 1993 means a Europe without boundaries; but the reality of the Spanish-French boundary in the Cerdanya suggests otherwise.

Still, border disputes in the Pyrenees are not the catalysts of military conflagrations or diplomatic entreaties as they are at other boundaries; and it is this relative "fossilization" of the boundary that requires explanation. That the Spanish–French boundary in this century has become less of a source of political tension than others in Western Europe, such as the Rhine, is due in large part to shifts of European geopolitical concerns. But the explanation also lies in the dual appearance of an undisputed boundary line and an accepted opposition of nationalities in the borderland. This book is concerned with the historical development of these two structural components of the nation-state—a national community within a delimited state territory—as they took shape in one section of the French–Spanish borderland between 1659 and 1868.

The dates are derived from the history of the political boundary itself. By the Treaty of the Pyrenees in 1659 and its addenda the following year, the French crown acquired the province of Roussillon and a portion of the Cerdanya valley. Political geographers call this the "allocation" of the boundary, the first step in a three-stage process of "allocation, delimitation, and demarcation."[2] The delimitation and demarcation of the Pyrenean border occurred more than two centuries later, when between 1854 and 1868 the Spanish and French governments agreed in the Treaties of Bayonne to mark an imaginary border line by posing officially sanctioned border stones. The French–Spanish boundary between 1659 and 1868 may have been stable, in the sense that no territories were exchanged between the two states. Yet in 1659 it was a boundary defined by the jurisdictional limits of specific villages. Much would happen before it became a delimited boundary defining national territorial sovereignty.

Modern definitions of territorial sovereignty focus on political boundaries as the point at which a state's territorial competence finds its ultimate expression. States are defined by their exclusive jurisdiction over a delimited territory; and the boundaries of territorial competence

---

2. These distinctions are shared by such disparate approaches as P. de Lapradelle, *La frontière: Etude de Droit International* (Paris, 1928); and S. B. Jones, *Boundary-Making: A Handbook for Statesmen, Treaty Editors, and Boundary Commissioners* (Washington, 1945).

define the sovereignty of a state. A recognized authority on international law, Charles de Visscher, wrote that

> the firm configuration of its territory furnishes the state with the recognized setting for the exercise of its sovereign powers. The relative stability of this territory is a function of the exclusive authority that the state exercises within it, and of the co-existence beyond its boundaries of political entities endowed with similar prerogatives. . . . It is because the state is a territorial organization that the violation of its boundaries is inseparable from the idea of aggression against the state itself.[3]

This idea of territorial sovereignty and the inviolability of political boundaries owes much to modern political nationalism. In the late nineteenth and twentieth centuries, territories and boundaries became political symbols over which nations went to war and for which citizens fought and died. Frederick Hertz, writing in 1944, evoked the political and ideological definition of territory as the codeword of political nationalism:

> The idea of the national territory is an important element of every national ideology. Every nation regards its country as an inalienable sacred heritage, and its independence, integrity, and homogeneity appear bound up with national security, independence, and honour. This territory is often described as the body of the national organism, and the language as its soul.[4]

This ideologically and politically charged idea of national territory is the final expression of territorial sovereignty as it developed historically in the west. Although the Greeks and Romans had their own ideas of territoriality, and the later middle ages witnessed the appearance of "a new limited territorial *patria*," such premodern conceptions of territory differed greatly from the tenets of modern nationalism.[5] This book is not about political nationalism as it developed in the later nineteenth century, but about its presupposition: the idea of national territorial sovereignty from the seventeenth to the nineteenth century. It considers the emergence of the notion of territory in the eighteenth century, the

---

3. C. de Visscher, *Theory and Reality in Public International Law*, trans. P. E. Corbett (Princeton, 1957), 197–198; see also his *Problèmes de confins en droit international public* (Paris, 1969).

4. F. Hertz, *Nationality in History and Politics: A Psychology and Sociology of National Sentiment and Nationalism* (London, 1944), 150–151.

5. On the emergence of territorial identity in the Classical world, see J. Armstrong, *Nations before Nationalism* (Chapel Hill, N.C., 1982), esp. 14–53 and 93–128; on the territorial state in later medieval Europe, see E. Kantorowicz, *The King's Two Bodies: A Study in Medieval Political Theology* (Princeton, 1981), esp. 232–272.

ways in which the French Revolution gave a national content to territorial sovereignty, and the politicization of territory in the nineteenth century. Unlike recent studies of the "invention of territory," it does so by focusing on the evolution of political frontiers and boundaries.[6]

Political geographers, following conventional usage, generally distinguish "boundaries" and "frontiers." The first evokes a precise, linear division, within a restrictive, political context; the second connotes more zonal qualities, and a broader, social context. Though the linear/zonal distinction draws its connotations in English from the American experience of the western frontier, similar distinctions are made in most modern European languages, where they too are colored by particular historical experiences.[7] The fact of this dualism has often misled theorists into perceiving an evolutionary movement, necessary and irreversible, from a sparsely settled, ill-defined zone toward an uncontested, nonsubstantial, mathematically precise line of demarcation. Such was the model that nineteenth- and early twentieth-century theorists of the frontier adopted, including the father of modern political geography, Friedrich Ratzel.[8] Applied to the historical experience of state formation in Europe, and in particular to the paradigmatic example of France, the model fails to explain much of anything. As a schema, it ignores two critical dimensions of political boundaries: first, that the zonal character of the frontier persists after the delimitation of a boundary line; and second, that the linear boundary is an ancient notion. As a historical description, the model falls dramatically short of the evidence.

On one hand, the persistence of a zone after the delimitation of the

6. See P. Alliès, *L'invention du territoire* (Grenoble, 1972). After critically assessing the notion of territory in contemporary legal thought (pp. 10–19), Alliès focuses on the construction of the administrative institutions of territorial sovereignty, tracing the development of a "homogenous space" of bureaucratic power in Old Regime France. See also J. Gottman, *The Significance of Territory* (Richmond, Va., 1973); and *Territoires,* no. 1 (Paris: Ecole Normale Supérieure, 1983).

7. On the French distinction of *frontière* and *limite*, see L. Febvre, *"Frontière*: The Word and the Concept," in P. Burke, ed., *A New Kind of History: From the Writings of [Lucien] Febvre,* trans. K. Folca (London, 1973), 208–218; and D. Nordman, "Des limites d'Etat aux frontières nationales," in P. Nora, ed., *Les lieux de mémoire,* 2 vols. (Paris, 1986), 2 (pt. 2): 50–59; and his "Frontiera e confini in Francia: Evoluzione dei termini e dei concetti," in C. Ossola, C. Raffestin, and M. Ricciardi, eds., *La frontiera da stato a nazione: Il caso Piemonte* (Rome, 1987), 39–55. On the Spanish distinction of *frontera* (or *marca*) and *limite,* see A. Truyol y Serra, "Las fronteras y las marcas," *Revista española de derecho internacional* 10 (1957): 107; and J. M. Cordero Torres, *Fronteras hispanicas: Geografía e historia* (Madrid, 1960).

8. Febvre, "Frontière," 212, citing F. Ratzel, *Politische geographie* (Paris and Munich, 1903); C. Vallaux, *Le sol et l'état* (Paris, 1910), chap. 10; and J. Brunhes and C. Vallaux, *La géographie de l'histoire* (Paris, 1921), 344 et seq.

boundary has long been noted by jurists and students of international law. The zonal character of the frontier is a political construction of each state independently and of two contiguous states together. The zone consists in the distinct jurisdictions that each state establishes near the boundary for the purposes of its internal administration—thus a military zone, a customs zone, and so forth. And the zone represents the area where contiguous states realize policies of international cooperation and friendship, or *bon voisinage*. Although forms of international cooperation often precede the delimitation—as was the case in the Pyrenean frontier—they are codified and given stature in international law as part of the delimitation proceedings.[9]

On the other hand, the concept and practice of a linear boundary is an ancient—perhaps the most ancient—part of the frontier, one that long preceded modern delimitation treaties of the eighteenth and nineteenth centuries. Techniques of delimitation were known to the Greeks and Romans, and the Treaty of Verdun in 843 involved 120 "emmisaries" who worked more than a year to determine the boundaries of the parcels distributed to the three heirs of Charlemagne.[10] Historians once argued that the medieval polity in France had no conception of precise territorial boundaries. The division of Verdun remained without significance, it was claimed, not only because of the complete absence of topographical maps, but also because the extensive fragmentation of authority and the growth of feudal jurisdictions soon became the rule in western Europe.[11] More recently, medieval historians have recognized that the extension of feudal relations from the tenth to the thirteenth centuries did not mean the disappearance of questions over boundaries. The territorial extent of a *seigneurie* could be largely ignored, but it could also be precisely delimited, especially in areas where the seigneurie took shape within the limits of the ancient gallo-roman divisions, or *pagi*. Moreover, the kingdom's boundaries were in general

9. Lapradelle, *La frontière*, pt. 2; I. Pop, *Voisinage et bon voisinage en droit international* (Paris, 1968).

10. On the "primitive" and "sacred" character of linear boundaries, see Lapradelle, *La frontière*, 18–19; and A. Van Gennep, *The Rites of Passage*, trans. M. B. Vizedom and G. L. Caffee (Chicago, 1960), 15–25. On Roman and Greek conceptions of the linear boundary, see M. Foucher, *L'invention des frontières* (Paris, 1987), 63–96; and Lapradelle, *La frontière*, 20–25. On the Treaty of Verdun, see R. Dion, *Les frontières de la France* (Paris, 1947), 71–85.

11. For example, R. Doucet, *Les institutions de la France au XVIe siècle*, 2 vols. (Paris, 1948), 1: 16; Lapradelle, *La frontière*, 29–31; and G. Dupont-Ferrier, "L'incertitude des limites territoriales en France du XIIIe au XVI siècle," in *Académie des Inscriptions et Belles Lettres: Comptes Rendus* (Paris, 1942), 62–77.

well-defined, marked by stones, rivers, trees, and sometimes man-made trenches, even if these borders were often disputed.[12]

Yet in the eleventh and twelfth centuries, the political boundaries between kingdoms were fundamentally similar in kind to feudal limits within the kingdom. Only in the later thirteenth century did the two become different. The word "frontier" dates precisely from the moment when a new insistence on royal territory gave to the boundary a political, fiscal, and military significance different from its internal limits. The "frontier" was that which "stood face to" an enemy. This military frontier, connoting a defensive zone, stood opposed to the linear boundary or line of demarcation separating two jurisdictions or territories. But from the sixteenth century onward, and especially in the later eighteenth century, the two words tended to overlap; and the notion of delimitation became one of finding the *limites de la frontière,* the "boundaries of the frontier."[13]

Yet the conception of a linear political boundary as it appeared in the early modern period was not identical to the border line that slowly emerged after the seventeenth century. Peace treaties of the sixteenth and seventeenth centuries sometimes included provisions for the delimitation and demarcation of boundary lines, but the Old Regime state was something less than a territorial one. The French monarchy continued to envision its sovereignty in terms of its jurisdiction over subjects, not over a delimited territory, relying on the inherited notions of "jurisdiction" and "dependency" instead of basing its administration on firmly delineated territorial circumscriptions.[14]

Thus the Treaty of the Pyrenees of 1659 named the Pyrenees Mountains as the division between France and Spain, and further stipulated that commissioners were to meet to define more precisely which were the Pyrenees. The commissioners used the word "delimitation" and claimed to seek the "line of division," but they resorted to ideas of "jurisdiction" and "dependency" when dividing up the villages of the

12. J.-F. Lemarignier, *Recherches sur l'hommage en marche et les frontières féodales* (Lille, 1945); P. Bonenfant, "A propos des limites médiévales," in *Hommage à Lucien Febvre: Eventail de l'histoire vivant,* 2 vols. (Paris, 1953), 1: 73–79; and B. Guenée, "Des limites féodales aux frontières politiques," in P. Nora, ed., *Les lieux de mémoire,* 2 (pt. 2): 11–33.

13. B. Guenée, "Les limites," in M. François, ed., *La France et les français* (Paris, 1972), 57–64; Febvre, *"Frontière,"* 210–211; and Foucher, *L'invention des frontières,* 104–110.

14. Febvre, *"Frontière,"* 213–214; and Lapradelle, *La frontière,* 35–37; the notion of "jurisdiction" is explored more fully below, chap. 1.

Cerdanya. Only in 1868 did the Bayonne commissioners "delimit" the boundary by establishing an imaginary border of two national territories and "demarcate" the division by means of boundary stones.

The history of the boundary between 1659 and 1868, then, can hardly be summarized as the simple evolution from an empty zone to a precise line, but rather as the complex interplay of two notions of boundary—zonal and linear—and two ideas of sovereignty—jurisdictional and territorial. The two polarities can be found at any given moment in the history of the boundary, although the dominant but hardly unilinear tendency was the collapse of separate jurisdictional frontiers into a single territorial boundary line. The French Revolution gave to the idea of territory a specifically national content, while the early nineteenth-century states politicized the boundary line as the point where national territorial sovereignty found expression.

NATIONS AND NATIONAL IDENTITY

The creation of the territorial state constituted one component of the modern nation-state; the emergence of national identity formed another. According to received wisdom, modern nations were built from political centers outward and imposed upon marginal groups or peripheral regions in a process of cultural and institutional "assimilation" and "integration."[15] National identity, in this view, is the expression of cultural unity and national consciousness consolidated within the political framework of a centralized state. The paradigmatic experience is, of course, the French one. Though an older generation of scholars saw in the French Revolution a formative period in the creation of French unity, more recent scholarship suggests that France only became a unified nation at a surprisingly late date. For only during the early Third Republic (1870–1914) did the French state create the road and railway

---

15. Examples include K. Deutsch, *Nationalism and Social Communication: An Inquiry into the Foundations of Nationality* (Cambridge, Mass., 1953); K. Deutsch, "Some Problems in the Study of Nation-Building," in K. Deutsch and W. Foltz, eds., *Nation-Building* (New York, 1963), 1–16; C. A. Macartney, *National States and National Minorities* (Oxford, 1934); and R. Bendix, *Nation-Building and Citizenship* (New York, 1964). On the resistance of mountain regions to "the establishment of the state, dominant languages, and important civilizations," see F. Braudel, *The Mediterranean and the Mediterranean World in the Age of Philip II,* trans. S. Reynolds, 2 vols. (New York, 1966), 1: 38–41; and C. Levi, *Christ Stopped at Eboli,* trans. F. Frenaye (New York, 1947), 37–38.

networks, policies of compulsory primary education, and the universal military conscription by which peasants became Frenchmen.[16]

The corollary idea is that peasants become national citizens only when they abandon their identity as peasants: a local sense of place and a local identity centered on the village or valley must be superseded and replaced by a sense of belonging to a more extended territory or nation. In the words of Arnold Van Gennep, the dean of French folklorists, nationhood is "the extension of real or symbolic love felt for the corner of land which belongs to the commune, to an entire valley, an immense plain, the steppe, and the great city like Paris or Vienna."[17] National identity means replacing a sense of local territory by love of national territory.

Focusing on how the nation was imposed and built from the center outward, and claiming that its acceptance meant giving up local identities and territories, this received wisdom denies the role of local communities and social groups in shaping their own national identities. This book argues that both state formation and nation building were two-way processes at work since at least the seventeenth century. States did not simply impose their values and boundaries on local society. Rather, local society was a motive force in the formation and consolidation of nationhood and the territorial state. The political boundary appeared in the borderland as the outcome of national political events, as a function of the different strengths, interests, and (ultimately) histories of France and Spain. But the shape and significance of the boundary line was constructed out of local social relations in the borderland. Most concretely, the boundaries of the village jurisdictions ceded to France were not specified in the 1660 division, nor were they undisputed among village communities. The historical appearance of territory—the territorialization of sovereignty—was matched and shaped by a territorialization of the village communities, and it was the dialectic of local and national interests which produced the boundaries of national territory.

16. H. Kohn, *Prelude to Nation-States: The French and German Experience, 1789–1815* (Princeton, N.J., 1967); E. Weber, *Peasants into Frenchmen: The Modernization of Rural France, 1870–1914* (Stanford, Calif., 1976); for critical reviews of Weber's use of modernization theory, see C. Tilly, "Did the Cake of Custom Break?" in J. Merriman, ed., *Consciousness and Class Experience in Nineteenth-Century Europe* (New York, 1979), 17–41; and T. W. Margadant, "French Rural Society in the Nineteenth Century: A Review Essay," *Agricultural History* 53 (1979): 644–651.

17. A. Van Gennep, *Traité comparatif des nationalités: 1. Les éléments extérieures de la nationalité* (Paris, 1922), 144.

In the same way, national identity—as Frenchmen or Spaniards—appeared on the periphery before it was built there by the center. It appeared less as a result of state intentions than from the local process of adopting and appropriating the nation without abandoning local interests, a local sense of place, or a local identity. At once opposing and using the state for its own ends, local society brought the nation into the village.

Benedict Anderson has recently described nations as "imagined communities." The nation-as-community is imagined (in the sense of created and invented, as opposed to fabricated and dissimulated) "because the members of even the smallest nations will never know most of their fellow-members, meet them, or even hear of them, yet in the minds of each lives the image of their communion." The definition usefully corrects the positivist conception of national identity as a product of "nation building," focusing our attention instead on the symbolic construction of national and political identities.[18] Others have emphasized in recent years the importance of distinction and differentiation in the development and expression of ethnic, communal, and national identities. In the French–Spanish borderland, it is this sense of difference—of "us" and "them"—which was so critical in defining an identity.[19] Imagining oneself a member of a community or a nation meant perceiving a significant difference between oneself and the other across the boundary. The proximity of the other across the French–Spanish boundary structured the appearance of national identity long before local society was assimilated to a dominant center. This study develops what might be called an oppositional model of national identity in a particular historical setting: the Cerdanya, divided between Spain and France in 1659.

18. B. Anderson, *Imagined Communities: Reflections on the Origins and Spread of Nationalism* (London, 1983), 15. Recent work that brings out the cultural and symbolic dimensions of the nation "imagined" and includes C. Beaune, *Naissance de la nation France* (Paris, 1986); and A. E. Smith, *The Ethnic Origins of Nations* (Chapel Hill, N.C., 1982). See also L. Hunt, *Politics, Culture, and Class in the French Revolution* (Berkeley, Los Angeles, London, 1984), chaps. 1–3; and M. Agulhon, *Marianne au Combat* (Paris, 1979), which focus on the political dimensions of the nation as a cultural construction; and the contributions to the collective volume, *Les lieux de mémoire,* ed. P. Nora, 2 vols. (Paris, 1986–1988), which generally do not.

19. Recent work emphasizing the oppositional character of identities includes R. Grillo, Introduction to Grillo, ed., *"Nation" and "State" in Europe: Anthropological Perspectives* (London, 1980); S. Wallman, "The Boundaries of 'Race': Processes of Ethnicity in England," *Man* n.s. 13 (1978): 200–217; the contributions to A. P. Cohen, *Symbolizing Boundaries: Identity and Diversity in British Cultures* (Manchester, 1986); and J. Armstrong, *Nations before Nationalism* (Chapel Hill, N.C., 1982).

THE CERDANYA: BETWEEN FRANCE
AND SPAIN

"Beautiful, fertile, and well populated, this land can be compared with any other." So wrote an anonymous but proud inhabitant of the Cerdanya in the early seventeenth century, describing his native land. The "land and county of Cerdaña" may be glimpsed momentarily through his eyes:

> Its shape is in the form of a ship, with its prow to the east and its stern to the west, although it turns a bit south in the form of a half-moon, but without losing its shape. The oars can be likened to the many valleys on all sides. Its length is seven large leagues, from the Tet bridge where the Cerdanya ends and the Conflent begins, to a little below the Arséguel bridge, a league and a half from the Seu d'Urgell. By that point, it is much less wide, with high mountains which can be likened to the sides of the ship.[20]

"A very fertile land," as he reiterated, the Cerdanya produced most of what was "necessary to human life"; indeed, natives and foreigners alike saw the densely populated valley, with its rich alluvial plain, its plentiful rivers, forests, and abundant pastures, as an oasis within the more forbidding ecology of the Mediterranean Pyrenees. (see maps 1 through 3.)

Most of the eighty or so settlements of the Cerdanya are situated at the juncture of the "ship" and its "oars," where a series of perpendicular valleys open onto the main valley floor. Their location assured the inhabitants, who numbered perhaps 8,000 in the early seventeenth century, the optimum use of the ecological resources necessary to the reproduction of their agro-pastoral way of life. The settlements were mostly nucleated villages, although in the southwestern part of the

20. Biblioteca de Catalunya (hereafter BC) MS 184, fol. 1v: "Descripción de la tierra y condado de Cerdaña," n.d., ca. 1610. These limits have varied historically: for the purposes of this study, I have taken the Perxa Pass as the northeastern limit of the valley, thus excluding the villages of Sant Pere dels Forcats, La Cabanasse, and La Llaguna, which nonetheless formed part of the eighteenth-century viguery (*viguerie*) of the French Cerdagne. Although the southwestern limit described by the seventeenth-century text has been generally accepted as the boundary of the canton (*comarca*), the eighteenth-century administrative district (*corregimiento*) of Puigcerdà extended beyond the Arséguel bridge and encompassed the valley of Ribes as well. Concerning the southwestern portion of the valley, this study focuses on the plain upstream from Bellver, roughly that part of the Spanish Cerdaña which became the province of Girona in Spain's administrative reforms of 1833. On the medieval and modern limits of the Cerdanya, see E. Balcells, "Vicisitudes historicas de las comarcas descritas (Alto Urgell, Alto Bergada, Cerdaña, y Andorra)," *Actas del 7e congreso internacional de estudios pirineaicos* (Jaca, 1974), 117–133; on the history of territorial divisions of Catalonia, see P. Vila, *La divisió territorial de Catalunya* (Barcelona, 1979), 27–63.

Map 1. France, Spain, Catalonia

plain—that which became, after 1660, the Spanish Cerdaña—the settlement pattern showed more dispersal.[21] The village communities were corporate groups, associations of "neighbors" (Catalan: *veïns*), with appointed judicial officers or bailiffs (*batlles*) and elected councillors (*syndics* or *consols*), holding land and usufruct rights in common.

21. P. Vila, *La Cerdanya* (Barcelona, 1926), 97–113.

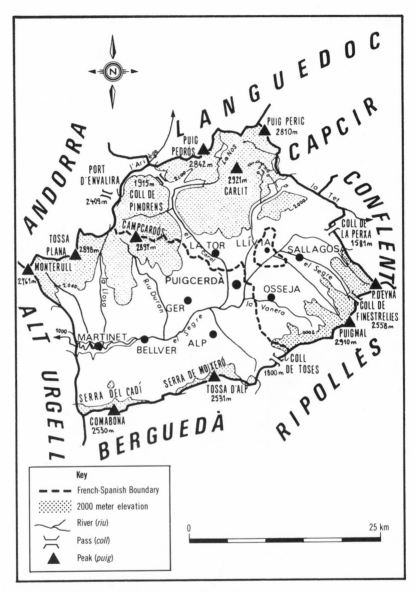

Map 2. The Cerdanya

These communities maintained a great deal of autonomy in the regulation of their public life, as the seigneurial regime was relatively weak in the Catalan Pyrenees, and the early modern state was a distant entity that interfered rarely in communal affairs.

The village communities were the cells of social life; grouped together, many of them formed more inclusive unities, often within the framework of the perpendicular valleys, such as Carol and Osseja. Resembling the federations or "valley–communities" of the central and western Pyrenees, these associations of villages and hamlets held land, pastures, and usufruct rights in common. Although the County of Cerdanya itself had no property in common, it nonetheless maintained in the early seventeenth century institutional and political expressions of a collective public life. The seventeenth-century description divides the valley into four "quarters" or districts. Each district sent militia levies, money, or provisions as requested by the General Council of Syndics. Representatives of the quarters met regularly according to ancient privileges and maintained the right of imposing a local tax. Elected every three years, the syndics had their obligations and ordinances: "almost like the ancient tribunals of Rome, they care well for the public good."[22] The Cerdanya, in fact, was in the early seventeenth century one of the most unified cantons (comarques) in all of Catalonia.

At the center of the plain, situated on a small rise, was the town of Puigcerdà, the political, administrative, economic, and cultural center of the valley. In Puigcerdà sat the royal law courts and administration, all under the authority of the veguer, the royal judicial officer in charge of the district. In Puigcerdà resided much of the local ruling class, the nobles, titled bourgeois, and large landowners who were increasingly drawn to the "town." The weekly markets and annual fairs brought peasants from all over the valley to buy and sell livestock and manufactured goods. Religious festivals were also the occasion for peasants to gather in Puigcerdà, where could be found the several churches and monasteries with properties, incomes, and seigneurial jurisdictions over many surrounding villages. Finally, as the principle fortified site in the district, the town afforded protection for villagers and townspeople alike, who found refuge within its walls.[23]

This small, self-contained, and relatively prosperous world was completely surrounded by a ring of mountains. The valley floor lies at an

22. BC MS 184, fol. 3v; S. Galceran i Vigué, L'antic sindicat de Cerdanya: Estudi socio-econòmic based en la historia inèdita dels segles XIV al XVII (Girona, 1973), esp. 65–69.

23. BC MS 184, fols. 24v and 26–33: "Descripción de Puigcerdan" by the Dominican monk, Joan Trigall, written in 1603. The most important religious institutions were the Collegiate Church of Santa Maria, who are "seigneurs of many villages with many vassals" and richly endowed with more than 80 benefices; the Augustinian monks of San Francisco; the Preachers of Sant Dominic; and the Nuns of Santa Clara.

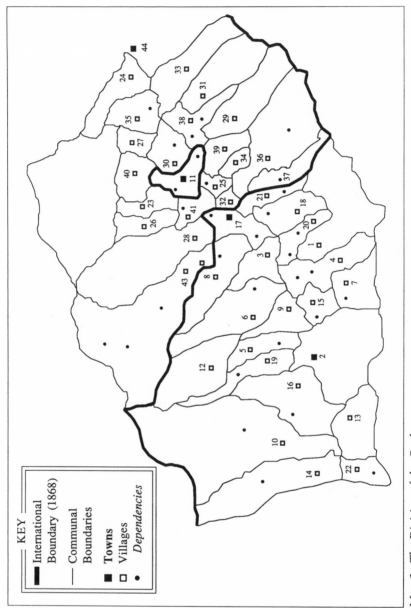

KEY

International Boundary (1868)

Communal Boundaries

■ Towns
□ Villages
• Dependencies

Map 3. The Division of the Cerdanya

SPANISH CERDAÑA

1 Alp
2 **Bellver**
3 Bolvir
   Aranser
4 Das
   Mossol
   Sanabastre
5 Ellar
   Ardovol
   Olopte
6 Ger
   Ventajola
   Saga
   Rigolisa
7 Grus
   Riu
   Les Pereres
8 Guils
   Saneja
9 Isobol
10 Lles
   Vilar
   Estoll
   Coborriu
11 **Llívia**
   Aja
   Sareja
   Gorguja
12 Meranges
   Estana

13 Montella
14 Musser
15 Prats
   Sampsor
16 Prullans
17 **Puigcerdà**
18 Queixans
19 Talltendre
   Orden
20 Urtx
21 Villalobent
22 Villec

FRENCH CERDAGNE

23 Angostrina
24 Bolquera
25 Càldegues
   Onzès
26 Dorres
27 Egat
28 Enveig
29 Err
30 Estavar
   Bajanda
31 Eyna
32 Hix/Bourg-Madame
33 Llo
   Rohet
34 Nahuja
35 Odello
   Via
36 Osseja
   Valsebollera
37 Palau

38 Sallagosa
   Ro
   Vedrinyans
39 Santa Llocaya
40 Targasona
41 Ur
   Flori
42 Vilanova
43 Carol Valley
   Carol
   La Tor de Carol
   Porta
   Porté
   Ques
44 **Mont-Louis**

average elevation of 1200 meters. To the north, the granite mass of Carlit Mountain (2921 meters)—on which were located the principal summer pastures used by the northernmost villages of the valley—separates the Segre, Ariège, and Tet River valleys. To the west, the mass of Puigpedros (2842 meters) divides the Cerdanya from the valleys of Andorra. To the south and east, a string of mountain peaks from the Serra del Cadi ridge to the Tossa d'Alp (2531 meters), Puigmal (2910), and Eyna (2705) form a watershed that separates the Cerdanya and the Ripollès, the valleys of Lillet and Ribes in Catalonia.

Although enclosed by high mountains, the Cerdanya was not as isolated as might first seem. The seventeenth-century author described the numerous passes in and out of the valley, the principle ones "very pleasant and delightful, with much water and crystalline springs, and easy to cross" in the summer months, and rarely impassable for people on foot even during the eight months when they remained snow-covered.[24] The autarky of Cerdanya was only relative. Grain, wine, oil, and livestock passed to, from, and through Cerdanya, linking the valley to both versants of the Pyrenees. Nor was the mountain range a barrier to the commerce of manufactured goods. In the late medieval period, the town of Puigcerdà was an important commercial and manufacturing center of cloth and linen, its merchants trading with French and Spanish towns and with northern Europe and the Levant. Although local industries were clearly in decline in the seventeenth and eighteenth centuries, the commercial role of Puigcerdà and the Cerdanya in a trans-Pyrenean economy remained an important one.[25]

The relative permeability of the Pyrenees, the openness of the Cerdanya to the world beyond its boundaries, was due in part to a distinctive geography. At both the eastern and western extremes of the Pyrenean chain, an otherwise well-defined ridge—what geographers call the "axial zone"—is shattered by a confusing topology of fault

24. Ibid., fol. 7v. Seventeenth- and eighteenth-century descriptions of these passes abound, many of them authored by military officials and engineers: examples include Roussel and la Blottière, *Légende de tous les cols, ports, et passages des Pyrénées (1716–1719)*, ed. J. Escarra (Paris, 1915); France. Ministère de la Guerre, Archives de l'Armée de la Terre (hereafter FMG AAT) MR 1084, no. 22: "Mémoire sur les communications de la Cerdagne française avec les pays de la domination française et espagnole," 1754; FMG AAT MR 1325 and 1327: "Guia de los caminos del Principado de Cataluña," n.d., ca. late eighteenth century; and France. Archives du Ministère des Relations Extérieures (hereafter FAMRE) Limites vol. 439, no. 91: "Premier mémoire sur la frontière du Roussillon," September 24, 1777, esp. fols. 20–37.

25. S. Galceran i Vigué, *La indústria i el comerç a Cerdanya* (Barcelona, 1978), passim; BC MS 184, fol. 25v.

zones and secondary ridges. In the east, the very idea of the Pyrenees as a dividing range of mountains loses its meaning as it approaches the Cerdanya. A "scar" or "transversal slash" cuts across the chain from southwest to northeast, creating a series of small basins—Conflent, Capcir, Cerdanya, Alt Urgell (Urgellet)—strung between the Roussillon and Urgell plains, between Perpignan and the Seu d'Urgell. Forming the upper basin of the Segre River valley on the southern or Spanish versant of the Pyrenees, the Cerdanya lies at the center of this transversal slash.[26]

In the early seventeenth century, the Cerdanya stood at the crossroads of the eastern Pyrenees. From Perpignan to Lérida, an ancient road, in use since Roman times, entered the Cerdanya at the Perxa Pass and left the valley through the narrow gorges of the Segre River. Less well-defined geologically, a second passage through the Cerdanya linked Toulouse and Barcelona, the Aquitain Basin, and the secondary depressions and coastal areas of Catalonia. The route into the Cerdanya from France crossed the Puigmorens Pass into the Carol Valley, and left the Cerdanya through the Alp Valley (via the Creu de Mayans or Tossa d'Alp Passes) before descending to Ribes, Berga, Ripoll, and Vic.[27] Mule drivers, peddlers, merchants, and smugglers passed back and forth along the dozens of mule and foot passages linking the two sides of the mountain chain. Neither the mountains surrounding the valley, nor the political boundary that was eventually to divide it, were impenetrable barriers.

If the ring of mountains surrounding the Cerdanya did not prohibit the movement of people or goods in and out of the valley in the early modern period, neither did it shelter the Cerdanya from the destructive forays of local bandits, so prevalent during the early seventeenth century in the Catalan Pyrenees.[28] But above all, it was the pillage and de-

26. P. Vilar, *La Catalogne dans l'Espagne moderne*, 3 vols. (Paris, 1962), 1: 199–343.

27. Ibid.; and L. Solé Sabaris, "Del paisaje de Cerdaña," *Ilerda* 19 (1955): 123.

28. On banditry in Catalonia and the organization of bandits into the competing factions of *nyerros* and *cadells*, see J. H. Elliott, *The Revolt of the Catalans: A Study in the Decline of Spain, 1598–1640* (Cambridge, Eng., 1963), passim; and X. Torres i Sans, "Els bandols de 'Nyerros' i 'Cadells' a la Catalunya Moderna," *L'Avenç* no. 49 (1982): 345–350. On the links of these factions to political groupings of royal officials and to important aristocratic families in the Cerdanya and the Catalan Pyrenees, see X. Torres i Sans, "Les bandositats de Nyerros i Cadells a la Reial Audiéncia de Catalunya (1590–1630): 'Policia o alto govierno?'" *Pedralbes* 5 (1985): 145–171; and N. Sales i Bohigas, "El senyor de Nyer sense els nyerros," in her *Senyors bandolers, miquelets i botiflers: Estudis sobre la Catalunya dels segles XVI al XVIII* (Barcelona, 1984), 11–101.

struction wrought by French troops which the Cerdans constantly feared, and the early seventeenth-century author, like his contemporaries, expressed an evident hostility toward the French, who "have entered this land thousands of times, doing great damage." Forming the frontier of France and Spain, of the Spanish Habsburg lands and those of the French Valois, the two counties of Cerdanya and Roussillon were under constant assault by French troops throughout the sixteenth and early seventeenth centuries. According to the local author, eyewitness to a familiar pattern of events:

> They take all the cattle and all that they can get, including men and even children as hostages, returning them at a great price; and if it were not for the fertility and wealth of the land, there would be no remedy against them; and if today the Cerdans are not rich, neither are they poor, but what is astounding is after all their suffering, they have anything at all. For if another land had experienced such encounters, I doubt if it would have resisted as well.[29]

The principal arenas of the Valois–Habsburg rivalries lay in Italy and later the Low Countries; but Catalonia and the Pyrenean frontier proved to be an important "theater of distractions" from the main battlefields. At each outbreak of hostilities, and often without waiting for formal declarations of war, the French directed troops toward the fortress of Salces, rebuilt by Charles V at the northernmost point of Habsburg territory; Perpignan, capital of the two counties; and Puigcerdà, the mountain stronghold and capital of the Cerdanya. During the century from Francis I to Henry IV, French troops raided and pillaged the Cerdanya more than a dozen times.[30]

The two counties of Cerdanya and Roussillon formed an integral part of Catalonia in the early seventeenth century, although they preserved a certain administrative autonomy within the Principality of Catalonia. Their distinctiveness stemmed in part from historical experience: in 1276, the two counties had been separated from Catalonia by James I to create a new kingdom for his son but were reunited to the principality in 1343. In 1463, Louis XI of France, in alliance with the Castilian monarchy against Aragon, conquered and annexed Roussillon and Cerdanya, returning them to Ferdinand of Aragon in 1493. Although a certain rivalry between the two counties and the principality

29. BC MS 184, fol. 3.
30. J. Sanabre, *El Tractat dels Pirineus i la mutilació de Catalunya* (Barcelona, 1961), 23–25.

existed in the early seventeenth century, there was no doubt that Roussillon and Cerdanya were part of Catalonia, and that Catalonia was part of Spain.[31]

Spain, in the early modern period, was formed from the dynastic union of Ferdinand of Aragon and Isabel of Castile in 1469, who began their joint rule in 1479. Within the constitutional ordering of the Spanish monarchy, Catalonia preserved its distinctive political administration, its own written and customary laws (the *constitucions*), monetary system, customs barriers, and fiscal administration. Catalonia also maintained its own linguistic and cultural identity distinct from Castile, and there was little attempt to impose Castilian culture on the principality.[32]

But which were Spain's boundaries in Catalonia, at the eastern end of the Pyrenees? In the early modern period claims were made for two distinct mountain ranges, both identified as "the Pyrenees." "The province of Catalonia," wrote the Perpignan doctor Lluis Baldo in 1627, referring to the principality and the two counties of Roussillon and Cerdanya, "begins at the crests of the Pyrenees Mountains and extends, across valleys and watersheds, toward Spain, of which it is the first region." But his Barcelona contemporary Salvador Fontanet thought differently, believing that Roussillon and Cerdanya were "outside the limits of Spain, which are the Pyrenees Mountains."[33] Lluis Baldo was referring to the Corbières ridge, which stretches from the Mediterranean, just south of Narbonne, southwest to the mountains north of Cerdanya. This was the range implied, although not named, in the Treaty of Corbeil of 1258, which established the boundary of France and Aragon (later Spain) until 1660. By the Treaty of Corbeil, Louis IX of France renounced claims to the counties of Roussillon and Barcelona, while James I of Aragon gave up his aspirations to the fiefs of the viscounties of Carcasonne and Narbonne.[34] It was only normal that Baldo and others should have named the Corbières as "the Pyrenees

31. V. Ferro, *El dret públic català: Les institucions a Catalunya fins al Decret de Nova Planta* (Vic, 1987), 16–17. For sources on the rivalry between the *comtats* and the *principat*, see below, chap. 3, n. 1.

32. J. H. Elliott, *Imperial Spain, 1469–1716* (New York, 1963), 17–43; and Elliott, *Revolt*, 1–48.

33. France, Bibliothèque Nationale (hereafter FBN) F01 48: printed pamphlet of Lluis Baldo, 1627; memoir of Salvador Fontanet, February 1627, quoted in Elliott, *Revolt*, 23. Seventeenth-century conceptions of natural frontiers are examined in chap. 1.

34. Vilar, *La Catalogne*, 1: 12; M. Sorre, *Les Pyrénées méditerranéennes* (Paris, 1913), 15–17; H. Guiter, "Sobre el tractat de Corbeil," *Revista catalana* 38 (1978): 13–15.

Mountains." But Fontanet was referring to the Albères range, which stretches from Cap Cerbère at the Mediterranean nearly due west to the Cerdanya. He claimed that the Pyrenees divided at once the County of Roussillon and the Principality of Catalonia, and that they ought to separate France and Spain as well.

In 1627 the debate over the location of the "true" Pyrenees, and over whether or not Roussillon ought to form part of France or Spain, had been ongoing for several centuries. It was later taken up by the negotiators of the Treaty of the Pyrenees, and eventually resolved once and for all. Article 42 of the Treaty declared that "the Pyrenees Mountains, which anciently divided the Gauls from the Spains, shall henceforth be the division of the two said kingdoms." Roussillon was ceded formally to France, but the case of the Cerdanya appeared more complicated. Historically united to Roussillon, it nonetheless appeared to lie on the Spanish versant of the Pyrenean chain. French and Spanish commissioners were named to designate the actual limits of France and Spain, and they spent much effort deciphering the specific meaning of a text that seemed to evoke both historical and natural frontiers. The result, after bitter struggle and compromise, was Spain's cession of sovereign jurisdiction over roughly half of the villages of Cerdanya to France, thus dividing the valley through its center into two parts, while leaving the town of Llívia under Spanish jurisdiction, where it remains today, completely surrounded by French villages.

How unique is the Cerdanya, a valley and a single ethnic and linguistic group divided politically between France and Spain? Catalan territories span the ill-defined Pyrenean mountain range as it meets the Mediterranean, just as provinces of Basque-speaking peoples span the chain as it nears the Atlantic. In both cases, an ethnic group was divided by the French-Spanish boundary. Yet there are no single valleys as neatly divided by a political boundary as the Cerdanya. In the central and eastern Pyrenees, however, and especially those valleys immediately west of Cerdanya, different permutations of ethnic, topographical, and political divisions can be found.

The Principality of Andorra, for example, situated next to the Cerdanya on the southern versant of the Pyrenees, is ethnically Catalan and enjoys the dual sovereignty of France and Spain. But in contrast to the Cerdanya, where France and Spain divided the valley territorially, the remarkable survival of the feudal division (*pariatges*) of 1276 gave the bishop of Urgell and the count of Foix (and his political descendants, the king of France and president of the Republic) shared sovereignty over the Principality of Andorra.[35] West of Andorra lies a clearly defined axial zone separating Pallars (Spain) and Couserans (France): here the topographical division coincides with distinct ethnic and linguistic groups and political membership.[36] West of Pallars, however, the Valley of Aran is geographically a part of France, linguistically and culturally related to Gascony, but politically dependent from the early medieval period on the comital dynasty in Catalonia—although continuously disputed in the later Middle Ages between the Aragonese and the counts of Comminges and, later, France.[37]

The Cerdanya, then, represents one permutation of a more general structure of intersecting relations among ethnicity, topography, and politics. Because the political division runs through the center of the plain, on one hand, the Cerdanya might be compared with large sections of France's northern boundaries, where no single topographical feature serves to mark the division between France and its neighbors. The existence of the Spanish enclave of Llívia in the French Cerdagne would also seem to link the valley to France's northern and eastern frontiers, where jurisdictional and territorial enclaves appeared as the norm, at least until the late eighteenth century; in the Pyrenees, Llívia was the only such enclave.[38] On the other hand, the northern and eastern boundaries were essentially mobile from the seventeenth to the nineteenth centuries and can thus be distinguished from the Pyrenean boundary—even if the evolution of territorial sovereignty more generally passed through the same stages.[39]

Topographical considerations would suggest the Alps as an inevitable site of comparison: a natural frontier, like the Pyrenees, the

35. J. Descheemaeker, "La frontière pyrénéene de l'Océan à l'Aragon" (Thèse de droit, Université de Paris, 1945); and the works cited by J. Armengol, M. Batlle, and R. Gual, *Materials per una bibliografia d'Andorra* (Perpignan, 1978).

36. L. Birot, *Etude comparée de la vie rurale pyrénéenne dans le pays de Pallars et de Couserans* (Paris, 1937).

37. J. Regla Campistol, *Francia, la corona de Aragon, y la frontera pirenaica. La lucha por el valle de Aran, siglos XII–XV*, 2 vols. (Madrid, 1951); and Ferro, *El dret públic*, 18.

38. On the problems surrounding the four extant groups of enclaves in western Europe—Llívia, Busingen (West Germany and Switzerland), Campione d'Italia (Italy and Switzerland), and Baarle (thirty-eight enclaves between Belgium and Holland)—see H. M. Catudel, *The Exclave Problem of Western Europe* (University, Ala.: 1979).

39. N. Girard d'Albissin, *Genèse de la frontière franco–belge: Les variations de limites septentrionales de la France de 1659 à 1789.* (Paris, 1970); F. Lentacker, *La frontière franco–belge: Etude géographique des effets d'une frontière internationale sur la vie des relations* (Lille, 1974); and L. Gallois, *Les variations de la frontière française du Nord et du Nord-Est depuis 1789* (Paris, 1918).

Alps contain many regions that resemble the Cerdanya—including one group of valleys, the Barcelonetta, divided between France and Savoy by the Treaty of Utrecht in 1714.[40] But the Cerdanya can be distinguished from the Alpine frontier, and brought closer to the experience of central and eastern Europe, by the existence of an ethnocultural group spanning the political division of two national states.

The possibility of comparisons between the political boundary and borderland in the Cerdanya and elsewhere in France and Europe suggest the utility of focusing on this single case in depth. For the Cerdanya is in many ways exemplary. Because the division of the valley into two parts occurred at a precise moment and has since remained in place (although it has changed its character dramatically), the Cerdanya can serve to demonstrate the long-term effect of a single political boundary. The political division of an ethnic group, neither French nor Spanish, makes the problem of nation building all the more salient. The division of the valley, moreover, occurred precisely at the moment when France and Spain began to move, at different speeds and in different ways, toward unified, territorial nation-states. The Cerdanya presents itself as a set of laboratory conditions in which to examine the dual process of state formation and nation building in France and Spain.

ORGANIZATION OF THE BOOK

This study traces the emergence of territory and identity in the Cerdanya valley since the seventeenth century from the double perspective of the states and of local society. "State" and "local society" are abbreviations for different configurations of social and political groups which, acting out of private or collective interests, constructed the boundaries of territory and identity in the Cerdanya. The "state" includes, at different historical moments or simultaneously, ministers and kings in Versailles and Madrid; provincial authorities in Perpignan and Barcelona; and local judicial officers, tax collectors, customs guards, and soldiers in the Cerdanya itself. "Local society" refers to the classes

---

40. P. Sopheau, "Les variations de la frontière française des Alpes depuis le XVIe siècle," *Annales de géographie* 3 (1893–1894): 183–200; S. Daveau, *Les régions frontalières de la montagne jurassienne: Etude de géographique humaine* (Lyon, 1959); R. Désorges, "Le Briançonnais, une région frontière de montagne" (Thèse de droit, Université de Grenoble, 1956); and R. Tardy, *Le pays de Gex: Terre frontalière* (Lyon, 1970).

of landless peasants and small property owners, many of whom survived by contraband trade, annual migration from the valley, or domestic industries such as knitting stockings; to wealthy landowners, most of them members of the privileged classes, resident in Puigcerdà and Llívia; and to the corporate village communities, which made up the institutional cells of social and economic life. Each of these social configurations had different interests and subsequently a different relation to the problem of the boundary, and each shaped its identities accordingly. If there is a single history to the boundary and the borderland, it must take into account this multitude of voices, many of which can barely be heard.

The distinction of "state" and "local society" in the Cerdanya might be thought of as a quadrant with two axes. A horizontal axis divides the political and civil orders. Intersecting this axis is a vertical one distinguishing the territorial and national identities of France and Spain. The quadrant is complicated, however, by the existence of the distinctive identity of Catalonia. Despite the decline of the institutional and juridical framework of the medieval Catalan state, Catalonia continued to represent a linguistic and ethnic unity, a unity that persisted well after the political division of Catalonia and the Cerdanya between France and Spain. After 1659, the Cerdans, political subjects of the French and Spanish kings and eventually citizens of France and Spain, remained Catalans: the ethnic and linguistic identity of the Catalans stood in no *necessary* relation to their adopted national identities, at once imposed from above and created from below within the political framework of the French and Spanish states.

This book is organized in a way designed to highlight the shifting relations of state and society over time. It is cast chronologically, treating the principal events of French and Spanish history and international relations as they shaped, and were occasionally shaped by, local life in the Catalan borderland. Chapter 1 examines the Treaty of the Pyrenees and the division of the valley; chapter 2 treats the frontiers of the Old Regime state. Both chapters consider the Cerdanya from the state's perspective, paying little attention—in the same way that the states paid little attention—to local life in the borderland. Chapters 3 and 4 shift perspectives, focusing first on the resistance of local society to the state, and second, on the uses made of the boundary and of political membership by the privileged classes and peasant communities of the valley. Chapter 5 draws these two perspectives together in an account of the French Revolution, suggesting that it was in this critical period that the perspectives of state and society began to merge. Chapter 6 considers the Spanish crises of the Old Regime in the borderland during the early nineteenth century, developing both the political implications (the politicization of the national boundary) and the local ones (the nationalization of local identities). Chapter 7 treats the formal delimitation of the boundary during the Treaties of Bayonne and the problem of economic and cultural difference in the valley. The conclusion of the book is followed by an epilogue that brings the history of the borderland up to the present, considering the implications of the development of Catalan nationalism in Spain and political and economic integration in France on national identities in the borderland.

1
The Treaty of the Pyrenees and the Division of the Cerdanya

Between August and November 1659, on an island in the Bidassoa River near Bayonne, the first ministers of France and Spain, Cardinal Mazarin and Don Luis de Haro, negotiated and signed the Treaty of the Pyrenees. The following June, in an elaborate ritual of royal alliance, the Spanish Infanta, María Teresa, was presented to her husband-to-be, Louis XIV. Although the ceremonies surrounding the negotiation and the alliance were not excessive by seventeenth-century standards, the symbolic encoding of space reveals much about contemporary ideas of boundaries. The Bidassoa River formed the division of France and Spain. The two ministers chose their meeting site on the Isle of Pheasants, which was declared neutral (see illustration 1).[1] Cardinal Mazarin explained the proceedings:

> Without further delay we had bridges built to link each side of our island [with the mainland], and plan to build equal lodgings, and a large room at the head of the island equidistant from the two lodgings, in which there will be two doors, one on his side and one on mine, by which we can enter, each holding rank in the chairs which will be prepared for us on each side of the room, which we will take care to build and to furnish, each one his own half.[2]

1. G. Gualdo Priorato, *Histoire du traité de la Paix conclue à Saint Jean de Luz entre les deux couronnes en 1659* (Cologne, 1675), 46.
2. *Lettres du Cardinal Mazarin où l'on voit le secret de la négociation de la Paix des Pyrénées* (hereafter *Lettres Pyrénées*) (Amsterdam, 1690), no. 8 (to War Minister Le Tellier, August 5, 1659).

25

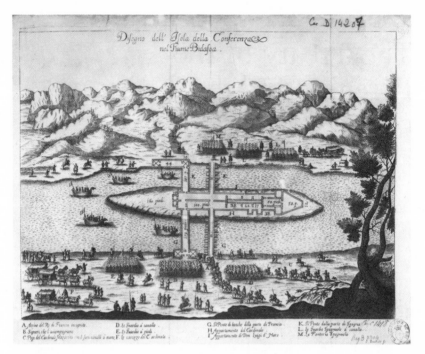

Illustration 1. "Isle of Pheasants (Isle of the Conference)," with the pavilion constructed for the negotiations between the French and Spanish ministers. This Italian engraving of 1678 shows the arrival of Cardinal Mazarin with his entourage in August 1659. Although it includes the French king traveling incognito, the young Louis XIV in fact only came to the Bidassoa the following June to celebrate his marriage to María Teresa, daughter of Philip IV, on the same site.

The symbolic equality still gave rise to competition between the two nations once the ministers and their entourages began to meet, and again during the marriage ceremony itself.[3] More important, the ceremonial symmetry belied French strength at the bargaining table, as France, victorious on the battlefields, dominated the negotiations.

Yet these rites of diplomacy did not simply mask power; they expressed a developed consciousness of the history and territorial boundaries of royal sovereignty. Mazarin and Haro knew that the Isle of

3. Ibid., no. 9 (to Le Tellier, August 10, 1659); and François Colletet, *Journal contenant la relation véritable et fidèle du voyage du Roy* (Paris, 1659), and *Entrevue et conférence de son Eminence le Cardinal Mazarin et Dom Luis d'Aro, le 13 août, 1659* (Paris, 1659), both in Houghton Library, Harvard University, FC 6 L9297 W 660 M vol. 7 (1) and vol. 14.

Pheasants was "five hundred steps" from the site where the Habsburg and Bourbon dynasties had consecrated a double marriage alliance in 1615. It was on these same banks that Louis XI had met Henry IV in 1463, and that Francis I was exchanged for his two children held hostage in 1526. In 1565, furthermore, Charles IX and his mother Catherine de Medici came to meet her daughter the Spanish queen near, though not on, the island.[4]

Indeed, meetings of royal persons and their first ministers at the frontier formed part of a long-standing European tradition. The seventeenth-century Italian historian of the Pyrenees Treaty, Gualdo Priorato, remarked that it "has always been the custom to build pavilions or buildings on the confines of each country." At least since the tenth century, kings and counts had met at their boundaries to settle their differences, seal alliances, or pay respect and hommage.[5] More generally—and more anciently—the primary function of kings in the Indo-European conception of royalty was not to govern, to wield power, but to take responsibility for the religious act of tracing boundaries. The philologist of Indo-European languages Emile Benveniste linked "the delimitation of the interior and exterior, the sacred and the profane kingdoms, national and foreign territories [sic]" to the regal function of establishing rules and defining "right."[6] The negotiation of the Treaty of the Pyrenees on the boundary of France and Spain in 1659 formed part of an archaic and enduring representation and function of sovereignty.

---

4. FAMRE MD 1970, fols. 8–9: "Cérémonial observé au Congrès des Pyrénées en 1659"; Mazarin, *Lettres Pyrénées*, no. 7 (to Le Tellier, July 30, 1659); see also J. Boutier, A. Dewerpe, and D. Nordman, *Un Tour de France royal: Le voyage de Charles IX, 1564–1566* (Paris, 1984), 96–98.

5. Gualdo Priorato, *Histoire du traité*, 46; Lapradelle, *La frontière*, 227–229, lists ceremonial encounters of kings at their frontiers since the tenth century; see also Lemarignier, *Recherches sur l'hommage en marche*, passim; and B. Guenée, "Des limites féodales, in Nora, ed., *Les lieux de mémoire*, 2 (pt. 2): 11–33. Similarly in rural society, representatives of peasant communities held ritual encounters at their boundaries, giving symbolic expression to their territorial extensions: see Dion, *Les frontières*, 23–32. The annual meeting of the syndics of Roncal and Barétous, valley communities on opposing sides of the watershed in the western Pyrenees, was described in the seventeenth century by the Béarnese-born Pierre de Marca, who was later to play a critical role in establishing French claims to the Cerdanya: see his *Histoire de Béarn* (Paris, 1640), 554. Its more recent political transformations are described by M. Papy, "Mutilation d'un rite: La junte de Roncal et Barétous et la crise du nationalisme français dans les années 1890," in J. F. Nail, *Lies et passeries dans les Pyrénées* (Tarbes, 1986), 197–223. For a similar but more prosaic example of boundary encounters in nineteenth-century Cerdanya, see below, chap. 4 sec. 3.

6. E. Benveniste, *Le vocabulaire des institutions indo-européennes*, 2 vols. (Paris, 1969), 2: 14–15.

In 1659, Cardinal Mazarin and Don Luis de Haro, as plenipotentiaries of their respective kings, did not simply respect a preexistent
boundary. Rather, by their presence they *created* the territorial division of the two kingdoms of Spain and France. For in their absence,
there was nothing to define the precise territorial boundaries of their
lands—no boundary stones, no line of soldiers, no customs guards. The
boundary line, as the permanent expression of territorial sovereignty,
only made its appearance in the nineteenth century. The seventeenth-
century state was not, strictly speaking, a territorial state: it was structured instead around "jurisdictions."

The exercise of jurisdictional sovereignty can be defined in three
different contexts. First, jurisdictional sovereignty was above all a relation between king and subject (*regnicole* in France). The early modern
state inherited the medieval practice of giving precedence to this political bond over the territorial one. Early modern "citizenship," defined
in relation to the king, was symbolically affirmed in the oaths of loyalty and allegiance by individuals and corporate groups.[7] Second, jurisdictional sovereignty involved a form of administration that gave
precedence to jurisdiction over territory. Generally conceived, sovereignty consisted of the exercise of authority within a wide range of domains: military affairs, justice, ecclesiastical policies, commercial and
economic activities, and taxation. Each of these jurisdictional domains
was an administrative circumscription with its own boundaries; these
boundaries often failed to coincide, and they remained distinct from the
political boundary of the kingdom.[8]

Finally, jurisdictional sovereignty meant that throughout the Old
Regime, kings and princes ceded or acquired, in war as in diplomatic
settlement, specific political jurisdictions and rights to domains—
among them fiefs, bailiwicks, counties, bishoprics, seigneuries, towns,
or even villages, mixing purely "feudal" forms of dominion with administrative circumscriptions of distinct origin.[9] Seventeenth-century
treaties rarely concerned themselves with specifying territorial bound-

7. B. Guenée, "Etat et nation en France au Moyen Age," *Revue historique* 237
(1967): 25–26; see below, chap. 2.

8. See, for example, A. Brette, *Les limites et divisions territoriales de la France* en
1789 (Paris, 1907); and the sources listed below, chap. 5, n. 1.

9. On the "feudal" nature of the state's "territory" in the early modern period, see
Lapradelle, *La frontière*, 37; Febvre, "*Frontière*," 213; and Clark, *The Seventeenth Century* (New York, 1961), 140–143; but cf. N. Girard d'Albissin, "Propos sur la frontière,"
Revue historique du droit français et étranger 47 (1969): 390–407, who correctly distinguishes jurisdictional circumscriptions from purely "feudal" forms of suzerainty.

aries in their major clauses. Thus articles 42 and 43 of the Treaty of the Pyrenees gave the County of Roussillon to France and the County of Cerdanya to Spain. But this customary language was complicated— and indeed, contradicted—by a more territorial conception of the kingdom's boundaries, one defined by the historical and natural frontier formed by the Pyrenees. Article 42 of the Treaty proclaimed that "the Pyrenees Mountains, which anciently divided the Gauls from the Spains, shall henceforth form the separation of the two kingdoms." This tension between the jurisdictional and territorial conceptions of sovereignty informed the lengthy and bitter diplomatic debates in the year that followed the Treaty itself.

FRANCE AND ROUSSILLON

Most of the 124 articles of the Treaty of the Pyrenees were not concerned with the Pyrenees at all, but with princely alliances, commercial agreements, and the cession of jurisdictions along the French frontier of the Spanish Netherlands and the Franche-Comté, where the major battles in the Bourbon–Habsburg phase of the Thirty Years War had been fought. French foreign policy, largely the creation of Richelieu, was intended to stop the perceived "encirclement" of France by a universalist Habsburg empire and to allow France to intervene in the affairs of the German states; the Count Duke of Olivares, Richelieu's contemporary and counterpart, was concerned to assure the unity of the two branches of Habsburgs, and with alliances along the eastern frontiers of France that would assure control over the strategic "Spanish road" to Flanders.[10] Throughout most of the seventeenth century, France and Spain fought their major military and diplomatic struggles far from the Pyrenees. But in the late 1630s, unrest in Catalonia brought the theater of war to the frontier of the two monarchies. In the two decades of warfare and diplomatic struggles that followed, France and Spain gave to their frontier its definitive, though still crudely defined, shape. Until the delimitation treaties of the 1850s and 1860s, there were to be no more permanent acquisitions of territory on either side of the Pyrenees.

10. G. Zeller, "Saluces, Pignerol, et Strasbourg: La politique des frontières au temps de la prépondérance espagnole," *Revue historique* 193 (1942–1943): 97–110; W. F. Church, *Richelieu and Reason of State* (Princeton, N.J., 1972), 294–300; and J. H. Elliott, *Richelieu and Olivares* (Cambridge, Eng., 1984), 119–125.

The Treaty of the Pyrenees ended a secular phase in the political formation of the French–Spanish frontier: it marked the final cession and acquisition of territories and jurisdictions by France and Spain along the Pyrenean frontier. The medieval states that flourished in the Pyrenees had rarely recognized the mountain range as a boundary; to the contrary, these polities had spanned both watersheds. The crowns of Navarre and Aragon, at the extremities of the chain where the "axial zone" of the Pyrenees was not itself well defined, both developed as trans-Pyrenean entities.[11] Since the thirteenth century, however, the monarchical ambitions of France challenged the kings of Castile and Aragon repeatedly for control of these Pyrenean states.

In 1258 the question of the eastern Pyrenees appeared settled when Louis XI renounced claims to the Counties of Roussillon and Barcelona, while the King of Aragon renounced his claims to the Viscounty of Carcasonne and the County of Toulouse—thus recognizing the Corbières ridge as the division of Aragon and France. In the early fifteenth century, another diplomatic settlement resolved the dispute between the king of Aragon and Philip IV of France over the Aran valley, situated on the French watershed but definitively attached to the crown of Aragon. The dismemberment of the kingdom of Navarre began in 1512 with the Spanish conquest of a portion of the territory and ended in 1620 with the French annexation of Béarn. The 1659 Treaty of the Pyrenees, which incorporated Roussillon and part of Cerdanya, completed this first phase in the historical *creation* of the Pyrenean frontier. By 1659, the idea of the Pyrenees as a mountain range that divided France and Spain had come into being.[12]

France's annexation of the Roussillon had distant origins in this process, yet it had not been *consistently* the object of French foreign policy. Between 1462 and 1492, during the revolt of the Catalans against the kings of Aragon, Louis XI twice conquered and annexed the two northern counties of Roussillon and Cerdanya, which he held as security against the payment of a promised sum of 300,000 écus. Though the money was never paid, Louis, on his deathbed, returned the counties to Aragon. During the wars between Charles V and Francis I in Italy, and later as the Valois–Habsburg rivalry shifted north to the

---

11. J. Regla, "La cuestión de los Pirineos a comienzos de la Edad Moderna," *Estudios de historia moderna* 1 (1951): 1–32; and the still useful, if not always reliable, J. E. M. Cénac Montcaut, *Histoire des peuples et des états pyrénéens,* 5 vols. (Paris, 1860).

12. R. Plandé, "La formation politique de la frontière des Pyrénées," *Revue géographique des Pyrénées et du Sud-Ouest* 9 (hereinafter *RGPSO*) (1938): 221–242.

Low Countries, Perpignan was besieged twice (1542 and 1597), but the French crown did not launch a full-scale invasion of the two counties until the events set off by the momentous year of 1640.[13]

Catalonia rebelled in 1640 in order to preserve its traditional liberties and institutions, as guaranteed by the terms of the union of the crowns of Aragon and Castile. The Count Duke of Olivares' attempt to engage the Catalans in the defense of Spain within the renewed warfare of Spain and France formed the catalyst of the revolt and the context of the French conquest and annexation of the Roussillon and Cerdanya.[14] The two counties had expressed a desire for independence from the principality in repeated political struggles during the 1620s but had no significant interest in being joined to the French crown. Richelieu himself wanted to establish Catalonia as an independent republic under French protection, but this proved impossible, and the political leadership in Barcelona signed treaties in 1640–1641 which instead placed the principality and the two counties under the suzerainty of the kings of France, "as their ancestors had done in the times of Charlemagne."[15] Only after a series of French failures to secure the coastal strongholds of Roses and Cadaqués from the Castilian armies did the French seriously consider the conquest and annexation of Roussillon. To underline the importance of the operations, Louis XIII himself assumed command of the 1642 siege of Perpignan, capital of Roussillon. The city fell in October 1642, and the acquisition was definitive. "That which is conquered by the sword," wrote the first minister after Perpignan's fall, "cannot be returned."[16]

Richelieu justified his conquest by the force of arms, but military prowess alone would not justify the annexation once the diplomats sat down at the bargaining table. After the failure of France and Spain to negotiate a peace settlement at the Münster conferences in 1648,

13. On the fifteenth-century revolt of the Catalans and the French intervention, see J. Calmette, *La question des Pyrénées et la marche d'Espagne au moyen age* (Paris, 1925); on the sixteenth century, see J.-L. Blanchon, "La Cerdagne devant la rivalté franco–espagnole au XVIe siècle," *Conflent* 29 (1965): 7–24.

14. The essential study of the origins of the 1640 revolt in Catalonia is Elliott, *Revolt of the Catalans*; for a detailed chronology of the events of warfare and diplomacy during the revolt, see J. Sanabre, *La acción de Francia en Cataluña en la pugna por la hegemonía de Europa, 1640–1659* (Barcelona, 1956).

15. Sanabre, *La acción,* 221–222; Richelieu's agent in Catalonia, Bernard Duplessis-Besançon, however, noted shortly after Catalonia's submission to Louis XIII that "the most solid advantage that the king can draw from the uprising in Catalonia is the conquest of the Roussillon": *Mémoires,* ed. H. de Beaucaire (Paris, 1892), 150; see also C. Vassal Reig, *Richelieu et la Catalogne* (Paris, 1935).

16. Quoted in C. Vassal-Reig, *La prise de Perpignan* (Paris, 1939), 69.

Hughes de Lionne was sent by the French court in the summer of 1655 to negotiate secretly in Madrid with the Spanish first minister, Don Luis de Haro.[17] In the midst of these talks, which ultimately failed, Lionne overstepped his authority and offered to exchange the Roussillon for Artois. The young Louis XIV soundly rebuked him and, in a memoir apparently penned by Cardinal Mazarin, he argued for a kind of "proprietary dynasticism":

> You cannot be unaware that the counties of Roussillon and Cerdagne are part of the ancient patrimony of my crown, that the Spanish crown used an illegitimate claim to deprive me of them, and since God enabled me to recover them by force of arms, I shall never renounce my rights.[18]

The construction of a historical discourse justifying conquest and annexation was an increasingly common practice in seventeenth-century statecraft, especially in France. The historical claims to annexed territories, however, did not end with vague incantations about ancient patrimonial rights. Beginning in the 1630s, such erudite researchers as Cassan, Bret, Dupuy, and Godefroy in France were given royal commissions to track down innumerable charters and titles—especially to those in frontier regions—many of them long forgotten in the Trésor des Chartes. The published results of this research evoked, as a matter of course, the tenets of "fundamental law" and the prohibition of alienating the royal domain. Historical erudition also devoted much effort to establishing dynastic claims through detailed researches into past royal alliances and into the customary laws of the territory in question which would give the French throne a claim of inheritance, as for example in the French apologist Pierre de Caseneuve's "Rights of France over Catalonia," commissioned in 1637 and published in 1644.[19]

17. J. Valfrey, *Hughes de Lionne: Ses ambassades en Espagne et en Allemagne; La Paix des Pyrénées* (Paris, 1881), 3–66; A. Morel-Fatio, ed. *Receuil des instructions données aux ambassadeurs et ministres de France*, vol. 11, no. 1: *Espagne, 1649–1700* (Paris, 1894), chap. 4. On the failure of France and Spain to reach a settlement at Münster, see J. Castel, *España y el tratado de Münster (1644–1648)* (Madrid, 1956); and Sanabre, *La mutilació*, 39–46.

18. FAMRE MD Espagne, vol. 54, fols. 4–6: Louis XIV to Lionne, August 16, 1656; and ibid., fols. 51–59: "Divers partis à proposer touchant le Roussillon," August 6, 1656, in which the king underscored his "total inflexibility" to never let go of the County of Roussillon, "if only for what the Spanish king holds in the Kingdom of Navarre." On notions of the right to the kingdom as a dynastic property right, see H. Rowen, *The King's State: Proprietary Dynasticism in Early Modern France* (New Brunswick, N.J., 1980).

19. Church, *Richelieu*, 349–371; G. Zeller, "La monarchie d'Ancien Régime et les frontières naturelles," *Revue d'histoire moderne* 8 (1933): 316–318; Caseneuve, *La Catalogne française, où il est traité des droits du Roy sur la Catalogne* (Toulouse, 1644);

It might seem that such historical erudition would serve only as a public justification of what was already accomplished. "The King having conquered Lorraine," wrote Dupuy about the Trésor des Chartes, "the Sieur Godefroy was sent to Nancy to consult the titles and charters of the country." But historical claims often informed and anticipated foreign policy objectives, even highlighting their significance. In 1630, for example, Dupuy wrote the "Rights of the King to the Crown of Aragon and Navarre," which established the possibilities of female inheritance in Aragon and a fifteenth-century marriage beween the house of Foix and the kings of Aragon.[20] More important, such a use and abuse of history was to enter into the diplomatic negotiations of the Treaty of the Pyrenees themselves, even to the point of redefining military strategies and giving new claims to the French crown.

In June 1659, Cardinal Mazarin and Don Antonio de Pimentel signed a preliminary peace treaty in Paris. The treaty left unresolved a number of minor points, and when in August 1659 Cardinal Mazarin and Don Luis de Haro began their negotiations on the Isle of Pheasants, there were "five or six" issues that remained to be discussed.[21] The most debatable of these turned out to be the status of the Conflent, the district covered by the upper Tet Valley between the plains of Roussillon and Cerdanya. The exact status of the Conflent had been left open by the Paris treaty: Mazarin insisted that it was a "dependent and inseparable annex" of Roussillon, but Pimentel refused to have the Conflent named in the text, "not having enough information on it as to whether or not it is dependent on Roussillon."[22]

That information was quickly forthcoming. In early July, Haro requested from the Council of Aragon's archives a number of charters, titles, and notarial acts. The viceroy of Catalonia, in turn, ordered a historical treatise to justify Spanish claims to the Conflent. "Written in forty-eight hours," this learned dissertation made a convincing argu-

and the collection of memoirs in FAMRE CP Espagne, vol. 20, fols. 371–432v: "Mémoires concernant les droits de la France sur la Navarre, le Roussillon, la Catalogne" (1644).

20. J. Dupuy, *Traité touchant les droits du Roy* (Paris, 1655), 1005–1018 ("Du Trésor des Chartes du Roy") and 164–190 ("Des droits du Roy sur l'Aragon et la Navarre").

21. Mazarin, *Lettres Pyrénées*, no. 14 (to Le Tellier, August 21, 1659); see also Marqués del Saltillon, "Don Antonio Pimentel de Prado y la Paz de los Pirineos," *Hispania* 7 (1947): 24–124; and J. Regla, "El tratado de los Pirineos de 1659," *Hispania* 11 (1951): 101–166.

22. FAMRE CP Espagne, vol. 38, fol. 84: "Convention particulier fait entre les plenipotenciares des Ses Majestés Catholiques et Très Chrétien fait le même jour que le traité de Paix" (June 24, 1659).

ment that the Conflent, far from depending directly on the County of Roussillon, was a dependency of the Cerdanya. Citing dozens of charters, donations, and documents from the great abbeys of the Conflent—Sant Miguel de Cuxa, Sant Marti de Canigó—as well as the researches of local scholars such as the Rousillonnais historian Andreu Bosc, the memoir seemed to leave little doubt.[23] Indeed, when Haro presented his case to Mazarin, the French minister, originally so optimistic about acquiring the Conflent, became less sanguine. As Hughes de Lionne explained in a letter to Mazarin,

> As to our claims to the Conflent: frankly speaking, the King doesn't have the least claim there, not only because we do not have any rights, but also because Don Luis could justly refer to the Treaty signed in Paris, which we are using as a basis for this one. The Treaty, in fact, gives us no reason to claim it, since Don Luis has shown me that the Conflent is not a dependency of Roussillon, and we are wrong to seek it.[24]

In matters of historical erudition, it was one small victory for the Spanish negotiator. But despite the strength of Haro's historical argument, Mazarin refused to give up his claim to the Conflent. In order to make a case, he temporarily abandoned history and turned to geography instead.

THE POLITICS OF NATURAL FRONTIERS

When his historical documentation fell short, Mazarin argued that whatever its historical relation to the Cerdanya, the Conflent lay on "this side"—the French side—of the Pyrenees, thus making a conscious appeal to the principles laid down in the Madrid negotiations of 1656.[25] In 1656, while many issues divided Hughes de Lionne and Don Luis de Haro, the two men concurred in a conception of the Pyrenean chain that divided France from Spain and Roussillon from Catalonia. The exact relation of the Cerdanya to the "dividing crests" of the Pyrenees was more problematic. Instructions to the French plenipoten-

23. Archivo de la Corona de Aragon (hereafter ACA) CA leg. 231, nos. 33, 35, and 27: Documents and memoirs on the status of the Conflent; France, Archives Nationales (hereafter FAN) 21 Mi 202, fols. 67–94: correspondence of Haro and Philip IV on the Conflent during September 1659; and Arxiu Històric Municipal de Barcelona (hereafter AHMB) MS 184, fols. 65–72: "Discurso hecho en que se prueba con evidencia que Conflent es parte del condado de Zerdaña y no del Rossellon, hecho en 48 horas."

24. FAMRE MD Espagne, vol. 61, fol. 128 (Mazarin to Le Tellier, August 31, 1659); see also Sanabre, El tractat, 59–64.

25. Lettres du Cardinal Mazarin (hereinafter Lettres Mazarin), ed. M. A. Chéruel, 9 vols. (Paris, 1879–1906), 9: 365–368 (to Lionne, October 15, 1659).

tiary defined Louis XIV's rights to all of Roussillon and Cerdanya "which is understood to include all the territories and towns, places, lands, and seigneuries which are on this side [*au deça*] of the Pyrenees Mountains, on the French side." At another point he proved willing to cede to Spain "the dependencies of the Cerdanya situated on that side [*au delà*] of the Pyrenees."[26] In either case, the Pyrenees were believed to form a division of France and Spain that might or might not coincide with the boundaries of the jurisdictions in question.

The idea that mountains ought to form the limits of adjoining polities was a widespread dictum of geographical discourse during the seventeenth century. In his *Science de la géographie*—"the first attempt in France to elevate geography to the rank of science," according to François Dainville—Père Jean François argued that mountains serve

> as very strong walls and ramparts between kingdoms, sufficient to stop the progress of a conquerer and the armies of the enemy. Such are the Pyrenees between France and Spain, the Alps between France and Italy . . .[27]

Of course, the idea of natural frontiers was not a seventeenth-century invention. Throughout the medieval period, French publicists tended to focus on waterways as natural frontiers. In particular, the "four rivers"—the Rhône, the Saône, the Meuse, and the Escault—were deemed to form the eastern boundary of France, although official inquests and testimony of local inhabitants suggested that they were only approximations.[28] By the seventeenth century, the focus was on mountains, which played the dual role of protecting a kingdom and of limiting the ambitions of a prince. "God created the Pyrenees," said a member of the Aragonese Cortes of 1684, "to free the Spaniards from the French." But natural frontiers could also serve to keep the ambitions of a prince or of a nation within bounds. An article in the semi-official *Mercure Françoise* (1624) urged France and Spain to remain at peace, pointing out that the

26. FAMRE CP Espagne, vol. 35, fol. 150: "Mémoire des divers parties que M. de Lionne peut proposer touchant le Roussillon"; FAMRE MD Espagne, vol. 53, fol. 15: "Instructions du Roy," June 1, 1656.

27. Quoted in F. Dainville, *La géographie des humanistes* (Paris, 1940), 280.

28. Dion, *La frontière*, 79–85; N. J. G. Pounds, "The Origin of the Idea of Natural Frontiers in France," *Annals of the Association of American Geographers* 41 (1951): 146–157; and A. Lognon, "Les limites de la France et l'étendue de la domination anglaise à l'époque de Jeanne d'Arc," *Revue des questions historiques* 18 (1875): 446–448. On the question of the Meuse, where "learned" and "popular" opinion diverged, see Guenée, "Des limites féodales," 18–28; and an anonymous memoir of 1537 in France, Bibliothèque Nationale (hereafter FBN) MS Dupuy 472: "Actes et mémoires pour les limites de France," fol. 22.

Ocean, the Alps, and the Pyrenees were limits which nature has placed between the two belligerent nations to keep each one shut up in its own country.

Thus it was geographers, wrote the anonymous author of an atlas at the time of Richelieu, who "study what are to be the confines of the kingdom," in order to limit the ambitions of the prince.[29]

Such geographic claims were often invoked in conjunction with historical ones. Indeed, the idea of natural frontiers carried more weight within seventeenth-century political culture when qualified as history. An oft-cited passage from Cardinal Richelieu's *Political Testament* argued that

> it was the goal of my ministry to restore to Gaul the limits that Nature has traced for her, to render all the Gauls to a Gallic King, to combine Gaul with France, and where the ancient Gaul had been, to restore the new one.[30]

Over fifty years ago, the French historian Gaston Zeller dismissed the text because, like Voltaire, he correctly doubted its authorship. Indeed, the apocryphal *Testamentum Christianum* (Lyon, 1643), from which the phrase is drawn, was the work of Labbé, the Jesuit geographer and royal publicist. Zeller thus dismissed the idea of natural frontiers under the Old Regime as "mere journalism," the claim of "a few isolated voices and chimerical spirits." Historians today tend to follow this line of argument, claiming that seventeenth-century ideas of natural frontiers were not only cloaked in history but were inevitably justifications for other, "more questionable motives." Natural frontiers were but "an intellectual pretext in search of strategic positions."[31]

Zeller and others since have failed to consider the problem of France's natural frontiers other than the Rhine. The case of the Pyrenees suggests how natural and historical frontiers formed a substratum of belief and an image that helped to shape the idea of a unified state in seventeenth-century France, and even entered into the formulation of the crown's specific policies. More than masking military strat-

29. The Aragonese Cortes is quoted in H. Kamen, "The 'Decline of Spain': An Historical Myth?" *Past and Present* 81 (1978): 45; the *Mercure Françoise,* and the atlas, in Dainville, *Géographie,* 343 and 352.

30. Quoted in Zeller, "La monarchie," 312.

31. Zeller, "La monarchie," passim; and his "Histoire d'une idée fausse," *Revue de Synthèse* 11 (1936): 115–131. On the authorship of Richelieu's *Testament,* see L. André, ed., *Le testament politique du Cardinal de Richelieu* (Paris, 1947), Introduction. Recent dismissals of the idea of natural frontiers include Clark, *Seventeenth Century,* 148; Vilar, *La Catalogne,* 1: 175–176; and most recently, F. Braudel, *L'identité de la France,* (Paris, 1986), 1: 289–292.

egy, the idea of natural frontiers could actually determine it. For once introduced into the discourse of diplomacy, the idea took on a life of its own and could even set the agenda that might then be justified by military rationale, as we shall see. Ideas about geography, history, and strategy coexisted and shaped each other in different contexts. They appeared in no set hierarchical order of importance. Thus the Treaty of the Pyrenees succeeded in giving to France a boundary that was as much the result of the treaty-making process as it was the crown's original goal—a boundary whose historical and geographical indeterminancy was matched by its military uselessness.

In thinking about France's boundaries, seventeenth-century ministers and military officials came to rely heavily on maps. Henry II seems to have been the first French monarch who saw the possibilities of maps for the business of governing, but French cartography lagged well behind Spanish technical virtuosity. Only in the early seventeenth century, under Henry IV and Cardinal Richelieu, did royal engineers and topographers begin to undertake the mapping of France's boundaries.[32] Royal and commercial cartography of the late sixteenth and seventeenth centuries helped to diffuse the idea of natural frontiers. Before then, the only boundaries drawn were crude dashed lines, while highly exaggerated rivers tended to function as delimitations, as in the *Theatrum Orbis Terrarum* of Ortelius (1570). By the seventeenth century, however, mountain ranges came to suggest the ideal of political divisions marked out by nature. Contours were still practically unknown, and would remain so until the nineteenth century, and mountains tended to be drawn obliquely from afar in the perspective "from horseback" (*à cavalière*), although the seventeenth century witnessed the use of hatch marks and shaded angles (see illustrations 2 and 3). Until the very end of the seventeenth century, maps generally failed to distinguish provincial and state boundaries, portraying them indistinctly with dotted or dashed lines. But mountains often doubled as political boundaries, and it was not unknown for publishers to highlight in color the mountain ranges that served to separate different territories.[33]

32. R. Mousnier, *The Institutions of France under the Absolute Monarchy, 1598–1789: Society and the State*, trans. B. Pearce (Chicago, 1979), 689–694; D. Buisseret, "The Cartographic Definition of France's Eastern Boundary in the Early Seventeenth Century," *Imago mundi* 36 (1984): 72–80; and M. Pastoureau, *Les atlas français, XVIe–XVIIe siècles: Répertoire bibliographique et étude* (Paris, 1984).

33. On the symbols of early modern cartographers, see F. Dainville, *Le langage des géographes: Termes, signes, couleurs des cartes anciennes* (Paris, 1964), esp. 272; for brief surveys of cartographic representations of mountains, see J. P. Nardy, "Cartographies de la montagne: De l'édifice divin au bas-relief terrestre," in *Images de la montagne:*

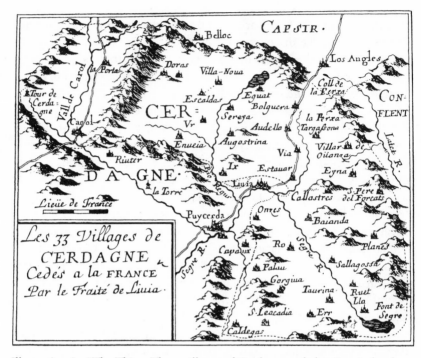

Illustration 2. "The Thirty-Three villages of Cerdagne ceded to France by the Treaty of Llívia," included in Pierre du Val's *Acquisitions de la France par la Paix* . . . (Paris, 1679). The representation of natural frontiers is typical of seventeenth-century cartographic techniques: the mountains are drawn as seen "from horseback"; the dotted lines are the same for provincial and for state boundaries; and the map shows a range of mountains dividing the Cerdanya, in the southeast, where in fact there are none.

Royal cartographers such as Nicholas Sanson and Pierre du Val, alongside many lesser-known figures, frequently portrayed France's boundaries as a set of natural frontiers, providing an image in the service of the Bourbon crown. More than propaganda, such maps were also used in classrooms, and especially by the Jesuits who trained so many future officers and statesmen. Identified as the natural frontiers of Gaul, the idea of the Rhine and the Pyrenees as France's ideal limits thus came to occupy a central place in the political culture of seventeenth-century France.[34]

In the mid-seventeenth century negotiations between France and

Catalogue et essais (Paris, 1984), 77–79; and J. Konvitz, *Cartography in France, 1660–1848: Science, Engineering, and Statecraft 1660–1848* (Chicago, 1987), 82–102.

34. P. Solon, "French Cartography and Bourbon Ambition," *Proceedings of the Tenth Annual Meeting of the Western Society for French History* (Lawrence, Kans., 1984): 94–102; on Jesuits and their use of maps and atlases, see Dainville, (Paris, 1980),

Spain, both kings and their ministers adopted what must have seemed a commonplace in asserting that the Pyrenees divided the two kingdoms. But although the idea was a central tenet of the Madrid negotiations in 1656, Pimentel and Mazarin seem to have entirely disregarded natural frontiers when, in the spring of 1659, they returned to the idea of jurisdictions. Article 36 of the provisional treaty, signed in June 1659, gave to Spain all dependencies and annexes of the Cerdanya, "in the mountains or wherever they may be found, although not named by the present treaty."[35]

But several months later, when Haro and Mazarin met on the Isle of Pheasants, the French minister reintroduced the notion of "dividing mountain crests" in defense of his failed historical proposition that the Conflent was a "dependency" of the Cerdanya. As the two ministers debated the issue, it became clear that they held very different ideas of where the Pyrenees lay. The problem, as Haro observed, was that the Conflent was over 125 leagues from both courts, and 80 from the site where the treaty was negotiated.[36] At the time occupied by French troops, its topography was not well understood. Mazarin, who appears to have first consulted a map only after three weeks of discussions, was less gracious about the reasons for their disagreement.

> I had Doctor Trobat come here this morning before going to the conference, and I examined the maps of Catalonia and the location of the County of Conflent and that portion of the Cerdanya which must remain ours. I find that the situation of the mountains is very confusing: that there are some that start at Leucate, and others between Roussillon and France, and between Conflent and Roussillon, which are not the height of the Pyrenees, which following the ancient geographers separated the Gauls from Spain. I cannot doubt that Don Luis . . . in obstinately saying that we shall only keep that which is on this side of the mountains, intends to pass off as mountains those which are not, and which strictly speaking are only hills.[37]

When the Spanish commissioner would not concede his vision of where the Pyrenees lay—and it appeared that the issue of the Conflent would

298–317. For a critique of the idea that maps had much influence at all on images of France before the late nineteenth century, see E. Weber, "L'hexagone," in Nora, ed., *Les lieux de mémoire*, 2 (pt. 2): 104–109.

35. Sanabre, *El tractat*, 51–67.

36. AHMB MS B 184, fol. 50: "Discurso geográfico, historico, y juridico sobre la división y limites de las Españas y de las Gallias," n.d., ca. September 1659.

37. Mazarin, *Lettres Pyrénées*, no. 82 (letter to Le Tellier, September 24, 1659). At a later stage in the negotiations, several manuscript maps were prepared for and used by Cardinal Mazarin and Pierre de Marca, the commissioner who negotiated at Ceret: see FBN MS 4240, fol. 346 bis: map of Llívia and the northern portion of Cerdanya; and below, n. 45.

Illustration 3. "Porton of Catalonia and Roussillon." This anonymous map of the late seventeenth century was copied from a military engineer's drawing of ca. 1685. The orientation from north to south suggests an idealized, Parisian perspective. The map exaggerates the narrow width of the valley, thus emphasizing its role as a passage across the Pyrenees, at the head of which the French military engineer Vauban ordered the construction of the fortress of Mont-Louis in 1679.

be left to commissioners—Mazarin insisted then on "certain words which I suppose will not be useless," namely that "the Pyrenees Mountains, which anciently divided the Gauls from the Spains," be used to define the limits of the two kingdoms (see text in app. A).[38]

In adding the phrase, Cardinal Mazarin was not neglecting the strategic interests of the French crown. Indeed, throughout the negotiations, royal ministers and military commanders were constantly preoccupied with military and strategic concerns; although they rarely described these publicly, they used strategic rationale to justify decisions made on the basis of other interests. For example, the French justified the conquest and annexation of Roussillon by stressing its strategic importance to the crown. Both the French and Spanish monarchies saw Roussillon and its capital, Perpignan, as a "key" to both offensive and defensive strategies. The French sought to acquire the Conflent since it was seen as a "key passage" to Perpignan, and its control could prevent the sacking of the city.[39] Mazarin tried to retain at least a portion of the Cerdanya because of its strategic role. The valley was "a passage" offering "free entry into Catalonia"; it was a rampart "that could guarantee possession of the Roussillon"; and it was a fertile plain, which could provide troops with sustenance. Pierre de Marca summed up the strategic importance of the valley:

> The French commissioners have seen the necessity of keeping the Conflent, the Cerdagne, and the Urgellet, not only to protect against the dangers of surprise, but also to solidify the new conquest of the Roussillon. The Cerdagne has the additional advantage of being able to supply an army, and to allow the penetration of the mountains on the side of Ribes and Tossa; the Urgellet gives [us] passages through these mountains via Oliana, to go south to Prats, Balaguer, and Lérida, or else toward Aragon, via the district of Tremp.[40]

Unlike geographers, when ministers and military strategists looked at mountains, they saw passages, not barriers (see illustration 3). Seventeenth-century strategists of mountain warfare went beyond the notion of a linear boundary, stressing the need to control key passes, and thus to conquer the mountain range itself. For France, the Pyrenees were not an effective barrier unless these passes were held. Nor was it simply a

---

38. FAMRE MD Espagne, vol. 61, fol. 232 (Mazarin to Le Tellier, October 7, 1659); see also *Lettres Mazarin*, 9: 365–368 (Mazarin to Lionne, October 15, 1659).

39. Sanabre, *Resistència*, 24–25; Sanabre, *La acción de Francia*, 589.

40. FBN MS 4309, fols. 111–122v: "Avis de Mgr. de Marca . . . touchant les limites des deux Royaumes," May 10, 1660; see also FBN MS 4309, fols. 6–8v: "Instructions aux commissaires," December 18, 1659.

question of keeping French access open: the entire pass had to be occupied, and with it the territory along its southern flanks. Such was the strategy of Duplessis-Besançon, Richelieu's agent in Catalonia, following the French–Catalan accord of 1641, when he sought to control the passes of the Pyrenees believed to form the boundary of Roussillon and Catalonia. Such was also Mazarin's strategy after 1652 when he ordered the conquest of the fortified towns on the southern flanks of the Albères and in the Cerdanya.[41] It was the conquest of natural frontiers which formed the basis of an effective military policy.

The strategic principle of conquering and controlling natural frontiers was neither isolated nor innovative. In 1462–1463, Louis XI effected the same policy in Roussillon and Cerdanya. As Zeller pointed out, French ambitions in the Alps during the sixteenth century included attempts to control the towns and fortresses of Saluces and Pinerola, on the Lombard side, thus giving France both a defensive and an offensive footing in Italy. As with mountains, so with rivers. Richelieu's attitudes toward the Rhine, for example, suggest the need to secure passages over the river, as at Breisach, with which to defend against the perceived Habsburg policy of encirclement.[42]

While negotiating the Treaty of the Pyrenees, the military concerns of the French crown remained implicit. More importantly, far from being masked by historical and geographical assertions, they were in fact contradicted by the adopted phrase that "the Pyrenees Mountains . . . shall henceforth form the division of the two kingdoms." The two ministers could agree to the phrase because they thought very differently about the topography of the Pyrenees. Article 42 of the treaty far from resolved their disagreement. It specified:

> If there be found any part of the County and Viguery of Conflent only, and not of Roussillon, on the side of the Pyrenees Mountains toward Spain, it shall remain of the Catholic Majesty; and likewise, if any part of the County and Viguery of Cerdanya only, and not of Catalonia, is found to be on the side of the Pyrenees Mountains toward France, it shall remain of the Most Christian King.

And in order to find out "which are the Pyrenees Mountains," commissioners appointed by the two crowns were to meet "in good faith" and to "mark the limits" of the two kingdoms (see text in app. A).

41. Sanabre, La acción, 547–576.
42. Zeller, "Saluces, Pignerol, et Strasbourg," 97–110; H. Weber, "Richelieu et le Rhin," Revue historique 239 (1968): 265–280; and L. Battifol, "Richelieu et la question d'Alsace," Revue historique 138 (1921): 161–200.

THE CERET CONFERENCES (SPRING 1660)

Between March 22 and April 10, 1660, the French and Spanish commissioners met in the village of Ceret, at the foot of the Pyrenees in Roussillon. The French commissioner Pierre de Marca, Archbishop of Orange, dominated the negotiations. Born in Béarn, a learned scholar and publicist of French ultramontane positions, Marca was appointed overseer in Catalonia through a special commission in 1644, where he served as a loyal and judicious servant of the French crown for over seven years. Assisted by Hyacinthe Serroni, Bishop of Toulouse, the French commission presented a forceful and well-prepared case at the conferences.[43] On the Spanish side, Don Miguel de Salvà i Vallgonera and Josep Romeu i Ferrer argued from a weaker position, both personally and politically, although the extent of their claims has, until now, been largely unknown.

At the center of the debate was the phrase from article 42 of the November 1659 treaty. Between their two copies the texts differed marginally in wording, but the negotiations were to make this difference substantial. The French text defined the Pyrenees Mountains as those "which anciently divided the Gauls from the Spains" while the Spanish text cited the Pyrenees as those "which commonly had always been the division of the Spains from the Gauls."[44] This discrepancy both reflected and informed the differing historical stances adopted by Marca and Salvà.

Pierre de Marca's claims drew first of all upon the "Ancients." In his opening discourse, he based the French claims on the territories historically occupied by the pre-Roman peoples (gens). Following Strabo, Pliny the Elder, and other Greek and Roman geographic treatises, Marca established that the "Cerretani" tribe was "in the Pyrenees" up to the limits of the "Basconi," the Basques. De Salvà countered with the same texts, emphasizing that the ancient Cerdans were to be found "in Spain." But the French commissioner rebutted with a text from

43. P. Torreilles, "Le role politique de Marca et de Serroni durant les guerres de Catalogne, 1644–60," *Revue des questions historiques* 69 (1901): 59–98; see also F. Gaguère, *Pierre de Marca, 1594–1662: Sa vie, ses oeuvres, son gallicanisme* (Lille, 1932).

44. French text: "les Monts-Pirénnées qui avaient anciennement divisé les Gaules des Espagnes"; Spanish: "les montes Pirineos que communemente han sido siempre tenidos per division de las Españas y de las Galias"; FAMRE CP Espagne, vol. 38, fols. 61–62 (Serroni to Mazarin, March 27, 1660); FBN MS 4309, fol. 26 (Marca to Le Tellier, March 27, 1660); see Regla, "El tratado," 120–121; and P. Torreilles, "La délimitation de la frontière en 1660," RHAR 1 (1900): 21–32.

Pliny, who divided the "Cerretani" into two groups: the "Juliani" had their capital of Julia Llívia (in the Cerdanya) and extended to the Seu d'Urgell, while the Augusti, with lands further to the south and east, were based in the Lérida plain. Since Strabo specified most of the Ceretans were Spanish ("Eas maiori ex parte tenent Ceretani Hispanica gens"), Marca deduced that the "Juliani" were in Gaul.[45]

The "Ancients" established one set of claims, the "Moderns"—medieval authors, such as Julian of Toledo—another. Marca thus turned from the historical anthropology of peoples to the administrative divisions of the late Roman empire. He cited authors who defined Gallic Septimania as including the entire Cerdanya as well as the Seu d'Urgell. By these texts, Marca argued that France had rights to what was commonly understood as the *vegueria ancha* ("wider viguery") of Cerdanya, including in addition to the valley itself the adjoining districts of Ribes and the Viscounty of Castelbò.[46] This was clearly a violation of the spirit if not the letter of article 42, as the Spaniards were quick to point out. Carried away by his own research, Marca had given shape to a new territorial claim at the same time that he legitimated it.

Don Miguel de Salvà insisted on the term "commonly" in the cited clause. He interpreted it to mean that which is "commonly held," that which "all those who live in these mountains understand as the dividing line [*linea divisoria*]." A little ethnography supported his position. Serroni, seeking to verify what was commonly believed, went to the Cerdanya and asked several shepherds, "Which were the Pyrenees?" He was told that the Pyrenees separated France from the Cerdanya, and that the other mountain chains were known by their proper names—Serra del Cadi, Puigmal, and so forth. Citing Pontificus Honorio on the truth of testimony by "the most simple and rustic," Salvà argued that "common belief" supported his position. As for the word "always," Salvà disagreed with the French derivation of the Latin *antiquitus*, "the most ancient," arguing that the Spanish *siempre* was understood as a continuous and uninterrupted state of more than a hundred years. His sources at this point were drawn from medieval canon law, and

45. *Marca hispanica sive limes hispanicus*, trans. into Catalan by J. Icart (Barcelona, 1965), bk. 1, chap. 12; FBN MS FR 8021, fols. 249–267: "Détermination des limites du Roussillon à la suitte du Traité des Pyrénées"; FAMRE Limites vol. 433, no. 4: "Mémoire sur les limites de la Cerdagne"; and the manuscript map of the Cerretani, "Confinium galliae narbonensis in qua regio sardonien sive Sardonia," by Joanne Rosset, addressed to Cardinal Mazarin, in FBN MS Colbert, vol. 479, fol. 336.
46. FAMRE Limites, vol. 433, no. 4: "Mémoire sur les limites de la Cerdagne."

his scholastic argumentation moved into the realm of metaphysical abstraction.[47]

In a more convincing argument, Salvà turned to the historical debates on the location of the "true Pyrenees." In the late fifteenth century, the Catalan humanist Joan Margarit had argued in reaction to Louis XI's attempts to annex the Roussillon that the northern Catalan county had always formed part of *Hispania*.[48] The debate about whether Roussillon and Cerdanya formed part of Spain continued beyond the context of French invasions. In the 1580s, the erudite lawyer from Vilafranca in Roussillon, Francesc Comte, wrote an unpublished chronicle of his native region, *Illustrations dels comtats de Rosselló, Cerdanya, i Conflent*. In this mythohistorical treatise, he offered a discussion of two shepherds, Jermens and Rosselló, the first guarding the herds of the French, the second of the "ancient Spaniards." Rosselló's arguments for the French origins and dependency and of the county of his name were solid enough, drawing as he did on such "modern" authorities as Florian de Ocampo and Ambrosi de Morales. But those of Jermens, founding his arguments with such "ancients" as Strabo, Pliny the Elder, Pomponius Mela, and Ptolemy, were more convincing. In the early seventeenth century, learned opinion—sometimes in defense of the autonomy of the "two counties" from the Principality of Catalonia—tended to affirm that Roussillon and Cerdanya were indeed part of Spain. Andreu Bosc's 1629 treatise in Catalan, *Sumari índex o epitome dels . . . títols d'honor de Catalunya, Rosselló, i Cerdanya*; Esteve de Corbera's *Catalunya illustrada* (written in the 1620s but published in Naples in 1678); and the Barcelona lawyer and chronicler Jeronio Pujades all agreed that the frontier of France and Spain lay along the Corbières, reaching the Mediterranean at Leucate.[49]

Salvà's history shared none of the mythico-fantastical elements of a Comte or a Bosc. But he could cite them to support his claim, as

47. AHCB MS B 184: "Discurso geográfico"; see also FBN MS 4240, fols. 246v–253: "Advis des commissaires espagnols touchant les limites," in Castilian, n.d.

48. J. Margarit, "De terra Ruscilionis an sit in Hispania," fol. 5v, quoted in R. Tate, *Ensayos sobre la historiografía peninsular del siglo XV* (Madrid, 1970), 142–143.

49. F. Comte, "Illustrations del comptats de Rosselló, Cerdanya, i Conflent," with a foreward by Esteva de Corbera, in Spain, Biblioteca Nacional (hereinafter SBN) MS 615; A. Bosc, *Sumari, índex o epítome . . .* (Perpignan, 1629); see J. S. Pons, *La littérature catalane en Roussillon au XVIIe et XVIIIe siècles*, 2 vols. (Toulouse, 1929), 1: 54, on the seventeenth-century debate over these boundaries.

well as such reputable sources as the official chronicler Zurita's *Anales de Aragon*. In advancing the argument that the "true Pyrenees" were the Corbières ridge, the boundary implied in the 1258 Treaty of Corbeil, Salvà was on weak ground indeed. The French annexation was already concluded; Mazarin upheld the clause of article 42, arguing that the commissioners were seeking "the ancient limits which divided the Gauls from the Spains, and not the historical frontier of the two crowns."[50]

As Salvà refused to concede on this point, the commissioners reached an impasse: Salvà's obstinancy was matched by Marco's blatant and unscrupulous manipulation of the historical facts. "In disputing with the Spanish commissioners," he wrote, "I invent many reasons to make them uncomfortable with things they held for certain." He even went so far as to claim that while France had no rights to the Cerdanya according to the articles of the 1659 Treaty, once *Spain* had introduced the historical definition of the Pyrenees as "anciently dividing the Gauls from Spain," this became the determining clause![51]

By early April, the two commissioners were even further from agreement than Haro and Mazarin had been in October, and more tenacious in defense of their respective arguments than in the "good faith" required by article 42. In fact, the November 1659 Treaty had provided for such a deadlock, and the commissioners deferred to their first ministers. Time was now of the essence. The royal marriage was to be celebrated shortly on the shores of the Bidassoa, and both kings insisted that it would not take place until all the issues were settled. In the weeks before the ceremonies of June 1660, as the two royal cortèges were steadily proceeding toward the limits of the two kingdoms, Don Luis de Haro and Cardinal Mazarin bitterly proceeded to come to terms.

The disgust of the Madrid court was so great that Haro went so far as to propose the indefinite adjournment of the royal marriage.[52] Mazarin quickly ordered Hughes de Lionne to Hendaye in order to keep the discussions going. The cardinal, suffering greatly from gout, himself arrived on the border in the middle of May and opened talks with Don Luis de Haro. With the cardinal on the scene, the discourse shifted

50. FAMRE CP Espagne, vol. 39, fol. 62 (Marca to Mazarin, March 27, 1660).

51. FBN MS 4309, fol. 96 (Marca to Mazarin, April 29, 1660); and FBN MS 4309, fols. 111–123: "Avis de Marca touchant les limites."

52. FAMRE CP Espagne, vol. 39, fols. 92–96 (Haro to Mazarin, April 12 and April 18, 1660).

levels: the debate moved away from historical and geographical claims to a series of reciprocal concessions. At first, the ministers ceded jurisdictions located at opposing ends of the Cerdanya valley: Mazarin renounced claims to the Urgellet and Ribas, while Haro reluctantly conceded the Conflent.[53] But the Cerdanya valley itself remained in dispute. "The commissioners are at a point of honor," wrote Baluze to Marca on May 25. "It is reduced now to a point of reputation," wrote Don Luis to his king. The French were no less intransigent than the Spaniards. "If we should stay here twenty years," wrote Mazarin, "this affair will not be settled without us holding on to part of the Cerdagne. It is useless to waste time if they claim to end this negotiation in any other way."[54]

And so it was. Day by day, between May 23 and June 1, concessions were made on each side. Now it was down to towns. Mazarin gave up his claims to Bellver and, after much hesitation, to the capital of the valley, Puigcerdà. On May 28, Haro agreed to cede a portion of the Cerdanya situated to the north of Puigcerdà, without naming villages, and he bitterly accused the French of duplicitously trying to mark the supposedly natural frontier in the middle of the plain.[55]

Having renounced all claims to the southern part of the Cerdanya, Mazarin still sought to retain a portion of the valley. Beyond the question of "honor," the French minister put forth a strategic rationale for his claims. He sought possession of the Valley of Carol, which contained the ruins of two fortified sites and controlled the Puigmorens Pass into the County of Foix. In addition, he sought French sovereignty of the villages along a "continuous passage" between the Carol Valley and the Capcir. In the end, this was the claim that resulted in the accord signed on May 31, 1660, as a clarification of article 42. Spain retained sovereignty over the County of Cerdanya, with the exception of

> thirty-three villages together with their jurisdictions, which should be composed of those in the said Valley of Carol and those found to be in the said passage joining Carol and Capcir; and if there are not enough villages in the said valley and the said passage, the number of thirty-three will be supplemented by other villages of the said county of Cerdanya to be found among the most contiguous.[56]

53. FAN 21 Mi 204 (Haro to Philip IV, May 24 and May 25, 1660).
54. FBN MS 4309, vol. 39 fol. 129 (Baluze to Marca); FAN 21 Mi 204 (Haro to Philip IV, May 24, 1660); *Lettres Mazarin*, 9: 614–616 (to Lionne, May 23, 1660).
55. FAN 21 Mi 204 (Haro to Philip IV, May 28, 1660).
56. FAMRE CP Espagne, vol. 39 fols. 158–169 (letters of Lionne to Mazarin, May 25 to June 1, 1660); FBN MS 4309, fols. 132–135 (patent letters of Louis XIV ratifying

The French commissioners, who by their political and personal powers had dominated the discussions, had long insisted on the natural and historical boundaries of France along the Pyrenees; they ended up with a frontier drawn through the center of the Cerdanya plain. The irony did not escape Haro, but he was unable to convince, and unwilling to be convinced, otherwise. The limits of these natural frontiers, then, were neither historical nor geographical but rather a compromise resulting from a bitter diplomatic struggle.

The outcome was to give France a foothold in the Cerdanya; hence, it could be argued, the French commissioners emerged victorious with France's military and strategic interests upheld. Yet those military interests, initially shaped by a conception of natural frontiers, had been reshaped in the debates themselves. Moreover, as we shall see, the military frontier in the Cerdanya only belatedly recognized the political division of the valley. Less an acquisition of strategic importance, France's "passage" through the Cerdanya was an empty justification of a series of failed historical and geographical claims, a face-saving victory of no real strategic significance.

THE LLÍVIA ACCORD (NOVEMBER 12, 1660)

But the irony was not yet over. The enumeration of those thirty-three villages constituted the final diplomatic confrontation of the Peace of the Pyrenees. The meetings between Don Miguel de Salvà and Hyacinthe Serroni, in Pierre de Marca's absence, began during the last days of July in Llívia, and it was the issue of Llívia itself which brought about the most discussion. On August 13, the French commissioner wrote to Mazarin:

> I have already been in this region fifteen days, where [Salvà] ... has disputed, as I had predicted, everything which was clear in the article as concerns the thirty-three villages; but with all that I have already succeeded in acquiring twenty-eight, which are effectively more than fifty, for the words 'villages with jurisdiction' which Your Excellency had added have won me ten, which are only counted as one, and I received four or five, which are counted as two.[57]

In fact, the twenty-eight "villages" represented a portion of the Cerdanya valley far greater than the "continuous passage" between the

the explication of article 42); FAN 21 Mi 205 (Haro to Philip IV, June 1, 1660). The text of the May 31, 1660, accord appears in app. A.

57. FAMRE CP Espagne, vol. 39, fol. 244.

Carol Valley and the Capcir. In the middle of these villages was the settlement of Llívia, ancient capital of the valley during Roman times and, in the seventeenth century, a large settlement of about 700 inhabitants. The French commissioner insisted that it be one of the villages given to France. Don Miguel de Salvà claimed that he had received assurances from Mazarin at the end of May that Llívia would remain Spanish, although in fact its status had never been formally discussed.[58]

But Salvà was determined at all costs to hold on to Llívia. To do so, he turned to the text of the May 31 accord, which stated that thirty-three *villages* could be ceded. (The Spanish text of the accord had translated the French word into a Castilian neologism, *villajes*.) The Spanish commissioner made his case based on a lexical difference of the French and Castilian languages. Claiming that Llívia was a *villa* (in Castilian), Salvà argued that it could not form one of the *villages* turned over to France. As Serroni explained to Mazarin,

> The name of *village* is altogether French; the Spanish language has no such word, and on this the Spanish commissioner has agreed. And when the Spaniards used the word [*villaje*] they borrowed it from us, and thus it ought to signify the same thing as in France, where under the name of *village* are included all the places which are not *cités*. And the same Spanish authors say that what is *ville* in France, is *cité* in Spain, and that which is *village* in France is *ville* in Spain. The Spanish commissioner admits that Llívia is not a *cité*: it should thus be included under the name of *village*, the French word, even though he insists on calling it *ville* in Spanish.[59]

Although Salvà asserted that Llívia was a *villa*, he could adduce little evidence to support the claim. Serroni gathered sworn attestations from the municipal council of Llívia that they had never held the privilege of *villa*, and added that the councillors did not participate in the "estates" of the realm, as did others in Spain who came from *villas*.[60] But Salvà only became more obstinate. Having lost the war, Roussillon, and a good part of the Cerdanya, Spain was determined to hold on to Llívia.

The discussions dragged on through September and October. The Spanish commissioner tried a dozen tactics and subterfuges. He offered

58. Ibid., fol. 265 (Memoir on Spanish claims, in Castilian, n.d.).

59. Ibid., fols. 263–264 (letter of September 10, 1660); see also FAN 21 Mi 17 (Philip IV to Spanish ambassador, Conde de Fuensaldaña, August 31, 1660).

60. FAMRE CP Espagne, vol. 39., fols. 264 (Serroni to Mazarin, September 10, 1660); ibid., fol. 275v (Serroni to Mazarin, October 8, 1660); and ACA CA 231, no. 33 (*consulta* of the Council of Aragon sent to Madrid, October 19, 1660). On the status of Llívia in the medieval period, see J. M. Font Rius, *Cartas de población y franquicia de Cataluña*, 2 vols. (Barcelona, 1969): 1: 301–368 and 637.

to exchange the thirty-three villages for certain territories in Flanders, but Serroni responded that "we are not accustomed in France to compromise." Salvà challenged the villages already ceded, claiming in his turn that "jurisdiction" ought to be understood as the "most ancient" and not the current "dependencies" of any given village. He reinterpreted the May accord, claiming that the word "contiguous" could be read as those villages next to the Carol Valley and tried to force Serroni to accept as French certain villages along the mountain flanks separating Andorra from Cerdanya. Finally, he left the Cerdanya, on the excuse of going to Barcelona to give his oath of loyalty to Philip IV.[61] These "metaphysics," as Serroni called Salvà's tactics, were matched by the directness of the French approach. Serroni offered to buy Llívia for 1,000 livres, which insulted the Spanish king and convinced him of the illegitimacy of French claims. The French commissioner then placed a garrison in Llívia. Although he argued that it was necessary to prevent uprisings against French authority by the peasants of the Conflent and Capcir, Serroni clearly intended to intimidate the Spanish commissioner.[62]

Once again, the discourse changed levels. In Paris, the Spanish ambassador raised the issue of Llívia directly with Cardinal Mazarin. Both courts instructed their separate commissioners to "remain firm," but by the end of October the two diplomats reached an agreement in Paris. The Count of Fuensaldaña admitted that Llívia was not a *cité* but that neither could it be a "village," since the French word translated as *aldea* in Spanish. Although not entirely convinced, Mazarin ceded, and Llívia remained part of the Spanish monarchy. "There is no further question of disputing the issue," wrote Mazarin to Serroni on October 29 and, on November 12, Serroni and Salvà signed the Llívia convention (text in app. A).[63]

Yet a final issue remained unresolved. In a letter of October 29, Mazarin instructed Serroni to include a clause by which Llívia, under Spanish sovereignty, would remain unfortified, "otherwise we shall have to fortify several gates on our side in order to guarantee passage and communication among the places which remain ours." In effect, the clause was inserted into the Llívia accord. In a memorandum to the

61. P. Torreilles, "La délimitation," 28–32; FAMRE CP Espagne, vol. 39, fols. 269–286, esp. fol. 277 (Serroni to Mazarin, October 15, 1660).

62. FAMRE CP Espagne, vol. 39, fol. 269v (Serroni to Mazarin, September 24, 1660); FAN 21 Mi 17 (Philip IV to Spanish ambassador, September 10, 1660).

63. FAMRE CP Espagne, vol. 39, fol. 286 (Mazarin to Serroni, October 29, 1660).

king, the Council of Aragon, supporting Don Miguel de Salvà's position, mistakenly argued that

> the French will not retain Llívia, and thus the condition of not fortifying Llívia is unimportant; in any case, it is not necessary. We are fortifying Puigcerdà, and the most important point is that [Puigcerdà] not remain French, since it is very important to fortify this site in order to have a foothold in the plain of Cerdanya.[64]

Writing to the viceroy of Catalonia in late November, Philip IV expressed his dissatisfaction. In May 1661, the king and Council of State ordered that the accord be renegotiated dropping the nonfortification clause. Although the French court apparently agreed to discuss the issue, six months later there was no sign of new negotiations. The question of Llívia passed away, and the king of Spain never formally ratified the Llívia convention of November 1660.[65]

The Llívia accord completed the division of the Cerdanya but it was not, strictly speaking, a territorial division. For even if the commissioners spoke of the "delineation of the frontier" and understood that "the line, which has to be almost mathematical, has necessarily to occupy a very narrow width,"[66] the conception of a dividing line remained vague and unspecific. The accord only specified that the boundaries of the villages ceded to France were to form the boundaries of the two kingdoms, except in the case of Hix. Although the village of Hix was situated on the northern bank of the Raour River, the village territory extended to the other side. Drawn to the possibilities of a natural frontier, the commissioners ended up dividing the territory of Hix between France and Spain:

> [T]he division of France and Spain will be taken as the said river following its natural course . . . until it enters and meets the territory of Aja, which shall remain of Spain. . . . This division does not separate the said territory of Hix from the village of Hix, as to that which concerns the seigneurie, property, fruits, pastures, or any other such relevant thing, so that this sep-

---

64. ACA CA 231, no. 33 (consulta of November 22, 1660).

65. ACA CA 231, no. 33: "Sobre la orden que VM ha resuelto en consulta de la junta de Estado," May 18, 1661; see also J. M. Guiler, *Unitat històrica del Pirineu* (Barcelona, 1964), 164–170. Spain's refusal to ratify the Llívia accord was frequently noted by Spanish officials in the following centuries. For example, in 1737, during a conflict between the intendancy of Roussillon and the captain-general of Catalonia over the taxation of disputed lands near Llívia, a certain Don José Ventura wrote that "it is common knowledge that Philip IV did not ratify it, which is also proven by the fact that this document was sought in 1703, and not found either in Barcelona or Madrid": Spain, Archivo Histórico Nacional (hereafter SAHN), Estado libro 673d, fols. 111–119.

66. AHCB MS B 184, fol. 2: "Discurso geográfico."

aration is understood as dividing Spain from France, and not the seigneurie or private property of the said territory, which shall always remain part of Hix.[67]

The final shape of the boundary was far from that which each side had envisioned during the course of the negotiations. Beginning in the fall of 1659, the ministers and commissioners of the two crowns had advanced a complex series of historical and geographical claims, which shaped and were shaped by their military and strategic interests. In the end, they settled on a boundary that was not a natural frontier, in which military considerations played a secondary role, and which ran counter to any conceivable history of the Cerdanya. The accord shattered the unity of the valley; far from the resolution of a diplomatic struggle, it marked the beginning of a lengthy historical process of constructing the political boundary of France and Spain.

CONSTRUCTING JURISDICTIONS IN THE BORDERLAND

The years immediately following the Treaty of the Pyrenees were marked by the normalization of juridical and political life in the borderland of France and Spain. There were two dimensions to this process of normalization, the one domestic, the other international. Domestically, each crown established its own set of political and administrative institutions: in Spain, this meant the reincorporation of that which remained of the County of Cerdanya into the institutions of the Principality of Catalonia;[68] in France, it meant the creation of new institutions for the annexed province of Roussillon. Internationally, the process of normalization involved the cooperation of the two crowns in sorting through the myriad claims and issues that had arisen after twenty years of warfare and political crisis in Catalonia. The Treaty of the Pyrenees outlined the settlements to be reached: article 55 allowed the Catalans—ecclesiastics and laity, nobles and commoners—to return to possession of their estates, annuities, and benefices in the Roussillon; article 57 specified the case of ecclesiastical benefices; while articles 58, 59, and 112 referred to the return of goods confiscated during the war.

67. See the complete text in app. A.
68. ACA CA leg. 220, no. 33. Instead of maintaining a separate governor and officers of the Royal Patrimony for the Spanish portion of the Cerdanya, the Council of Aragon, following the opinion of the Real Audiència and arguing the dictates of economy, urged Philip IV to incorporate the district into the institutions of the principality, which was done.

The treaty mandated the appointment of commissioners to resolve all disputes over properties, benefices, lands, and rents, many of which were in litigation or had already been reassigned by the French king. New commissioners thus met between 1665 and 1668 in the Catalan border town of Figueres to sort through the continuing claims, as well as a host of other issues concerning the problems of local relations in the borderland.[69]

Neither in their joint conferences nor in their separate policies did France or Spain seek to specify further the territorial dimensions of the French–Spanish boundary. It is true that when the governor of Puigcerdà ordered the hamlet of Sant Pere de Cedret to furnish its allotment of soldiers, the protests of the municipal councillors of La Tor de Carol reached the Figueres commissioners.[70] Sant Pere was not mentioned in the Llívia accord as one of the villages ceded to France, but the commissioners decreed it to be a "dependency" of La Tor de Carol—an issue that was to arise again, periodically, until the Treaties of Bayonne delimited and demarcated the boundary in 1866 and 1868. But this question of "dependency" did not concern a territorial boundary line, and neither commissioner appeared inclined to establish one.

To the contrary, the commissioners became embroiled in instances of contested jurisdictional sovereignty. Foremost among these was the fact that the viceroy of Catalonia and others in the Spanish court continued to use the title of "Count of Roussillon" when referring to Philip IV—long after the Spanish king had given up that jurisdiction. In 1661, the French court instructed the new French ambassador in Spain to request that the Spanish court drop the title. And over the next few years, the Council of Aragon politely advised Philip IV to stop calling himself Count of Roussillon.[71] The debate over the use of a title reveals the fundamental preoccupation of both monarchies: territorial boundaries remained unimportant compared to the boundaries of jurisdictional competency in the borderland.

69. Documentation on the Figueres conferences from the Spanish side is dispersed throughout ACA CA, including legs. 208, 234–235, 255–256, 314, 316, and 322. For the French side, see ADPO C 1368–1393, and the detailed analytical inventories by B. Alart, *Inventaire sommaire des archives départementales antérieures à 1790*, vol. 2 (série C) (Paris, 1877), 236–244; a chronological summary of the conferences may be found in C 1368, the first part of which was published in *Centre d'études et de recherches catalans* (hereinafter *CERCA*) 7 (1960): 151–159.

70. ADPO C 1361: "Information faite par le juge de Cerdagne," September 29, 1663.

71. Morel-Fatio, *Receuil des instructions*, 186–188: "Mémoire pour servir d'instruction" to the archbishop of Embrun, June 10, 1661; see also ACA CA leg. 231, no. 28 (consulta of May 20, 1662); and ACA CA leg. 6, no. 61 (consulta of February 17, 1662).

Since jurisdictional sovereignty was defined primarily as a relation between the king and his subjects, the ritualized oath taking was an important affirmation of royal sovereignty in each kingdom. The newly annexed communities of the Cerdagne gave their solemn and symbolic oath in June 1660—thus before the final accord—"to be a good vassal of the King our Most Christian Lord . . . and not to take any rewards from any foreign prince without the express consent of His Majesty."[72] This ritual affirmation was matched by the royal policies of consolidating that loyalty among the privileged classes of the provinces, the *gens du bien* on whose political loyalty depended that of the other inhabitants. Article 55 allowed the French crown to "prescribe the place of abode . . . appurtenances, and dependencies" of all those who proved "not acceptable to His Majesty," which in the Cerdanya meant several noble landowners and at least one holder of an ecclesiastical benefice.[73]

In the years following the Treaty of the Pyrenees, the French monarchy affirmed its jurisdictional sovereignty within four domains: the administration of justice, ecclesiastical affairs, economic and commercial activities, and finances. The boundaries of these jurisdictions failed to coincide, nor were they coterminous with the "division of France and Spain" stated by the Llívia accord.

The most important of these prerogatives was the administration of justice. In December 1659, Cardinal Mazarin ordered an archival inquiry to examine "the mode of justice and the form of government" practiced in the two counties of Roussillon and Cerdanya during Louis XI's fifteenth-century occupation.[74] It seems unlikely that these were helpful, for in the end the French crown did not rely on such historical precedents. Rather, maintaining a delicate balance between innovation and the preservation of local Catalan institutions, the French crown managed to incorporate Roussillon and part of the Cerdanya into a highly centralized polity while assuring the relative noninterference of the state in the relations of everyday life.

By the edict of Saint Jean de Luz in June 1660, Louis XIV abolished the administrative and judicial institutions of the Counties of Roussillon and Cerdanya, the royal Council of Catalonia, and the courts

72. FBN MS 4240, fols. 145v–150v: "Instruction . . . touchant la division du Royaume," December 10, 1659; and J. Sanabre, *La resistència del Rosselló a incorporar-se a França* (Barcelona, 1970), 94.

73. Regla, "El tratado de los Pirineos," 134–146; ACA CA leg. 317 and leg. 231, no. 36 (case of Don Pedro Rubi y de Sabater); and ADPO C 1390–1 (case of Pere Pont and the office of *chanoine archdiacre* of the Cerdanya).

74. FBN MS 4240, fols. 145v–150v: "Instruction."

of the Royal Patrimony at Perpignan. In their place, he instituted a streamlined Sovereign Council of Roussillon, which acted as an appellate court and the crown's central political authority of the newly annexed province. Offices of the Sovereign Council were held by appointment and were not for sale, suggesting the way in which the French monarchy experimented with new forms of state building on the periphery. The viceroy's jurisdiction over Roussillon was replaced by that of a military governor, a post staffed by the Noailles family. But the key institution in the crown's control of the province was the intendancy. Serving as the first president of the Sovereign Council, the intendant was appointed directly by the crown and held a virtually unlimited competency in fiscal, administrative, and political affairs.[75]

At the upper levels of the royal administration, the French crown introduced certain innovations; by contrast, political administration at the local level was marked by a profound continuity. The crown took over the office of *viguier* (*veguer* in Catalan) of the Cerdanya, the political and judicial officer at the head of the *viguerie* (*vegueria* in Catalan) and responsible directly to the viceroy. Previously situated in Puigcerdà, the veguer was gallicized and shifted to Sallagosa, making the office directly responsible to the intendant. (Throughout the Old Regime, the office of viguier of the French Cerdagne was held almost exclusively by the family dynasty of Sicart, one of the few families loyal to the French monarchy.) The crown also took over the royal jurisdictions and the prerogatives of "high justice" with regard to the seigneurial jurisdictions of the Cerdanya, which included the privilege of appointing bailiffs (batlles), local judicial officers directly responsible to the political administration.[76] In this way, the crown established a direct line of

75. The text of the edict appears in Sanabre, *La resistència*, 91–92. On the Sovereign Council, see G. Clerc, "Recherches sur le Conseil Souverain de Roussillon, 1660–1790: Organisation et compétence," 2 vols. (Thèse de droit, Université de Toulouse, 1973); and P. Torreilles, "L'organisation administrative du Roussillon en 1660," *RHAR* 1 (1900): 263–273. For comparative material on the establishment of royal justice in the "annexed provinces" during the later seventeenth century, see Marquis de Roux, *Louis XIV et les provinces conquises* (Paris, 1938) 161–182. On the intendancy in Roussillon, see the late eighteenth-century memoir of Pierre Poeydavant, an employee in the intendancy, published as "Mémoire sur la province de Roussillon et pays de Foix," ed. E. Desplanques, *Bulletin de la société agricole, scientifique, et littéraire des Pyrénées-Orientales*, no. 35 (hereinafter *BSASLPO*) (1901): 322–325, and 51 (1910): 137–177.

76. On the offices of veguer and batlle in Catalonia before the Bourbon reforms of 1714, see J. Lalinde Abadia, *La jurisdicción real inferior en Cataluña* (Barcelona, 1966), passim; and Ferro, *El dret públic Català*, 120–125; on the French reorganization of these offices, see Poeydavant, "Mémoire sur la province," *BSASLPO* 35 (1901): 329–331; and E. de Teule, *Etat des juridictions inférieures du comté de Roussillon avant 1790* (Paris, 1887). On the French control of seigneurial jurisdictions, see the dossiers and memoirs

command in the province, while at the same time preserving the local institutions of justice. To strengthen this tutelage, in 1664 the French crown prohibited meetings of the local syndics, elected representatives of the village communities and the valley as a whole, unless a royal officer were present—a policy the viceroy in Catalonia had already taken on his side of the boundary.[77]

Yet the jurisdictional competency of the French courts, far from extending throughout the territory of the newly annexed province, was limited by the persistence of Spanish seigneurial jurisdictions that penetrated into the French Cerdagne. For example, the College of Priests of Santa María of Puigcerdà maintained their seigneurial rights and privileges over the communities of Bolquera, Palau, and others. Other religious communities of Puigcerdà not only owned tithes and annuities but administered justice in half a dozen communities of the French Cerdagne. Lay seigneurs, including such grandees as the Duque de Hijar, retained the prerogatives of "high justice" over Estavar, Bajanda, and Callastres. Similarly, nobles from Puigcerdà, members of the ruling class such as Pastors, remained seigneurs of French villages until the Revolution.[78]

The second dimension of jurisdictional sovereignty was the exercise of control over religious affairs, and here again the problem of boundaries was at issue. The boundaries of royal authority over religious practice failed to correspond to the limits dictated by the Treaty of the Pyrenees. For not only did the Capcir remain under the authority of the bishop of Alet, in Languedoc, but the jurisdiction of the bishop of Urgell—and the archbishop of Tarragona in Spain—extended across the boundary to the thirty-three villages of the Cerdanya ceded to France. The issues surrounding ecclesiastical administration were far from resolved in the Figueres conferences; even the most important demand of the French crown—that the bishop of Urgell appoint a native French subject (regnicole) to administer justice in the villages—went unheeded and was to remain an issue, as we shall see, until the 1730s.[79]

in ADPO C 1282: "District de la viguerie de Cerdagne qui comprend la Vallée de Carol," 1767; and ADPO C 1274 (criminal jurisdictions in the French Cerdagne, 1763).

77. Sanabre, La resistència, 96. More generally, on the preservation and strengthening of the institutions of the rural community in the construction of the absolutist state in France, see H. L. Root, Peasants and King in Burgundy: Agrarian Foundations of French Absolutism (Berkeley, Los Angeles, London, 1987), esp. 22–44.

78. A list of seigneurial jurisdictions may be found in Sahlins, "Between France and Spain," app. 2.

79. A. Marcet, "La Cerdagne après le traité de Pyrénées," Annales du Midi 93 (1981): 141–155; and below, chap. 2. On French religious policies in the annexed provinces, see Marquis de Roux, Louis XIV, 203–249.

The third dimension of jurisdictional sovereignty—the control of economic and commercial affairs—pointed to the lack of congruence between the territorial extension of the monarchy and its economic (dis)unity. The province of Roussillon, like others incorporated relatively late by the French monarchy, retained the status of a *province réputée étrangère*. This meant that customs dues and royal taxes on the passage of merchandise in and out of the province were raised twice: once on the frontier of Languedoc and Roussillon, and then on the boundary of France and Spain. In fact, customs duties were often higher on the former. Colbert abolished certain internal customs barriers when he established the *Cinq Grosses Fermes* in 1664, but the barriers between Roussillon and France were not among them. And in the same way, Catalonia maintained its own customs frontier with Castile, so that both these zones represented a space distinct from the "limits of the two monarchies."[80]

There were other questions raised during the Figueres conferences concerning royal jurisdiction over economic affairs. Indeed, the first issue discussed in January 1665 was the grain trade in the borderland. The question had been raised the year before when the governor of Puigcerdà prohibited the exit of grains from the Spanish side of the valley in order to assure the provisioning of the town of Puigcerdà. Responding to an inquiry by the Roussillon Intendant Macqueron, Le Tellier ordered similar prohibitions levied on the exit of grains from France until an accord could be reached. But the commissioners failed to reach such an accord, and the issue of grain trade in the borderland was to preoccupy both states periodically over the next two centuries.[81]

So too did the issue of taxation, the fourth dimension of jurisdictional sovereignty. The "limits of France and Spain" as suggested by the Llívia accord reproduced local boundaries in the Cerdanya but also divided a great many properties and incomes on both sides of the border. Declarations made to the Figueres commissioners suggest an important feature of property ownership in the borderland: that nearly

---

80. Poeydavant, "Mémoire sur la province," *BSASLPO* 51 (1910): 102–104; on Colbert's reforms and French economic policies toward the annexed provinces, see Marquis de Roux, *Louis XIV*, 184–202; P. Dockès, *L'espace dans la pensée économique du XVIe au XVIIIe siècle* (Paris, 1969), 46–51; and J. F. Bosher, *The Single Duty Project: A Study of the Movement for a French Customs Union in the Eighteenth Century* (London, 1964), 1–24. On the Spanish customs frontiers, see H. Kamen, *Spain in the Later Seventeenth Century* (London, 1980), 15–16.

81. ADPO C 1361 (Figueres conference of July 13, 1664); ADPO C 1359 (conference of August 12, 1664); ADPO C 1369 (conference of January 31, 1665).

four times as much "Spanish" property was held in the French Cerdagne than vice versa.[82] Although the issue of property was not to preoccupy the two states until the eighteenth century, the dimensions of the problem were already revealed during the Figueres conferences. The question of property ownership across the boundary suggests the essential permeability of the frontier as established by the Llívia accord. Indeed, the continuity of social relations in the borderland was written into the accord itself, which guaranteed the freedom of passage among French villages and between Llívia and Puigcerdà, "without the possibility of their [inhabitants] being bothered by the officers of the two kings under any pretext." The only qualification was that such freedom "cannot be allowed to serve as a pretext for crimes," and that each crown could arrest on its own territory known or suspected criminals. At the Figueres conference, following numerous complaints of unwarranted seizures of livestock, the commissioners underlined the need to maintain this freedom of passage.[83]

But there were many aspects of continuity within local life in the borderland which went unaddressed by the Figueres commissioners or by either of the crowns. One was the character of the "neutral road" linking Puigcerdà and Llívia. Puigcerdà also petitioned the commissioners to assure the possession and enjoyment of its usufruct rights and properties located in the villages ceded to France—including its canal and the Carlit Mountain as well as the possession of the chapel of Belloch and its usufruct rights in the Bolquera Forest.[84] These local concerns went completely ignored by the Figueres commissioners—in marked contrast to the care with which the French and Spanish commissioners would treat such issues two centuries later during the delimitation conferences at Bayonne.

CONCLUSION

During the negotiations leading up to and following the signing of the Treaty of the Pyrenees, French and Spanish diplomats and commissioners spoke of the "delimitation" and the "delineation" of the "limits of France and Spain," implying in their language—as in the rites of di-

82. ADPO C 1365 (Figueres conference of January 24, 1665); and below, chap. 4 sec. 2.
83. ADPO C 1361 (Figueres conference of March 26, 1665).
84. AHP MA 1660-2, fol. 68–69: "Copia del memorial de Puigcerdà per sos interesos . . . sobre la división que esta fa dels llochs de Cerdaña," ca. 1665.

plomacy themselves—a linear boundary separating two contiguous territorial polities. Article 42, naming the Pyrenees as the "separation of the two kingdoms," seemed to affirm this vision. But the linear boundary never emerged from the negotiations: indeed, it was only to appear in the delimitation accords of 1868. Although the word was the same in the seventeenth and nineteenth centuries, much would happen before the fixing of the modern border line separating the national territorial states of France and Spain.

Natural frontiers, history, and military strategy all entered into the negotiations of 1659–1660, but not in a predetermined hierarchy of importance. Claims that the Treaty of the Pyrenees ought to reproduce natural and historical frontiers were not simply masks for military concerns; in particular, the idea of natural frontiers had a special appeal to seventeenth-century diplomats and statesmen. The appeal was so great, in fact, that the French cartographer Pierre du Val's map of the "Thirty-Three Villages Ceded to France," included a 1679 volume commemorating the conquests of Louis XIV, traced a line of mountains where there were none (illustration 2).

The division of the Cerdanya through the center of the plain, and the anomalous position of Llívia as a Spanish town surrounded by French villages, was the outcome of a process that constantly inverted the relation of interests and ideology. Natural frontiers, far from disguising strategic and military concerns, in the end determined such interests. By achieving the division of the Cerdanya, the French had won for themselves a military position that was to prove, in the next six decades, completely devoid of utility.

2
The Frontiers of the Old Regime State

Maps, both in manuscript and published, of the Principality of Catalonia, the two Counties of Roussillon and Cerdanya, and the eastern French–Spanish borderland during the later seventeenth and eighteenth centuries, reveal much about the evolving conceptions of territories and boundaries under the Old Regime. The first printed map to portray Catalonia without the County of Roussillon or the French Cerdagne appeared only in 1701, more than forty years after the Treaty of the Pyrenees. Indeed, until the later eighteenth century, the norm in maps of Catalonia was to include the "Two Counties."[1] At the same time, printed maps dating as early as two decades after the treaty already distinguished the "two Cerdanyas" by name. Such was the case of two popular maps of Catalonia, both extensively reprinted in the later seventeenth century: F. de Wit's *Accuratissima Principatus* (1670), specifying the division of "Cerretania Gallica" and "Cerretania Comitatus"; and *La Catalogne* (1675), by Nicholas Sanson, "the govern-

1. I. Colomer Preses, *Els cent primers mapes del Principat de Catalunya, segles XVI–XIX* (Barcelona, 1966), passim. French maps of Roussillon (without Catalonia) appeared much earlier. The first seems to be the one included in a printed memoir of the Perpignan doctor Lluis Palau in 1627. Jean Boisseau's *Nouvelle description du comté de Roussillon* (1639) anticipated the French conquest of the province, and the map of Nicolas Sanson, *Comté de Roussillon* (1660), commemorated it: see N. Broc, "Géographes et naturalistes dans les Pyrénées catalanes sous l'Ancien Régime," in *Trois Siècles de Cartographie des Pyréneés* (Lourdes, 1978), 55–73.

ment's first official cartographer."[2] In both cases, as more generally in printed maps of the seventeenth century, dotted lines marking the separation of France and Spain were indistinguishable from those separating administrative divisions within Catalonia itself. Old Regime cartography never accurately represented these divisions: most maps of the Cerdanya leave one or two villages on the wrong sides of the boundary. This was true even in the first French geodetic map survey by the Cassini clan, completed in the second half of the eighteenth century.[3]

The representation of the two Cerdanyas as pertaining to distinct territories in seventeenth-century maps is significant since it anticipated in theory what later appeared in practice: a territorial division within the military frontier. The military frontier dominated all other boundaries throughout the French and Spanish struggles that lasted until the 1720s. Initially, the control of the military frontier ignored the political division of territories but, over the course of several decades, the territorial division made its appearance at the heart of that frontier. Maps of the Cerdanya drawn up by the military establishments in France and Spain throughout the Old Regime reflect both the geographical unity of the valley and its incipient territorial partition. On the one hand, manuscript maps conceived for military purposes under the Old Regime generally failed to represent the political division of the valley, highlighting instead the ring of mountains that enclosed it. French maps of the period tend to show French domination implicitly by portraying the valley viewed from Paris (see illustration 3). On the other hand, the military establishment in France—and especially the Corps of Royal Engineers—was instrumental in creating detailed maps of the French–Spanish borderland which portrayed, albeit inaccurately, the location of the political boundary. The maps of Roussel and La Blottière, completed during the War of the Pyrenees (1718–1721) and printed in the 1730s, served the military establishment throughout the Old Regime.[4]

2. Konvitz, Cartography in France, 2; Colomèr Preses, El cent primers mapes, nos. 13–18 and 25; and his Els mapes antics de les terres catalanes (Granollers, 1967), no. 45. On Sanson's map and earlier ones of the Pyrenees, see H. Lapeyre, "La cartographie des Pyrénées avant Sanson," Annales du Midi 67 (1955): 261–268.

3. On the generations of Cassini cartographers and the national map survey, see Konvitz, Cartography in France, 1–31. On the representation of political boundaries in European cartography of the late seventeenth century, see J. A. Akerman, "Cartography and the Emergence of Territorial States," Proceedings . . . Western Society for French History: 84–93.

4. A. Fraelich, "The Manuscript Maps of the Pyrenees by Roussel and La Blottière," Imago Mundi 15 (1960): 94–104; on La Blottière's detailed memoirs on the Pyre-

Besides portraying the appearance of a territorial boundary within the military frontier, the military cartography of the Pyrenees revealed the jurisdictional boundaries of the Old Regime state. Once again, the maps anticipated in theory what emerged in practice only after 1722, with the end of the period in which the military frontier dominated. An anonymous map of the Cerdanya, probably from the 1680s, depicts one dimension of jurisdictional sovereignty: the dominion of the king over specific villages, not territories. The key to the map noted that "the villages marked blue are of France, those marked red are of Spain . . . the valley of Andorra is not marked with any color." A second map, probably drawn up in 1679 on the orders of the military commander Durban, soon to become governor of the newly constructed fortress of Mont-Louis, illustrates a transition from jurisdictional to territorial sovereignty. Though it retained the different colors of French and Spanish villages, it nonetheless included a simple boundary line, albeit inaccurately drawn.[5]

The division of France and Spain made its most evident appearance in maps composed under official aegis with the expressed goal of resolving disputes over territorial sovereignty. Such was the case of the earliest maps of France's boundaries, drawn during the late fifteenth century.[6] The earliest example of a map designed to represent a contested section of French–Spanish boundary in the Cerdanya was done by a military engineer, Cheylat, in 1703, following a seizure on disputed lands of two mules loaded with salt heading into France (illustration 4). Yet most maps portraying disputed boundaries make their appearance after the middle of the eighteenth century, when the French state began to demonstrate for the first time a consistent policy of delimiting its territorial boundaries.

THE MILITARY FRONTIER (1660–1722)

The location and character of the late–seventeenth century military frontier in the Cerdanya was a function of the state of war between

nean frontier, revealing the military interests behind the maps, see above, Introduction, n. 24. On the corps of royal military geographers, see the older but still useful C. Berthaut, *Les ingénieures géographes militaires, 1624–1831,* 2 vols. (Paris, 1902).

5. FMG AAT J 10 C 355. Durban himself was well aware of the territorial division, since during the 1680s he persisted in forays across the boundary as a way of asserting French domination of the entire valley: see below, chap. 2 sec. 1.

6. F. de Dainville, "Cartes et contestations au XVe siècle," *Imago Mundi* 24 (1970): 99–121.

Illustration 4. "Portion of Cerdagne with Limits of France and Spain," drawn up by a French official in 1703 following a disputed seizure of salt brought to the Carol Valley from the Cardona mines in Catalonia. The map represents a new concern with territorial sovereignty which appeared within the later seventeenth-century military frontier and took shape during the eighteenth century: the need, according to the French intendant commenting on that seizure, "to place boundary stones to signify the limits of sovereignty of the two crowns."

France and Spain during that period, as France established its political and economic hegemony over a declining Spain. The two "belligerent nations" were engaged in active hostilities during the War of the Devolution (1667–1668), the later phases of the Dutch War (1673–1678), the unofficial hostilities of 1683–1684, and the War of the Augsburg League (1689–1697). While Louis XIV's policies of conquest and annexation extended France's northern and eastern frontiers—encroaching on Spanish possessions in the Low Countries, acquiring the Franche-Comté, most of the Duchy of Lorraine, and important parts of Alsace—he treated the contiguous frontier of France and Spain in Catalonia as a "theater of distractions" from the main battlefields.[7]

The French crown developed no formal plans to annex permanently the Principality of Catalonia; yet the two states challenged each other repeatedly for control of the mountain valley of the Cerdanya. Every time that France and Spain were at war—and occasionally both before and after war was declared officially—one or the other army occupied the entire valley of the Cerdanya.[8] The establishment of the military frontier along the Pyrenees, whether controlled by Spain or by France, deliberately erased the formal division of the valley as outlined in the Llívia accord of November 1660. As a military engineer was to write in the 1830s, the treaty was "torn up by the sword."[9] This may be a general description of warfare, but it had a particular application in seventeenth-century Cerdanya.

The military frontier meant that, however ephemeral each occupation and reannexation of the valley, the army in control reconstituted the institutional and geographical unity of the County of Cerdanya. Military occupation entailed royal administration of the entire valley. In 1667 and again in 1673, for example, the Spanish crown reconquered the valley and took over the administration of patrimonial revenues and royal dues for villages throughout the Cerdanya. These dues included the various forms of land rent and indirect levies owed to the crown as seigneur. The French state collected these revenues from

7. For an overview of Louis XIV's policies, see L. André, *Louis XIV et l'Europe* (Paris, 1950); F. Soldevila's *Historia de Catalunya*, vol. 3 (Barcelona, 1963), 1066–1094, contains a solid narrative of French actions in Catalonia.

8. Arxiu Capitular de la Seu d'Urgell (hereafter ACSU), Pleitos: "Informació del temps que lo enemich ha occupat lo officilat de Serdanya," November 10, 1700; and I. Lameire, *Théorie et pratique de la conquête dans l'ancien droit:* vol. 2. *Les occupations militaires en Espagne pendant les guerres de l'ancien droit* (Paris, 1905), 37–43, 48–73, and 647–671.

9. FMG AAT MR 1349: "Mémoire sur l'enclave de Llívia par M. Loreton Dumontet, capt. adjudt.," May 7, 1836.

Spanish villages when it controlled the Spanish Cerdaña in 1678 and again in the 1690s. Much the same pattern developed with respect to ecclesiastical administration: when the Spaniards were in control of the valley, the bishop of Urgell took advantage and visited the villages ceded to France, which he had been otherwise forbidden to do. And after each French invasion and occupation of the Spanish Cerdaña, military officials placed clergy loyal to the French king in the convents of Puigcerdà, as occurred in 1678, and confiscated clerical revenues on both sides of the boundary.[10]

The definition of the military frontier reconstituted the political and administrative unity of the valley. There was, of course, a strategic rationale for treating the valley as a unity. The Cerdanya continued to play its role as a breadbasket. The relative wealth of the Cerdanya surprised French military commanders from Lieutenant General Roger de Rabutin in 1654, himself born in the fertile Burgundian plains, to the Duc de Noailles, governor of Roussillon in 1678.[11] The valley played a consistent role as a winter garrison, a constant concern of both French and Spanish commanders. And the Cerdanya was the central passage between France and Spain in the eastern Pyrenees, the "key" of both and a passage worth fighting over.

Taking control of the Cerdanya meant reconquering its capital, Puigcerdà. In the first two decades following the Treaty of the Pyrenees, Spain dominated because it was able to hold on to Puigcerdà. The court in Madrid, the vice-regency of Catalonia, and the Catalan political elite all decried the lack of defense of the Catalan frontier and emphasized in particular the need to fortify the walls of Puigcerdà, "the key of Spain."[12] Under the leadership of the Viceroy Castel-Rodrigo, the Spanish government did manage to begin work on the walls, adding new bulwarks and half-moons "a la moderna." Hampered by the lack of funds, which made the obstacles of cold weather decisive, the work was

10. ACA RP Maestre Racional, nos. 256–257: "Procuración real de los condados de Rossellon y Cerdaña," 1653–1700; ACSU reg. 798: "Processu Visitation," 1674 and 1678; FMG AAT A1 611, no. 69 (intendant to War Minister Louvois, June 4, 1678); FMG AAT A1 1289: "Etat des biens confisqués," 1693; SBN MS 2406, fols. 57 and 242 (petitions of the College of Santa María of Puigcerdà, 1689); and ACA CA 337 (consulta of January 26, 1689).

11. *Mémoires de Messire Roger de Rabutin, comte de Bussy*, 2 vols. (Paris, 1696), 2: 552–553; FMG AAT A1 611, no. 62 (Noailles to Louvois, May 31, 1678).

12. ACA CA leg. 231, no. 13 (consulta of Council of Aragon, August 1662); ACA CA leg. 317 (draft of a consulta on the need to fortify Puigcerdà, n.d., but ca. 1667); ACA CA leg. 330 (consulta of the Council of Aragon, August 14, 1675); ACA CA leg. 337 (printed letter of the Diputació, July 27, 1685).

never completed; yet the restored fortifications of Puigcerdà did withstand a minor French offensive in 1667. Pushed on by the Council of Aragon and local political and military officials in the Cerdanya, Spain went on to reconquer the thirty-three villages ceded to France, returning them to Louis XIV only after the Treaty of Aix-la-Chapelle in May 1668.[13]

Unlike the Spanish crown and Catalan elites, the French administration in Paris and Perpignan remained fundamentally unconcerned with the military frontier in the newly acquired province of Roussillon. Although French armies might seek to occupy the Cerdanya valley during actual campaigns, the crown itself utterly neglected the fortification of the province during the first two decades after the annexation. The numerous troops garrisoned in Roussillon were hardly intended to protect the frontier. Their concern was the repression of local uprisings—notably the "Angels' revolt" against the newly imposed and universally detested salt monopoly between 1663 and 1672, and the political conspiracies of 1674.[14] The French government's true concern was manifest at the diplomatic level, as French ambassadors attempted to shift the geopolitics of acquisitions, exchanging the hard-won gains of the Pyrenees Treaty for the remainder of the Spanish Low Countries. Although such offers were made until 1677, Spanish diplomats never seriously considered the possibilities, choosing instead to concentrate their resources on the distant and costly provinces of Flanders and Hainault.[15]

The balance of power that gave Spain temporary domination of the Cerdanya shifted abruptly in 1678. Spanish troops had controlled the valley and a portion of the Roussillon since the renewed outbreak of hostilities in 1675, although the French armies carried out seasonal

13. SBCM Aparici 5 (1–4–1–3), fols. 160–161 and 260–261 (correspondence of the viceroy of Catalonia, 1665). On the reconstruction of Puigcerdà, see also BC MS 2371, fol. 45: "Discurso general hecho por el Mre de Campo, D. Ambrosio Borsano," n.d., ca. late 1680s. On the Cerdanya during the War of the Devolution, see P. Torreilles, "Troubles et Guerres en Roussillon," *RHAR* I (1900): 321–328; FMG AAT A1, nos. 225–227: "Affaires de Cerdagne," 1667–1668; and ACA CA leg. 321, esp. the correspondence of the governor of Puigcerdà (letters of March 17 and July 12, 1667), and the consulta of October 23, 1667.

14. A. Marcet, "Une révolte antifiscale et nationale: Les Angelets du Vallespir, 1663–1672," in *Actes du 102e congrès national des sociétés savantes* (Paris, 1978), 1: 35–48; Marcet, "Les conspirations de 1674 en Roussillon: Villefranche et Perpignan," *Annales du Midi* 86 (1974): 275–296.

15. P. de la Fabrègue-Pallarès, "L'affaire de l'échange des Pays-Bas Catholiques et l'offre de retrocession du Roussillon à l'Espagne," *BSALSPO* 68 (1953): 47–63 and 115–133.

raids into the Cerdanya plain.[16] But in early May 1678, the Duc de Noailles, governor of Roussillon, commanding some 12,000 troops, undertook the costly and surprisingly difficult siege of Puigcerdà. The remarkable efforts of the Spanish forces—some 2,000 strong, supported by 600 Catalan peasants—staved off thirty-three separate assaults over four weeks, holding off the French seventeen days after the walls had been breached.[17] The governor of Puigcerdà finally capitulated, and the French armies subsequently occupied the entire plain, taking oaths of loyalty from all the villages.[18] The Treaty of Nijwegen in October 1678 returned the conquered district and the razed fortifications of Puigcerdà to Spain. The walls of Puigcerdà were not rebuilt until 1875, and then only as a defense against the Spanish Carlists. During the rest of the seventeenth century, as during the Revolutionary and Napoleonic wars, France remained firmly in control of the military frontier of the Cerdanya.

On France's northern and eastern limits, the year 1678 and the Nijwegen Treaty marked a shift toward the consolidation of a new kind of military frontier. The "politics of open doors on neighboring countries" gave way to the "politics of the barrier." The idea was Vauban's, the architect of France's new frontier. Long before the Maginot Line, Vauban had built his "iron frontier" consisting of two lines of fortified sites. The idea was to abandon the most advanced fortresses and towns, relinquishing more distant outposts in the interests of a more enclosed space, while never abandoning France's European aspirations. Vauban first gave expression to this new politics of the frontier in a memorandum of 1673:

> The king ought to think a little about squaring his field. This confusion of friendly and enemy fortresses mixed together does not please me at all. . . . Preach the squaring, not of the circle, but of the field; it is a good and beautiful thing to be able to hold one's accomplishment in both hands.[19]

Thus in addition to reforming the double line of fortifications, the new military frontier required the "purging" of enclaves within France. The

16. ACA CA leg. 232 (dossier on the "Guerra con Francia, 1674–76").

17. FAMRE CP Espagne, vol. 32, fols. 537r–544r; FMG AAT 611, nos. 18, 32, 43, 45, 61, and 63 (journal of the Roussillon Intendant Camus de Beaulieu); and the printed account of the siege in FAMRE MD 12, fols. 140–147.

18. SBCM Aparici 5 (1–4–1–3), fol. 435 et seq. (petitions of the governor of Puigcerdà); AHP MA 1678–1683, fols. 1–3 (copy of the capitulation accord).

19. Quoted in Girard d'Albissin, *Genèse de la frontière franco–belge*, 160; see also G. Zeller, *L'organisation défensive des frontières du Nord et de l'Est au XVIIe siècle* (Paris, 1928), chaps. 3 and 5; and H. Chotard, *Louis XIV, Louvois, Vauban et les fortifications du nord de la France* (Paris, 1890).

"politics of reunions" undertaken by provincial "chambers" in Douai, Besançon, Brisach, Metz, and Strasbourg during the 1680s incorporated into France dozens of "dependencies and annexes" forming enclaves on French territory—even if France was forced to return most of these by the Treaty of Ryswick (1697). The consolidation of territorial sovereignty served not only to guarantee the military frontier but also to assure the free passage and communication among French towns and cities.[20]

For almost thirty years, Vauban directed the construction of a barrier system controlling the important passages into France, reconstructing strategically located feudal ruins and building military fortresses along the frontier of the Spanish Low Countries, Luxembourg, the Franche-Comté, throughout the Alps, and all along the Pyrenees. In Roussillon, the French military frontier took shape in the late 1670s with Vauban's reconstruction of the walls of Perpignan and a half dozen other fortified sites on the province's frontiers.[21] The French Catalan Trobat, who was to become intendant in the 1690s, wrote in 1670 about the need to make the French presence felt in the Cerdanya:

> If we leave this district without building anything, we abandon to the enemy during wartime a district seven leagues long and two leagues wide, of which half is His Majesty's property; and in times of peace, this district, which belongs to the King, remains more devoted to Spain than to His Majesty because Puigcerdà, which the enemy will certainly rebuild, dominates the valley.

But Vauban had different plans. Rejecting proposed sites that would simply oppose the fortress of Puigcerdà, he opted for a site in the Capcir, outside the Cerdanya itself. At the confluence of two passes, the site protected the Aude and Tet Valleys—the roads to Toulouse and to Perpignan. "The enemy," wrote Vauban, "will almost never know what is going on behind our backs."[22]

The fortress and citadel of Mont-Louis was completed between 1679 and 1682 at a cost of over one million livres, an impressive testimony of the French crown's strength and ability to translate intention

20. Girard d' Albissin, *Genèse*, esp. pp. 227–235; and L. André, *Louis XIV*, 167–175 and 187–205.

21. P. Torreilles, "L'oeuvre de Vauban en Roussillon," *BSALSPO* 42 (1901): 181–288.

22. FMG AAT 611, no. 122 (Trobat to Louvois, September 14, 1678); ADPO C 2042: "Mémoire de Vauban pour bastir une place dans la plaine de Cerdagne," March 17, 1679.

into reality with remarkable speed (illustration 5).[23] The military citadel failed nonetheless, to live up to expectations. Vauban himself later proved willing to concede "these 43 [sic] lousy villages . . . and even raze Mont-Louis," given that Vilafranca would protect the plain and Perpignan. Throughout the eighteenth century, dozens of military memoirs attested to the facility with which enemy troops could pass into France along a variety of nearby roads unprotected by Mont-Louis.[24] If the primary rationale for its construction was soon discredited, the fortress did play an important secondary function on the French frontier. For in times of peace as during war, Mont-Louis dominated Puigcerdà and the entire Cerdanya valley.

That domination, in the eyes of the Duque de Bournonvilla, viceroy of Catalonia, involved the inability to protect the Spanish king's "vassals." In August 1680 he wrote to the Council of Aragon that the French

> are bringing in more workers to the fortifications of Mont-Louis, in order to finish it as quickly as possible. The poor inhabitants of the Cerdaña villages which are of the dominion of Your Majesty are being treated as if they were vassals of France. . . . We are forced to show our weakness since we can do nothing to protect these poor vassals.[25]

In effect, the military governor of Mont-Louis made his presence felt in Puigcerdà throughout the 1680s, a decade of relative peace between France and Spain. Durban would periodically enter Spanish territory in a symbolic demonstration of his power, collect "contributions" from the villages of the Spanish Cerdaña, and threaten to level any fortifications begun on Puigcerdà. In 1681, he ordered Puigcerdà's water supply cut; diplomatic entreaties failed to reestablish the town's canal, which was only rebuilt in 1687 after Puigcerdà received a royal "establishment" by a privilege of Louis XIV.[26]

23. Torreilles, "L'oeuvre de Vauban," 230–236; *France, Archives de l'Inspection du Génie* no. 81 (hereinafter FAIG) (maps, correspondence, and fiscal accounts of Vauban); and M. A. Paillisse, "Mont-Louis, place forte et nouvelle" (Mémoire de maîtrise, Université de Montpellier, n.d.).

24. Vauban quoted in J. F. Pernot, "Les chevauchées des ingénieurs militaires en France au XVIIe siècle, ou la maturité de voyages d'études politico-administratifs," in *La découverte de la France,* 342–343; see also Vauban's memoir in FAIG 4.1.1, and map 1. Eighteenth century reevaluations of Mont-Louis include FMG AAT MD 1084, no. 50: "Observations sur le mémoire de Vauban en 1700"; and FMG AAT MD 1084, no. 22: "Mémoire sur la communication de la Cerdagne française," 1754.

25. ACA CA leg. 334 (letter of August 3, 1680).

26. ACA CA leg. 336 (consulta of April, 1684); AHP Asequia (letters and memoirs of 1681–1682 concerning the canal); and ADPO C 1413 (letter of Louis XIV, December 17, 1687).

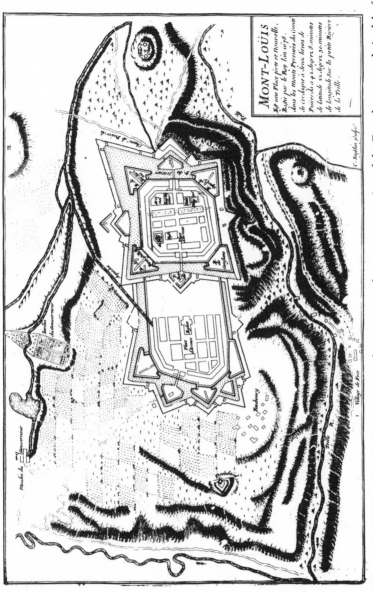

Illustration 5. Mont-Louis. An anonymous late seventeenth-century engraving of the French town and citadel of Mont-Louis, across the Perxa Pass from the Cerdanya, in the Capcir. Completed in 1682 under the direction of Vauban, the fortress definitively shifted the balance of power between France and Spain in the Cerdanya valley.

The failure of the government of Charles II to challenge French domination of the Cerdanya during the 1680s, despite constant emotional pleas from the Catalans and the Council of Aragon, had domestic causes as well. The standing committee of the Catalan Corts, the *Diputació,* was unwilling to share in the financial burdens of fortifying Puigcerdà, "the only shelter of this Province."[27] And the disagreements among the Councils of War, State, and Aragon and the viceroy of Catalonia—a failure of the conciliar system of government—left Puigcerdà at the mercy of the French.[28] Charles II's own inclinations helped the situation very little. The Spanish king was more concerned with his reputation than with Puigcerdà. He sought instead to raise money for a ten-day incursion into Roussillon which, he wrote, "could result in some credit to my Royal arms."[29] It was only the tireless efforts of Villahermosa, viceroy of Catalonia during the later 1680s, which made the Cerdaña a focus of some attention. But, since the "French troops are so superior, they would slaughter our own, and ruin whatever work we began," the viceroy considered other sites—notably Bellver and Montellà—beyond the shadow of Mont-Louis.[30] Delays and indecisions proved decisive, for during the spring of 1690 French troops again occupied the valley, which they were to hold until the Peace of Ryswick of 1697.

The struggles of the 1680s, a period when France and Spain were *not* formally at war, brought an increasing awareness by both French and Spanish military officials of the territorial division of the valley. For the first time since the Treaty of the Pyrenees, these officials became aware of the existence of the political boundary, and even, for the first time, of the two Cerdanyas.

In the first two decades following the treaty, neither French nor Spanish officials had a name for the portion of the valley ceded to France: most often, it was designated as the "district adjoining" Cerdanya. By the late 1680s, however, the terms "French" and "Spanish" Cerdanya had come into common usage. Don Borsano urged Charles II "to keep the two Cerdanyas, raising once again the walls of Puig-

27. SBCM Aparici 5 (1–4–1–3), fol. 80 (consulta of the War Council, May 1683).

28. SBCM Aparici 5 (1–4–2–4), fols. 59–70 (junta of members of the Councils of War, State, and Aragon on whether to fortify Puigcerdà). On the conciliar system of government and its failures, see J. Lynch, *Spain under the Habsburgs,* 2: 93–94; and J. H. Elliott, *Imperial Spain* (New York, 1963), 167–178.

29. SBN MS 2399, fols. 31–33 (Charles II to viceroy, September 12, 1689).

30. SBN MS 2402, fol. 80–83 (viceroy to Charles II, September 1, 1689); see also SBN MS 2403, fols. 331–334 (report of the Neapolitan Quarter Master Borsano to the viceroy on other possible sites, November 9, 1689).

cerdà," even if he mistakenly identified the number of villages in each one:

> If Puigcerdà were fortified, the enemy would be forced to abandon the contributions of the 25 villages of our Cerdaña . . . and the 23 villages of the French Cerdagne, which could then be placed under Spain's obedience.[31]

The Spanish seem to have used the specific terms before the French, for during the 1690s, French military account books refer to the "upper" and "lower" Cerdanyas, corresponding to the division instituted by the Treaty of the Pyrenees. The linguistic distinction nonetheless, corresponded to administrative realities during the French occupation of the 1690s.[32] Previous military frontiers ignored the political boundary, as the state in control took over the ecclesiastical and civil affairs in the entire valley. But the military frontier of the 1680s and 1690s was increasingly built around a territorial distinction. This territorialization of the military frontier was even more visible during the episodes of the Spanish War of Succession fought in the Cerdanya.

"The Pyrenees are no more," proclaimed Louis XIV—at least according to Voltaire—when his grandson the Duc d'Anjou left France to reign as Philip V, the first Bourbon king of Spain.[33] In fact, despite the Bourbon family alliance, the political boundary of Spain and France was to become a reality during the War of the Spanish Succession—even in the Cerdanya, where there was no natural frontier that the boundary could duplicate. It is significant that the first printed map of the French–Spanish boundary, done in 1701 by Nicholas de Fer, royal geographer but "more a popularizer than a topographer," was published at precisely the moment when the two Bourbon crowns allied themselves in preparations for the War of the Spanish Succession.[34]

The struggles of France and Spain against the Grand Alliance—the Habsburg Empire, England, and the United Provinces, soon joined by Savoy and Portugal—raged throughout Europe from 1701 until the peace settlements of Utrecht and Rastatt in 1713–1714. Within Spain,

31. SBN MS 2403, fol. 333 (Borsano to viceroy, November 9, 1689).

32. ADPO C 153, 157, and 170 (military expense accounts administered by the viguier of the Cerdagne, 1690–1697); ADPO C 184 (militia requisitions, 1697). The 1690 occupation also differed from previous occupations in that it was not a continuous occupation but coincided with spring campaigns. In this scheme, and with the southern end of the valley fortified, Puigcerdà became superfluous and was razed: see FMG AAT A1 1106, no. 159 (Trobat to Louvois, June 30, 1691); and SBCM Documentos (3–1–1–38): "Raisons qui portent à conserver les postes de la viguerie de Puigcerdà," n.d., ca. 1691.

33. Voltaire, *Le siècle de Louis XIV* (Paris, 1752), chap. 27.

34. *Images de la montagne*, 4.

this European struggle took shape as a series of civil wars: the peripheral provinces of Catalonia, Valencia, and Aragon, which initially accepted Philip V, rebelled in favor of the Archduke Charles against the two Bourbon monarchies of Spain and France.[35]

Once again, the Cerdanya was a site in which the major European powers confronted each other. After rising in favor of the Habsburg heir in October 1705, the Cerdanya was occupied by Austrian troops for two years. Reconquered in September 1707 by the Bourbon armies led by the Duc de Noailles, the Cerdanya became a base of military operations and a strategic garrison for the armies of "Their Two Majesties."[36] In point of fact, French officials and personnel maintained strict control of the military administration in the Cerdanya from 1707 to 1713. Puigcerdà became the center of operations, displacing Mont-Louis, while French commanders displaced their Spanish counterparts in the critical positions of authority.

Until the fall of Barcelona and the withdrawal of the Bourbon armies in 1714, French and Spanish officials continued to compete for jurisdictional supremacy within their shared military frontier, just as the overall military administration of the valley, although controlled by the French, increasingly recognized a territorial division of the French and Spanish Cerdanyas. For example, property confiscations between 1707 and 1714 were organized around the existence of the "two Cerdanyas," but the French administration appropriated the revenues to reward Bourbon loyalists and to help maintain the armies of "Their Two Majesties." In 1710, petitioning the Council of Castile, the Castilian military commander Don Antonio Gandolfo sought the position of "Adjudicator of Confiscations in the Spanish Cerdaña"; the position was ultimately given to a Roussillonnais.[37]

35. H. Kamen, *The War of Succession in Spain* (Bloomington, Ind., 1969); and Soldevila, *Història,* 2: 1096–1126.

36. FMG AAT A1 1891, nos. 219, 302, and 304 (reports of the intendant to the War Ministry, September and October 1705, on the Austrian occupation of the valley); FMG AAT A1 2053, no. 182 (letter of the Duc de Noailles, September 15, 1707); and below, chap. 3 sec. 1, on the local experience of these occupations.

37. SAHN Estado leg. 383 (Gandolfo to Council of Castile, January 23, 1710). Previous debates between French and Spanish officials over confiscations in 1668, 1674, and 1684 reveal how the French turned the exercise of a traditional prerogative of a sovereign at war into a means of political domination and source of income during times of peace, as the French crown reluctantly returned the confiscated properties of Spaniards in the French Cerdagne even after peace settlements were reached. These debates suggest how the two crowns came to recognize the different territories inscribed in the heart of the military frontier. On the earlier confiscations, see ACA CA leg. 232 (consultas of July 1668 and July 1669); and ADPO C 1411 (notebooks of conferences held between French and Spanish commissioners concerning confiscations, March–April 1687).

Spanish military officials challenged French control in other contexts. The court in Madrid, on the insistence of Gandolfo, nominated in 1712 a Spanish governor of Fort Adrien, a citadel built in 1707, adjoining Puigcerdà. The intendant of Roussillon, opposing such an infringement of jurisdictional supremacy, wrote:

> Fort Adrien was constructed by the orders and with the troops of the French King. Gandolfo, Brigadier of the troops in Cerdanya for several years, currently is in command of Puigcerdà and the Cerdanya. But he only has this post because the Count of Tournon, Quarter-Master General, has absented himself. . . . The garrison is entirely French. . . . The commander of Fort Adrien, M. d'Olive, is French. . . . This post, though situated in the Spanish Cerdaña, has always been considered a French post, and has always been guarded by French troops. . . .[38]

Such squabbles over jurisdiction seemed insignificant to the Duc de Noailles, and we might be inclined to accept his judgment. But they are important in revealing how the two monarchies thought about and competed for sovereignty in the heart of a military frontier that, for once, they shared. Jurisdictionally, there were two civil administrations in the Cerdanya, two administrations for the sequestered properties in each Cerdanya, and separate but equal mistreatment of the two Cerdanyas in the states' demands for money, firewood, grain, and men.[39]

The territorial division within the heart of the military frontier made its most dramatic appearance during the final episode of military struggles between Spain and France under the Old Regime. The War of the Pyrenees (1718–1721) intersected with news of the Marseilles plague in Catalonia (1720–1721) and the creation of a "sanitary cordon" that divided the Cerdanya valley into two parts. Although its existence was ephemeral, the new boundary line created significant disputes between France and Spain over their respective territorial extensions.

The War of the Pyrenees was the western theater of the Italian War, in which the Quadruple Alliance of France, England, the Empire, and Savoy opposed Spain and the court of Philip V, who refused to concede the Spanish king's succession to the French throne. It was a war of public opinion as much as of military strategy. As the Maréchal Duc de

38. FMG AAT A1 2406, no. 15 (intendant to war minister, August 14, 1712); see also the further correspondence with the Ministry of War, nos. 30–35, 76, 226–227, and 288.

39. For examples, see ADPO C 275: "Contributions, 1707–8"; ADPO C 283: "Ordonnance . . . modérant les taxes dans la Cerdagne espagnole," 1711; SAHN Estado leg. 383, no. 16: "Règlement des troupes dans les Cerdagnes françaises et espagnoles," 1709.

Berwick, commander of the French forces, explained in a pamphlet circulated in Paris, Madrid, and Barcelona, France was declaring war on the Spanish court, not the Spanish nation, "in order to guarantee the two nations a solid and durable peace." In order to embarrass Madrid, Berwick sought to take control of the Spanish Cerdaña. This he achieved in July 1719, again preferring the use of propaganda over force, since he promised to restore to the Catalans their privileges abolished by Philip V.[40] But the French control lasted less than six months, as a reorganized Spanish army reconquered the valley, occupied the villages of the French Cerdagne, and began to work on the fortifications of Llívia and Puigcerdà.[41]

Philip V renounced his claims to the French throne in June 1720, and negotiations for a treaty of alliance between Spain and France resulted in the secret Treaty of Madrid, signed on March 27, 1721.[42] Yet between the renunciation and the treaty, a series of developments unrelated to the military struggles was to reshape the military frontier in a fundamental way. When the news of the Marseilles plague reached the Cerdanya at the end of July 1720, the Spanish army in control of the valley erected the "sanitary cordon," which—despite much debate between the two crowns—was to define formally the territorial boundary of the two states.[43]

In control of both Cerdanyas, Spain established the sanitary cordon at the limits of the Cerdanya and the Capcir, along the passages into France. This was the customary location for such a preventative guard of the frontier, erected countless times when news of plague had arrived during the sixteenth and early seventeenth centuries. But this time, the "sanitary intendancy" erected its quarantine houses on French territory. The Madrid Treaty of Alliance declared that control of the French Cerdagne be returned to France, although Spain resisted doing so, despite diplomatic pressure, until September 1721. Only then did the Spanish military authorities move back the sanitary cordon so that it

40. ADPO C 602: "Lettre circulaire de M. le Maréchal de Berwick, envoyée sur les frontières," January 20, 1719; and ADPO C 603 (Berwick to the Roussillon intendant, August 10, 1719); see also H. Aragon, *La campagne de 1719 des armées de Roussillon et d'Espagne: Lettres inédites du Maréchal de Berwick et du Maréchal le Blanc, Ministère de la Guerre* (Perpignan, 1923), esp. 28 and 92–93; and AHP MA 1715–1719, fol. 40 (July 10, 1719); and fol. 43, (September 1, 1719).

41. ACA RA papeles no. 8 (Castelrodrigo to the Real Audiència, January 13, 1720); FAMRE Limites 433, no. 83 (viguier of the French Cerdagne to intendant, June 7, 1720).

42. SAHN Estado leg. 3369, no. 36: "Otro articulo secreto y separado del Tratado de Alianza."

43. AHP "Morbo de Marzella," passim; AHP MA 1720–1730, fols. 8v–9 (August 4, 1720); ACA RA, vol. 7, fols. 264–265 (consulta of November 24, 1721).

coincided with the approximate division dictated by the Treaty of the Pyrenees.[44]

By January 1722, the Spanish military authorities had constructed a line of twenty-seven wooden barracks, most of them on the valley floor; the number was increased to thirty-two in April.[45] Spanish troops, assisted by local peasants and townsmen, staffed these barracks, preventing all persons and goods from crossing into Spain. This first physical demarcation of the division of the valley, the line drawn by the Spanish troops, inevitably challenged French territorial sovereignty.

In November 1721, the viguier of the French Cerdagne sent a lengthy memoir to the French foreign minister detailing a dozen different points at which the Spanish military line "violated" French territory. Sicart reported that the lieutenant general of the Spanish troops claimed that Spain was only recovering usurpations of Spanish territory it had lost under Charles II. But French military officials claimed that the barracks usurped portions of French territory at Sant Pere de Cedret, Sant Pere de Senilles, Enveig, Ur, and Palau. The French sought to justify their territorial claims in a series of memoirs which, citing medieval charters and royal donations, proved that the lands on which the Spanish barracks were built were in fact part of France.[46] The immediate disputes remained unresolved until the formal withdrawal of the Spanish troops in June 1722, and the two states disputed these very same boundaries until the delimitation accords of 1868. Not only had a conjunction of circumstances in 1721—the war and the plague—given a concrete dimension to the territorial character of the military frontier emerging since the Treaty of the Pyrenees, but the disputes over the territorial boundaries of France and Spain posed during these years were to plague both states for the next century and a half.

ECCLESIASTICAL AND FISCAL FRONTIERS

The movement from a military frontier, encompassing the entire Cerdanya valley, to a territorial boundary dividing it into two parts, culminated in the sanitary cordon of 1721. But that division was

44. ADPO C 743 (correspondence between the intendant and the foreign minister, July–September 1721).

45. AHP MA 1720–1723, fol. 42 (January 2, 1722); AHP "Morbo de Marzella," no. 50: "Lista de la barracas de Sanidad echas en el condado de Cerdaña," April 9, 1722.

46. FAMRE Limites, vol. 459, nos. 7, 8, 10, and 12 (memoranda and correspondence concerning the barracks "constructed in French territory," November 1721); and no. 35 (correspondence of March, 1722); see also FAMRE Limites, vol. 446: "Mémoire sur la nécessité de terminer toutes les contestations qui concernent les limites," ca. 1721.

ephemeral, and the end of the episode brought back questions of jurisdictional sovereignty—over the administration of justice, taxation, and religious affairs. Between 1722 and the outbreak of the revolutionary conflicts in 1793, the Pyrenean frontier of France and Spain enjoyed a period unmarked by military struggles—the longest such period since the thirteenth century. In the first decades after dismantling the military frontier, France and Spain turned their struggle for European hegemony into one for jurisdictional supremacy in the borderland. The issues, as the Roussillon Intendant Jallais noted in 1732, concerned the exercise of sovereignty:

> There is no greater jurisdiction for sovereigns than that of exercising justice over their subjects, and that of taxing them and their lands as they see fit within the extent of their kingdoms.[47]

But justice and taxation were not uncontested prerogatives where the multiple jurisdictions crossed and contradicted each other, nor where the properties of separate subjects were inextricably intertwined. These prerogatives were especially complicated by the sovereigns' relative disregard for their territory, the "extent of their kingdoms." The issues of justice and taxation focused on the nationality of people, not territory; and the prerogatives of taxation within the kingdom were tempered by a fiscal system relying as heavily on taxing people as on taxing properties and lands.

The French intendancy's successful attempt to wrest control of ecclesiastical affairs in the thirty-three villages (twenty-four parishes) of the French Cerdagne from the bishop of Urgell represents a paradigmatic instance of the conflicts raised over the jurisdictional frontiers of France and Spain—and a remarkable instance of the "territorialization" of the frontier during the early eighteenth century. At the Figueres conferences between 1665 and 1668, one of the principal demands of the French commissioners was that the bishop of Urgell in Spain name a "general vicar and official" who was a French regnicole, a native-born subject of the French king, "with the power to assign benefices, convoke synodal assemblies of the priests of his jurisdiction, and impose necessary fines" for the thirty-three villages.[48] The French crown insisted that all judicial decisions concerning both spiritual and tem-

47. ADPO C 2046: "Observations pour répondre au mémoire envoyé par M. de Sartine," 1732.
48. ADPO C 1391: "Mémoire des entreprises commis par le Sr. Evêque d'Urgell contre la jurisdiction du Roy," 1663.

poral affairs had to be reached within the kingdom, under the ultimate jurisdiction of the Sovereign Council of Roussillon, and following French custom and law in church–state affairs. When the bishop refused to name a French general vicar, the French crown prohibited its subjects from appearing before foreign tribunals, forbade the bishop from visiting his parishioners in France, and gave full jurisdiction over ecclesiastical matters in the "adjacent Cerdagne" to the French viguier.[49] In practice, however, the French failed to enforce its demands. Even when France remained in control of the valley during the joint occupation of Bourbon forces between 1707 and 1714, Don Ramon Mossas, a subject of the Spanish king who resided in Puigcerdà, served as "official" for the villages of the two Cerdanyas.[50]

Pastoral visitation rights and the nationality of the general vicar were issues subsequently raised, and eventually resolved, only after the withdrawal of French troops and the end of the military frontier in 1715. The Spanish bishop renewed his demands to visit the French Cerdagne and asked that the priests of the French parishes be allowed to attend synodal assemblies in Urgell. Though the court in Paris granted his first demand, it firmly insisted that a French subject, a regnicole, administer justice. As the intendant of Roussillon—protesting that the general vicar of the Spanish Cerdaña, based in Puigcerdà, "claims to administer all kinds of justice" in the French Cerdagne—wrote in 1723:

> It is not right that French subjects be judged by foreigners who know neither our laws nor our rules, and who do not know the royal ordinances which must be followed in any given instance. . . . It would be dangerous were subjects of the king obliged to have recourse to foreign judges.[51]

Both the Roussillon intendant and the bishop of Urgell shared a growing concern in the late 1720s with the moral "disorder" of the clergy in the French Cerdagne. The intendant insisted that France had control over both "spiritual" and "temporal" affairs, while the bishop claimed to retain his "spiritual" jurisdiction—ultimately without success. In 1730, in exchange for visitation rights to the French Cerdagne, the bishop of Urgell finally consented, decreeing that both the long-desired "official" and an ecclesiastical court be named for the

49. ADPO C 1391: "Mémoire . . . par le Sr. Evêque d'Urgell"; ADPO C 1393 (edict of the Sovereign Council, August 3, 1669).

50. ADPO C 1354 (correspondence of bishop of Urgell and intendant, October–November 1721); ACSU leg. Cerdanya (intendant to bishop of Urgell, October 14, 1721).

51. ADPO C 1354: "Mémoire pour M. d'Andrézel," 1761; see also ACSU leg. Cerdanya (Cardinal de Fleury to bishop of Urgell, October 27, 1728).

French Cerdagne.[52] Beginning in the early 1730s, the intendancy took over the prerogative of nominating priests for the vacant posts in the French Cerdagne, preferring French subjects who had done their theological studies in Perpignan; and in the later 1740s, the French administration required that parish priests be non-Jansenists, "loyal subjects of the king," and able to "speak and understand French adequately."[53]

In fact, the French crown gained control over the "spiritual" affairs within the ecclesiastical frontier at about the same time that it solidified its hold over the "temporal" domain of the Spanish clergy in the French Cerdagne. The monarchy's attempt to tax their revenues and properties—estimated in 1726 at some 20,000 livres per year—consisted in extending the "free gift" (*don gratuit*) to the "exempt clergy" of the Roussillon. In truth, the two euphemisms were equally misleading. In 1724, when the intendancy drew up tax rolls including the revenues of Spanish clerics, the lower clergy—the priests and secondaries of the Cerdagne parishes—agreed to pay the tax and complained only when the tax assessment was increased in 1731. But the higher clergy, including the College of Priests of Puigcerdà and the bishop of Urgell, refused. The intendant of Roussillon ordered the revenues of the Spanish clergy seized for noncompliance, and the subsequent debates among ministers of both courts and the clergy of both provinces over the reciprocity of clerical taxation in France and Spain reveal more general, if implicit, philosophies about the nature of sovereignty and taxation.[54]

The Spanish clergy, whose spokesman was the bishop of Urgell, admitted that they had paid the "free gifts" in 1694 and again between 1701 and 1710, but insisted that both occasions were war-time levies, when French troops had controlled the borderland.[55] They further argued that neither the assessments of the *donativo gracioso* paid by the

52. ACA RA, vol. 137, fols. 227–231 (consulta of September 15, 1725); FAMRE MD FR 1747, fols. 258 et seq. (intendant to the Paris court, ca. 1728); ACSU leg. Cerdanya: "Advertencias del Obsipo de Urgel al Rdo Pedro Grau, rector de Hix y official de la Cerdanya françeso" August 1, 1730; and the (incomplete) registers of the ecclesiastical court in the French Cerdagne beginning in 1737, in ADPO Bnc no. 327.

53. ADPO C 1355 (lists of parish priests of the French Cerdagne, 1731–1785, but incomplete); see below, chap. 3 sec. 3, for a discussion of the problems of clerical gallicization and resistance during the eighteenth century.

54. ADPO C 868 (dossier on the *don gratuit* in the French Cerdagne, esp. "Députés du clergé dans la Cerdagne française," September 16, 1731); more generally, see M. Marion, *Dictionnaire des institutions de la France au XVIIe et XVIIIe siècles* (Paris, 1923), see "clergé" and "Don gratuit."

55. ACSU leg. Cerdanya (bishop to intendant, November 21, 1725). In fact, the 1701 tax was not a wartime levy: see ADPO C 866: "Etat des revenues," 1701.

Catalan clergy to Philip V at his accession, nor their "free gifts" of 1714 and 1715, had included the revenues in the Spanish Cerdaña of the great convents and churches of Roussillon. The Roussillon intendant turned the issue around and used the idea of reciprocity in France's favor: after all, the Roussillon clergy had paid the *quarto* and *excusado* taxes raised since 1660 on the Catalan clergy.[56] In 1732, the bishop of Urgell explained that these taxes were "graces" accorded by the Holy See to the Catholic king on a portion of the church tithes. The quarto and excusado were based on a cleric's revenue, not his person ("aunque las personas que perciben no son afectas, lo son los diezmos"). What was asked of the Spanish clergy was a personal tax, no less than

> a right of capitation, or a purely personal subsidy and charge, which cannot be levied on clergy who are said to be vassals of the Catholic King . . . and [therefore] not part of the Church of France.[57]

This distinction—between "personal" taxes levied on individuals and "real" taxes raised on lands "within the extent of the kingdom"— was an essential feature of the Old Regime tax system in France and Spain. Although the bishop of Urgell continued to insist until the late 1730s that the "free gift" was a personal tax, the French interpretation prevailed, as the intendancy was largely successful in taxing the Spanish clergy's revenues in the French Cerdagne. But when it came to the question of taxing lay properties in the French Cerdagne, the French crown paradoxically insisted upon the nonexemption of Spanish subjects from its own "personal" tax, the *capitation*.[58]

The royal ordinance of 1695 that created the first capitation clearly defined it as a "personal charge" to be levied on "subjects of the French king." Unlike other provinces less recently annexed to the French crown, and which tended to pay the *taille*, the Roussillon's capitation became an important source of royal revenue, amounting to about a third of the direct tax burden of the province and constituting between 20 and 30 percent of royal revenues from the French Cerdagne communities—second only to the *vingtième*. During the first capitation

56. ADPO C 868 (bishop of Urgell to intendant of Roussillon, April 2, 1732); and ACSU leg. Cerdanya (bishop to Spanish ambassador in Paris, April 28, 1732).

57. ADPO C 1354 (bishop of Urgell to intendant of Catalonia, March 17, 1732). On the nature and administration of the quarto and excusado taxes, see Elliott, *Revolt of the Catalans*, 92–93.

58. P. Sahlins, "An Aspect of Property Relations in the Borderland: Nationality, Residence, and the 'Capitation' Tax in the Eighteenth-Century French Cerdanya," in *Primer congrès d'història moderna de Catalunya*, 2 vols. (Barcelona, 1984), 1: 411–418.

in the Roussillon (1695–1701), no Spanish proprietors with land in the French Cerdagne were taxed; but after 1712, when the capitation was reintroduced, a change in the mode of levy raised the question of Spanish exemptions for lands in the French Cerdagne.[59]

Beginning in 1729, the Roussillon paid the capitation by subscription (*abonnement*): instead of a proportional tax on persons, it became a tax levied within the province on the basis of the customary tax rolls. The intendant and the viguiers distributed and assigned the tax burden to the communities and privileged groups of the province; the communities themselves assigned, under official tutelage, the imposed sums based on land ownership and revenue in the community. In 1729, the tax rolls excluded Spanish nationals with lands in the French Cerdagne, in keeping with the juridical definition of a "personal tax."[60] The issue, debated among Spanish landowners, French officials, and the Paris court until the late 1770s, was whether to tax Spaniards who subsequently acquired property in the French Cerdagne, whether by sale, inheritance, or gift.

As the Roussillon intendant explained in 1775, the practice was not to impose the capitation on Spaniards for their properties in the Roussillon itself. But in the Cerdagne,

> since many families are intertwined, their kinship ties and alliances make changes in property ownership more frequent, and all the more noticeable. We have thus considered that it was just to oblige Spaniards to pay the capitation.[61]

The intendants of Roussillon and the viguiers of the French Cerdagne had judged the capitation as "a mixed tax, half real and half personal." Since the 1730s, a customary set of procedures had developed—in accordance with a general ruling of 1713—whereby Spaniards who acquired land in France were taxed, but only if they resided in the French Cerdagne and farmed their estates themselves.[62] Spanish proprietors periodically challenged these criteria of taxation. The case of a particular Puigcerdà noble's claims for exemption are especially rele-

59. Marion, *Dictionnaire*, see "Capitation;" S. Mitard, *La crise financière en France à la fin du XVIIe siècle: La première capitation* (Rennes, 1934); ADPO C 793–794: "Arrêts, Ordonnances sur la Capitation."

60. Mitard, *La crise financière*, 128–131; ADPO C 804 (tax rolls of the nobility); ADPO C 843–846 (tax rolls of the French Cerdagne communities); and ADPO C 863 (list of individual exemptions).

61. ADPO C 864 (letter to intendant-general of finances, June 28, 1775).

62. Sahlins, "Property Relations in the Borderland," 414–416; and ADPO C 864 (viguier to intendant, May 2, 1744).

vant here, since they led to a series of decisions by Paris ministers upholding the juridical fiction of personal taxation in France.

Don Ignaci de Pera's petition in 1773 stated that his properties in the French village of Sallagosa had originally formed part of his family patrimony but were alienated in a form of sale that allowed the original owner to recuperate the property (called "venta a carta de gracia"). A century and a half after the sale, including nearly a half-century of litigation, Pera regained possession of these lands and, as a subject of the Spanish king residing in Puigcerdà, claimed exemption. The intendant admitted that Pera was Spanish and that the property was in France but claimed that he should not be exempted because the juridical nature of the capitation had changed after 1729. The case established French law, as Turgot, d'Ormesson, and Necker issued a series of decisions in the 1770s which lasted as long as the Old Regime.

The French ministers upheld the juridical definition of the capitation as a personal tax, even if this was a fiction in Roussillon and other provinces where it was levied by subscription. In doing so, the government ruled against the interests of French subjects—the French Cerdagne communities that had to pay the burden of subscription, including the exempted portions. Necker argued that not even the king could change the juridical nature of the tax; the capitation remained a poll tax, and "a foreigner not living in the kingdom can never be made to pay personal taxes."[63]

French insistence on the capitation reveals the limits of the territorialization of the state under the Old Regime. The French government gained territorial control of one jurisdictional frontier, ecclesiastical justice and taxation in the French Cerdagne, but it failed to secure—or even seek—the territorialization of the fiscal frontier, taxing the *properties* of the French Cerdagne.[64] By contrast, the Spanish state undertook a remarkable program of political reforms that, far more than the French, suggests an earlier and more radical movement toward the ter-

63. ADPO C 864 (memoir of Don Ignace de Pera, 1775); Necker to intendant, January 24, 1779, citing letters written by Turgot on May 30 and June 28, 1775.

64. The French fiscal administration did undertake, out of financial necessity, a number of other reforms, but none of these were directed toward consolidating the territorial character of state administration. Thus during the Spanish War of Succession, the crown "alienated" its patrimonial and seigneurial revenues in the French Cerdagne—leasing out the royal domain of Angostrina, Nahuja, Santa Llocaya, Dorres, as well as its indirect taxes on merchandise (lleudes) raised throughout the district: see Sahlins, "Between France and Spain," 241 and table 2.3. Over the course of the eighteenth century, the French monarchy also introduced a number of direct taxes—most notably the *dixième* and the *vingtième,* and their periodically increased rates of levy.

ritorialization of sovereignty. In France, it would take a Revolution—and then some—before the government would establish a single land tax on all the property within a state's territory through the survey and mapping of a land record or cadastre.

The introduction of the Spanish *catastro* in Catalonia formed part of Philip V's ambitious legislative program known as the *Nueva Planta*. The military, political, and fiscal reorganization of the political institutions of Catalonia was at once a retributive punishment for Catalonia's support of the Habsburg heir during the War of Succession, and an attempt to build an administratively unified and centralized government in Bourbon Spain. The Nueva Planta gave a new shape to the jurisdictional division of the kingdom, replacing the Catalan veguerias with Castilian-modeled administrative districts (corregimientos); it reformed the procedures and replaced the personnel of the *Real Audiència,* reorganized municipal government, and imposed the new system of direct taxation.[65]

Like the French system, the catastro made a distinction between "personal" and "real" taxes.[66] In practice, however, the administration of its "real" dimensions raised questions about the boundaries of fiscal sovereignty, touching off lengthy debates between France and Spain. The first dispute arose in 1717 with the repartition of properties in the Spanish Cerdaña: the formal valuation and description of these properties for the purposes of establishing tariff rates included lands situated in the villages and hamlets of Santa Eularia, la Vinyola, and Sant Pere de Cedret, as well as lands on the confines of Angostrina and Llívia and that part of the village of Hix situated on the southern side of the Raour River.

The problem, as a memorandum or *consulta* of the Real Audiència explained, was that the 1660 Llívia accord "did not specify or describe the limits and territory of each village." Two hamlets in question—Santa Eularia and la Vinyola—were not even mentioned in 1660 since they were at the time abandoned, but resettled in 1671. Don Francesc de Pastors, a local noble and seigneur of Enveig in the French Cer-

65. For general surveys of the provisions of Nueva Planta, see G. Anes, *El Antiguo régimen: Los Borbones* (Madrid, 1981), and the collected essays of J. Mercader i Riba, *Felip V i Catalunya* (Barcelona, 1968).

66. On the establishment and administration of the catastro, see Mercader i Riba, *Felip V,* 153–169; and M. Artola, *La hacienda del Antiguo régimen* (Madrid, 1982), 238–249. The crown continued to farm out its patrimonial revenues until 1808. For a list of these—the *leudes* of Puigcerdà and Llívia—their price of adjudication, and tax farmers, see P. Vilar, *La Catalogne,* 3: 450–455.

dagne, argued the case for Spain: those hamlets were dependencies of the Enveig Castle, which itself had depended, in military–administrative terms, on Puigcerdà. Puigcerdà's "jurisdiction" over the two hamlets was, according to Pastors, the same thing as its "territory"—even if the two hamlets were not on a contiguous terrain.

In fact, the idea of "territory" was used exclusively in local terms. The memoir, in referring to state territory, spoke instead of royal dominion. But this was still distinguished from royal jurisdiction. Concerning Hix, whose village territory was divided by the Raour River, forming the boundary of Spain and France, the Real Audiència stated that "it is not incompatible for lands to be of the dominion of one crown, and of the jurisdiction of the other," but concluded only that the two crowns ought to negotiate "a clear and durable settlement which will uphold their separate rights in these confines."[67]

Nothing was resolved in 1717, and over the next fifteen years complaints arose concerning these same lands.[68] The questions only found their partial resolution in 1732, when Spanish engineers and geometers, following instructions of the intendant of Catalonia Sartine, began a land survey (arpentaje) of the village communities for the purposes of administering the catastro. When they got to the enclave of Llívia, however, this domestic policy became an affair of state, for the engineers, producing a map of the communal territory, gave a finite shape and a territorial dimension to the communal boundaries that served as the international border line.

Between 1717 and 1732, the boundaries of Llívia had remained undefined. The Spanish state only considered the extension of the territory as a sum total of specific parcels of land that were taxed. The catastro, furthermore, only included owners of lands in Llívia if the principal properties forming their estates—the "body of the estate"—were located firmly within the municipality. Landowners from villages other than Llívia were considered "foreign landowners" (terratenientes estrangeros), whether they were foreign to the community or foreign to the kingdom. Until 1732, then, the catastro exempted those French landowners with estates outside Llívia, even if they held pieces of land within the enclave.[69]

67. ACA RA vol. 123, fols. 57v–66 (consulta of May 26, 1717).
68. See, for example, ACA RA, vol. 137, fols. 132–135v (consulta of April 12, 1725), as well as the disputes that arose when Spain erected its "sanitary cordon" (above, chap. 2 sec. 1).
69. See the registers of the catastro for Puigcerdà and Llívia in AHP and AMLL.

The 1732 survey and the resulting map were the first ones done for the town of Llívia, probably since Roman times (illustration 6). Yet there was no shortage of medieval and more recent boundary markers, including engraved rocks, stone crosses, and other signals. These had been placed—and displaced—following earlier disputes among communities, and occasionally between communities and seigneurs. The eighteenth-century engineers followed medieval descriptions of village territories as well as royal court decisions antedating the Treaty of the Pyrenees which mentioned boundary stones. In the Spanish survey these were connected by imaginary, straight lines. As a result, the survey came to include about 114 jornals of land belonging to French subjects—although the survey admittedly left out another 300 jornals also in dispute.[70]

The Roussillon Intendant Jallais and the French Cerdagne Viguier Sicart not only opposed this "usurpation" of French properties but argued more generally against a territorial division of the two kingdoms. The French position was that the boundaries of Llívia ought to be based on seigneurial jurisdictions, and thus on the descriptions of parcels of lay and ecclesiastical seigneurs. The Spaniards accused the viguier of acting in his own interest, since he himself owned several parcels within the confines of Llívia. But Sicart retorted that "to draw straight lines would result in a great prejudice to the king's jurisdiction, and is against the interests of his subjects since it reduces the extent of their own territory."[71]

Unable to reach a settlement concerning the territorial boundaries of Llívia and the neighboring French villages, the two intendants of Roussillon and Catalonia agreed to leave things as they stood prior to the land survey. French nationals whose "bodies of estate" lay firmly within the territory of a neighboring village were exempted from the Spanish catastro for the portion of their estates in Llívia, paying taxes on these lands to France. Conversely, Spaniards from Llívia whose estates extended beyond the village territory paid the catastro for these lands, but were exempted from land taxes in France.[72] The solution, in fact, favored the French state, since there were approximately twice as many proprietors of the French Cerdagne with lands in Llívia as there were Spanish landowners from Llívia with properties in neighboring French villages (table 1). (The direction of property ownership within the con-

70. ADPO C 2046 (Sicart to Jallais, January 18, 1732).
71. ADPO C 1046: "Observations pour répondre," 1732, articles 44–48.
72. ADPO C 2047, nos. 108–109 (memoir of the viguier Sicart, May 17, 1752).

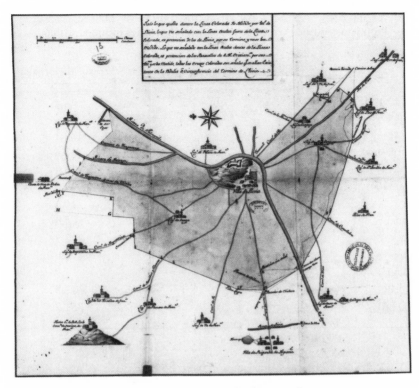

Illustration 6. Llívia Cadastre with Disputed Sections. This "Description or Map of Llívia" and the surrounding French villages was drawn up by Spanish geometers in 1732 during disputes between French and Spanish officials over payment of the Spanish land tax (*catastro*) in the borderland. Although the total acreage of disputed properties was relatively small (see table 1), the incident focused attention of the principles of territorial sovereignty.

fines of Llívia and the twelve neighboring communities reversed the direction of ownership in the Cerdanya more generally, where Spaniards held nearly four times as many properties in the French Cerdagne as the French did on the Spanish side of the boundary—see below, chap. 4 sec. 2 and table 4.) The two officials resolved their deadlock by asserting the primacy of a proprietor's residence and nationality, not the territory in which his lands were situated. The result was a set of fiscal frontiers of France and Spain less territorial than jurisdictional in nature.

In the second half of the eighteenth century, doubts were cast on the wisdom of the 1733 settlement. In 1764, it was the Spanish intendant

TABLE ONE   ESTATES ON THE BOUNDARIES OF LLÍVIA, 1762
Proprietors of the French Cerdagne with Properties included within
the boundaries of Llívia

Village	No. Proprietors	Acreage (in jornals)
Angostrina	15	52.4
Vilanova	4	14.2
Onzès	4	120.1
Targasona	12	67.6
Estavar	6	16.4
Sta. Llocaya	5	29.0
Ur	4	16.1
Eguet	2	4.9
Ro	2	8.1
Err	1	6.0
Palau	1	.9
Odello	1	13.8
TOTAL	58	351.7

Llívia Proprietors with Lands in Neighboring French Villages

Village	No. Proprietors	Acreage (in jornals)
Onzès	10	87.7
Ur	1	3.1
Estavar	2	8.1
Ro	2	5.3
Err	1	9.4
Sta. Llocaya	3	60.5
commune unknown	2	16.6
	21	188.5

SOURCE: ADPO C 864: "Estado o Aransel de las Tierras de Campo, y Prado, citas
en el termino de la Villa y Baylia de Llívia, contenidas en el Libro del R1. Apeo del año
1762 . . . que no pagan ni han pagado jamas en ningun tiempo el dho Real tributo qe.
corresponde a dichas tierras." The figures are given in the decimal system, although in
the table they are given as *jornals* and *ares* of arable land and meadows (16 are = 1
jornal = .9 acres).

who attempted to impose the catastro on a total of 138 French properties previously exempted, arguing for territorial sovereignty—although using a jurisdictional idiom.

> It is certain that any prince who has the right of sovereignty and dominion also has the right to impose taxes on all the lands which make up his estates, even if the owner of these lands are vassals of another prince, as long as one does not challenge the privileges of their persons.

But his French counterpart was not to be swayed.[73] Only in the 1780s did both the French and Spanish crowns and their provincial officials concur in the need to delimit their respective territories.

THE REPRESSION OF SMUGGLING: COOPERATION AND COMPETITION

After centuries of conflict, Spain and France allied themselves in the early eighteenth century on the continent and in the New World. The Bourbon monarchs affirmed their dynastic unity in a series of treaties: 1722, 1733 (sometimes known as the "First Compact"), and the "Family Compacts" of 1743 and 1762.[74] Concerned primarily with diplomatic and commercial affairs, these treaties were supplemented by a series of formal accords designed to assure the cooperation of Spanish and French government and military personnel in the Pyrenean borderland—the only shared frontier of the two states on the continent following the peace settlements of Utrecht and Rastatt.

As early as 1722, when the two states had not yet resolved their territorial disputes in the Cerdanya, the Paris and Madrid courts each agreed to return all "thieves, assassins, and deserters" who took refuge in the opposing kingdom. In 1765, a formal convention between the two Bourbon crowns agreed to return "deserters and criminals" who found refuge "in the lands of one dominion or those of another."[75] But it was their concern to repress smuggling that, above all, led the two

73. ADPO C 2048 (intendant of Barcelona to intendant of Roussillon, trans. into French, September 14, 1764; see also the letter of Sicart to the intendant, January 17, 1765).

74. V. Palacio Atard, *El tercer pacto de familia* (Madrid, 1921); W. N. Hargreaves-Mawdsley, *Spain under the Bourbons, 1700–1833: A Collection of Documents* (London, 1973), nos. 27, 37, 41, and 48.

75. On the negotiations of 1720–1722, see Archivo General de Simancas (hereafter AGS), Estado leg. 4682; ACA RA, vol. 3, fol. 54, and vol. 5, fol. 162; and FAMRE Limites, vol. 459, nos. 10 and 35 (correspondence of 1721–1722); a copy of the 1765 accord may be found in ADPO Bibliothèque 4880: "Convention entre le Roi et le Roi d'Espagne," September 29, 1765.

governments to cooperate—and to compete—in the borderland, and in the process to define a territorial boundary line.[76]

The frontier province of Roussillon was a privileged land of contraband trade, especially during the later eighteenth century, when markets began to expand in Catalonia and in France. Everything and anything of value could be smuggled, from sheep and mules to salt, tobacco, cloth, eau-de-vie, and currency. Of course, it was not always in the interest of the two crowns to cooperate in the repression of smuggling. During the eighteenth century, the Cerdanya was the most important land passage for Spanish gold and silver from the New World to France, accounting in the 1780s for some 40 million piastres a year. Although Spain strictly forbade the "extraction of bullion" from the kingdom, this liquid currency of gold and silver was important to French merchants as payment for merchandise and finished goods sold in Spain, and France had little interest in suppressing this illegal trade.[77]

The two states and their provincial administrations did cooperate in suppressing the illegal tobacco trade, for which each crown held a royal monopoly. Most of the tobacco came from Lombardy; discharged along the coast in Roussillon, it was expedited overland, toward Languedoc and, through the Cerdanya, into Catalonia.[78] By the 1770s, contraband trade in tobacco was growing to such a degree that it posed threats of "social disorder" as much as a loss of fiscal revenue. In May 1773, a band of 140 smugglers "took" the town of Puigcerdà while officials stood by helplessly, and the smugglers emptied the prisons of convicted colleagues. In May 1780, more than 60 smugglers attacked the governor of Roussillon while in the Conflent, and two years later an equally large group fought a battle with royal troops in the same district. This "banditry" troubled the Spanish authorities as well and led the two governments to sign accords in 1768, 1774, and 1786 establishing a special regime to police contraband trade in the borderland.[79]

76. M. Defourneaux, "La contrabande roussillonnaise et les accords commerciaux franco–espagnols après le Pacte de Famille, 1761–1786," in *Actes du 94e congrès national des sociétés savantes,* Pau, 1969 (Paris, 1971), 1: 147–163.

77. France, Archives Nationales (hereafter FAN) F(12) 1889, nos. 61–63 (correspondences of the Farmers General, June–July 1785); see also FMG AAT MR 1084, no. 41: "Mémoire sur la frontière des Pyrénées," 1777. On the periodic prohibitions in Spain concerning the export of currency, see Matilla Tascon, *Catalogo de la colección de ordenes generales de rentas,* 2 vols. (Madrid, 1962), ordinances of 1720 and 1757; and SAHN Hacienda, Libro 8039, fols. 328–331 (ordinance of 1787).

78. Defourneaux, "La contrabande roussillonnaise," 53–54.

79. FAMRE CP Espagne, vol. 570, fols. 332–334 (correspondence concerning the events of May 1773); ADPO C 1034 (*procès-verbal,* May 8, 1786); on the other ordi-

In agreeing jointly to repress smuggling, the French and Spanish states gave a formal expression to the territorial boundaries of their respective sovereignties. The 1768 convention explicitly allowed authorities of each crown to transgress the limits of France and Spain:

> The troops and guards of the two crowns may move freely beyond their reciprocal boundaries to stop the smugglers, provided that they mutually return the citizens arrested on the lands of one or the other power.

The accord acknowledged the existence of a boundary by stating its permeability; but in practice, it posed the contradictory goal of assuring the prerogatives of territorial sovereignty. Thus article 11 of the 1774 agreement specified that customs employees and soldiers of each state could "unite" and "mutually assist" each other on both territories while pursuing smugglers.[80] Competition and cooperation were built into the logic of the accords; both point to the way in which the states advanced a notion of territorial sovereignty in the context of a national customs boundary.

As in the case of its fiscal boundaries, the Spanish monarchy instituted a national economic frontier well before France. In 1717, as part of the political and administrative reordering of Bourbon Spain, Philip V suppressed nearly all the internal customs lines of the monarchy, making Catalonia a "united province" of the kingdom, along with Valencia, Aragon, Navarre, and the Basque territories. Only the *bolla*, Catalonia's own customs duty and tariff, remained in place until abolished by Charles III in 1769. Across the boundary, while the intendancy periodically wrote memoirs extolling the virtues of reforming the customs frontier and abolishing the customs posts between the provinces of Roussillon and Languedoc, Roussillon remained a province réputée étrangère until the last decade of the eighteenth century.[81] A grow-

---

nances, see Defourneaux, "La contrabande, 156–163; the texts can be found in G. F. Martens, *Recueil des principaux traités d'alliance, de paix, de treve ... depuis 1761 jusqu'à présent*, 6 vols. (Göttingen, 1791–1806): 1: 1–11 (Jan. 2, 1768); 6: 149–155 (Dec. 27, 1774); and 6: 227–234 (Dec. 27, 1786).

80. ADPO C 1041 (Comte de Mailly to the intendant, July 11, 1773); FAMRE CP Espagne, vol. 572, fol. 279 et seq. (Comte de Mailly to Duc d'Aiguillon, November 22, 1773, including various drafts of the accord).

81. Mercader i Riba, *Felip V*, 158–163; J. Muñoz Perez, "Mapa aduanero del XVIII español," *Estudios geográficos* 16 (1955): 747–797; FBN MS 221419, fols. 106–109: "Mémoires concernant les moyens pour faciliter le commerce en Languedoc et Roussillon"; and ADPO C 1014: "Observations sur le projet envoyé a M. le Controlleur Général," October 1761. On the more general movements during the seventeenth and eighteenth centuries to abolish internal customs, see P. Dockès, *L'espace dans la pensée économique*, 184–187. The definitive study of French proposals to reform the customs system during the eighteenth century is Bosher, *The Single Duty Project*, passim.

ing protectionism nonetheless formed part of the French monarchy's construction of a national economic territory, further evidence of the territorialization of sovereignty.

The policing of the grain trade through the Cerdanya provides the best example of the movement from a customs boundary to a territorial boundary line. Over the course of the eighteenth century, each state established an elaborate system of administration to ensure against the fraudulent extraction of primary, subsistence goods—firewood, cattle, and especially grain. Spain, by edicts of 1727, reiterated in 1732, 1737, and 1771, required all inhabitants within an eight-league-wide zone to carry declarations, or *guias*, with their merchandise. On the French side, the parallel system of *acquits à caution*—which required peasants to pass by one of the customs posts before transporting goods or livestock to other villages, French or Spanish—proved to be one of the more onerous aspects of living near the border, as complaints in the *cahiers de doléances* of 1789 so dramatically reveal.[82]

More generally, both states became increasingly protectionist over the course of the eighteenth century, forging distinct political economies in the Cerdanya borderland. Although the valley was self-sufficient during what the French intendancy called a "common year," most years in the later eighteenth century, marked by early frosts, excessive rain, or unusual cold spells, fell dramatically short.[83] During such years of shortages, the French Cerdagne sought its grain from Spain. In May 1777, for example, the Roussillon intendant requested rye for the French Cerdagne from his counterpart in Catalonia. The Baron de Linde agreed, claiming it gave him

> great satisfaction . . . to assist by allowing the export of rye to the villages of the French Cerdagne; I too am sincerely interested, as you are, in following the dictates of humanity, neighborliness, and the other ties which bind these two glorious monarchies.

Yet five years later, when a drought destroyed most of the crop in the Spanish Cerdaña, while that of the French side of the valley enjoyed a

82. BC Follets Bosom, nos. 6838 (edict of 1727), 6840 (1732), 6835 (1737); and the protests of Puigcerdà concerning these edicts in AHP Correspondencia "A," fols. 347, 361, et seq. On French complaints over the *acquits à caution* in 1789, see E. Frenay, ed., *Cahiers de Doléances du Roussillon* (Perpignan, 1979), 288–333.

83. P. Rosset, "Culture et élévage dans la viguerie de Cerdagne à la fin de l'Ancien Régime," *Revue d'histoire moderne et contemporaine* 31 (1984): 131–142; additional material on the police of the grain trade may be found in ADPO C 1024: "Sortie des grains," 1777–1781; ADPO C 1027 (bureau receipts of Sallagosa and La Tor de Carol); and ADPO C 1065–1069 (reports on the harvests, 1770–1789).

slight excess, the French intendant refused a similar request.[84] For by the 1780s, the French crown was prohibiting almost annually the export of subsistence primary goods, as part of a politics of unifying its national economic space. In 1785, Louis XVI finally ordered the suppression of the duties raised on merchandise leaving or entering Roussillon from Languedoc.[85] This internal reorganization of the customs frontier formed part of and capped the more general evolution toward a territorial boundary line.

TERRITORIAL BOUNDARIES OF THE OLD REGIME STATE

In *The Law of Nations, or Principles of Natural Law* (1758), the Swiss jurist Emmerich de Vattel wrote:

> It is necessary to mark clearly and with precision the boundaries of territories in order to avoid the slightest usurpation of another's territory, which is an injustice, and in order to avoid all subjects of discord and occasions for quarrels.

Vattel, following Christian Wolff, was among the first theorists of international law to identify territorial boundaries as the point at which sovereignty found expression.[86] This interest in the definition and maintenance of territorial sovereignty coincides with a new consciousness of territory that came to displace the older notion of jurisdictional sovereignty. This new awareness was visible in the language of local political and administrative officials, concerned to prevent territorial violations precisely where they occurred, on the periphery. But it was also the concern of the French Ministry of Foreign Affairs, which in the second part of the eighteenth century developed a coherent policy of "establishing and fixing the limits of the kingdom."

In the four decades before the Revolution, the French state began to

84. ADPO C 2049 (intendant of Catalonia to intendant of Roussillon, May 6, 1777; see also SAHN Estado libro 67, fols. 152–154 (municipal council of Llívia, July 1778, claiming 40,000 *quarteras* of grain were "sent down to the town of Perpignan . . . and to most of the County of Foix"); ADPO C 1068 (intendant of Roussillon to intendant of Catalonia, September 28, 1782).

85. ADPO C 1012: "Lettres patentes du Roi," September 1785; Bibliothèque Municipal de Perpignan MS 203 (memoir sent by the intendant of Roussillon to the comptroller-general, November 7, 1785); and Bosher, *The Single Duty Project,* 105–128.

86. E. de Vattel, *The Law of Nations, or Principles of Natural Law,* trans. C. G. Fenwick (Washington, 1916 [1758]), bk. 7, chap. 2; C. Wolff, *The Law of Nations,* trans. J. H. Drake (Washington, D.C., 1934 [1749]).

define its territorial sovereignty. According to an anonymous memoir of 1775,

> Of all the areas of administration, that of boundaries has been the most neglected in France. All of the kingdom's frontiers consisted in conquered provinces. The causes of the present disorder and uncertainty include the haste with which these peace treaties were drawn up, and the absolute ignorance of the plenipotentiary ministers concerning the nature and extent of the districts which they ceded.

The Chevalier de Bonneval, a royal engineer stationed at Besançon, explained how seventeenth-century treaties had involved "a victorious power dictating terms to a vanquished one," and there had been little concern with, or knowledge of, territorial extension. Partly influenced by Bonneval's persistent prodding, the ministry began to rectify the failures of past treaties.[87] Part of this effort involved building up an important map collection of all of France's frontiers. More significantly, by 1775 the Ministry of Foreign Affairs had taken over the formal jurisdiction of boundary matters from the War Ministry, despite the latter's extensive protest and continuing influence. Within the Foreign Affairs Ministry, the French government established a Topographical Bureau for the Demarcation of Limits, creating permanent "commissioners" to negotiate "treaties of limits" with France's neighbors.[88] During the 1770s and 1780s, the French government began the arduous process of delimiting its territory, signing nearly two dozen "treaties of limits" with the neighboring polities making up the Holy Roman Empire, the Swiss cantons, and the Kingdom of Savoy. In 1785, France signed its first delimitation treaty with Spain—although the accord only concerned the western Pyrenean frontier.[89]

The "rationalization" and "purification" of the frontier in the later

87. FAMRE Limites, vol. 7: "Observations concernant les limites du Royaume," 1775; for the earlier treaties, see the table drawn up in 1749, "Pouvoirs, Ratifications, et autres actes . . . par rapport à des reglements de limites entre les provinces et les états étrangers limitrophes du royaume, 1648–1749."

88. Konvitz, *Cartography in France*, 34–35; FAMRE Limites, vol. 7: "Conservateurs des limites," including the ordinance of January 31, 1773, creating an inspector "charged with the operation and affairs regarding exchanges and concessions of territory from sovereign to sovereign, and with regulations of the limits of the states and possessions of His Majesty with those of neighboring states." On the shift of jurisdiction under the Duc de Choiseul, when he held both portfolios of war and foreign affairs, see FAIG 4.3.1.1, no. 24: "Mémoire sur les affaires des limites (écrit vers 1770)."

89. J. F. Noël, "Les problèmes de frontières entre la France et l'Empire dans la seconde moitié du XVIIIe siècle," *Revue historique* 235 (1966): 333–346; and P. Sahlins, "Natural Frontiers Revisited: France's Boundaries since the Seventeenth Century," forthcoming in *American Historical Review*.

eighteenth century has long been noted by geographers and historians. The social geographer Vidal de la Blache saw in the process the French state's need to "carefully close its territory, as a peasant would enclose its field."[90] The simile is not altogether out of place in eighteenth-century discussions about the need to "fix the limits" and to "place boundary markers" on its frontiers—to define a linear territorial boundary. Not that eighteenth-century ministers thought the state's territory similar *in kind* to that of a peasant, for the monarchy defined territorial sovereignty precisely in opposition to the boundaries of jurisdictions and of private properties. "The dividing line of two territories," wrote the Chevalier de Bonneval in 1748, "must be drawn without reproducing the boundaries of attachment to a bailiwick or a *chatellenie*."[91] In 1772, the French foreign minister was advised that the boundary line

must be determined by markers . . . or by a ditch which functions as an outline. . . . It is better to cut the possessions of persons into two parts, under different dominations. The bad feelings would be only momentary, and the communities can remedy the situation by exchanging their goods so that they only have possessions under the same sovereign.[92]

But if state territory differed in kind from a peasant's possessions, the idea of a "peasant enclosing his field" does approximate the eighteenth-century monarchy's vision of "closing" its territory. Just as seventeenth-century writings spoke of "enclosing" belligerent states or nations within clearly defined natural frontiers, so did eighteenth-century experts, advisors of the Ministries of War and Foreign Affairs, continue to think about "closing" the territory of a state. Commissioners sent to negotiate treaties must seek to "purge the kingdom of foreign enclaves," and to

close the state as far as the nature of the district permits, so that in case of war it may easily be defended, and in times of peace, it can be easily protected against desertion and fiscal fraud, removing all hope of retreat from wrongdoers who break the law with the idea of an assured asylum. That is what happens when territories are confounded.[93]

90. Quoted in Zeller, "Saluces, Pignerol, et Strasbourg," 109.
91. FAMRE Limites, vol. 7 (memoirs of April 22 and April 29, 1747, and February 22, 1749); the second memoir is reproduced in Lapradelle, *La frontière,* app. 3.
92. FAMRE Limites, vol. 7: "Mémoire sur les frontières du Royaume," May 26, 1772.
93. FAMRE Limites, vol. 7 (Bonneval memoir of April 22, 1747).

The delimitation treaties of the late eighteenth century suggest a new stage in French state building, and in the European balance of power. Vauban's "iron frontier," consolidated in the 1680s, guaranteed France's annexations and conquests, but it was still a frontier designed for military purposes. A century later, the concerns of the state had changed dramatically. The era of annexations was over: the Treaty of Vienna in 1738 gave the province of Lorraine to Stanislaw Leczinski of Poland, and it passed to France upon his death in 1766. Lorraine, along with Corsica, bought by Louis XV in 1768, were the last of France's acquisitions under the Old Regime. As the Marquis d'Argenson, foreign minister under Louis XV, wrote in his memoirs: "This is no longer a time of conquests. France must be satisfied with its greatness and extension. It is time to start governing, after spending so much time acquiring what to govern."[94] The "treaties of limits" served the French state's domestic policies of political consolidation. As the Chevalier de Bonneval explained:

> The motives which engage us to undertake such accords are, in general, to suppress enclaves from which are born an infinity of inconveniences, to give to commerce a greater facility of communications, by land or river, to determine as clearly as possible the limits, so as to destroy all disputes between frontier inhabitants, and to give the government the means to take efficient measures against desertion and smuggling.

But the foreign policy implications of France's ultimately unsuccessful program of bureaucratic and administrative reform suggest the relative weakness of the French state in the eighteenth century, compared to a century earlier. France's boundaries became the sites of micropolitical contention between the French court and its undistinguished neighbors—two-bit bishoprics and petty principalities, with which it was forced to negotiate directly, as equals.[95]

The eighteenth-century delimitation treaties reveal the widespread preoccupation with tracing natural boundaries, especially rivers, which were deemed to be "fixed" and "permanent." "It is good to take streams, rivers, watersheds, or finally straight lines, when none of these [others] exist" as the boundaries of territories, reads a memoir sent to the foreign minister in 1772. And it was commonplace to name natural boundaries, both specifically and generally, in the prefaces and ma-

94. D'Argenson is quoted in Sorel, *L'Europe et la Révolution française*, (Paris, 1885), 1: 319–320.

95. FAMRE Limites, vol. 7 (memoir of April 22, 1747); Sahlins, "Natural Frontiers Revisited."

jor clauses of the "treaties of limits" during the 1770s and 1780s.[96] Such an insistence on natural boundaries received its most elaborate expression in the Pyrenees. In particular, the military establishments became the principal voices urging both governments to trace the French–Spanish boundary along the mountain crest.

The French War Ministry's position—eventually adopted by the French–Spanish commission named to delimit the boundary in 1785— was that only natural boundaries could be used to establish "clear, unchangeable, publicly known, and forever enduring demarcations."[97] The idea of natural frontiers, however, had undergone a significant transformation since the seventeenth century. For one, the dominant vision of the Pyrenees in the eighteenth century was no longer one of a conqueror. "Natural frontiers" (frontières naturelles) gave way to "natural boundaries" (limites naturelles). The principles of mountain warfare suggested less the control of passages across the Pyrenees than "mastering the heights" of the mountain range. The boundary established by the Treaty of the Pyrenees thus fell short of this military ideal. The boundary line "left the crest and summit of the mountains" and had been established instead on the basis of "such vague antiquities as the limits of Narbonese Gaul." The Comte de Varennes, in a military reconnaissance of the Cerdanya during 1777, failed to see why France had sought the thirty-three villages across the Perxa Pass: "The conqueror dictated terms to the conquered, but without great utility."[98] The French–Spanish boundary would be rectified by taking the watershed, without exceptions, as the proper limit.

Second, eighteenth-century military officials reinterpreted the Treaty of the Pyrenees in the spirit of an enlightened emphasis on Nature. Hume, Rousseau, Montesquieu, Turgot, and others believed that nature had established the correct boundaries of states in general and France in particular, although none of their commentary on natural boundaries included any explicitly political claims. As for the Pyrenees, though they may have been considered in the seventeenth century the *historical* division of ancient peoples, their most important quality in

96. FAMRE Limites, vol. 7: "Mémoire sur les frontières"; on the naming of rivers in the eighteenth-century delimitation accords, see D. Nordman, "L'idée de frontière fluviale en France au XVIIIe siècle," in *Frontières et contacts de civilisations* (Neuchatel, 1979), 84–88.

97. Quoted in Girard d'Albissin, "Propos sur la frontière," 398.

98. AIG 4.3.2.2: "Memoire sur les limites de la frontière en Roussillon," 1775; see also FMG AAT MR 1084, nos. 41 and 43: M. de la Varenne, "Itinéraire et remarques locales et militaires sur la frontière," November 1777.

the eighteenth century was the role of the mountain as a "natural barrier" defined by the idea of a watershed.[99] Shorn of historical baggage, the Pyrenees presented themselves as the classical and unequivocal example of a natural boundary.

Negotiations toward the delimitation of the Pyrenean frontier began in 1784, although the courts in Paris and Madrid had made overtures as early as 1761. The Caro–Ornano Commission was named after the French and Spanish negotiators, both military officers. With their respective royal engineers and cartographers, they began their work at the mouth of the Bidassoa River, and signed a formal convention on August 27, 1785, delimiting the Basque and Navarese Pyrenees.[100] It was the commission's only achievement: its further work was interrupted by the outbreak of the Revolution in France. Yet this single accord was doomed to failure because of the commissioners' insistence on natural boundaries.

The preamble to the "Definitive Treaty of Limits between France and Spain" summarized the states' concern and intimated the cause of their failure:

> The most Christian and Catholic Kings, animated by the desire to unite beyond the ties of blood and friendship which link them so tightly, and wishful to see their subjects enjoy the advantages of this great harmony, have decided to destroy and abolish the foundation of quarrels and discussions among their respective frontier inhabitants [by] drawing a dividing line which separates and divides all the lands between the two powers, the property of the valleys, and the sovereignty of the two kings.

The French and Spanish commissioners had been instructed to "follow the mountain crests and watersheds, unless there are contrary titles or visible inconveniences in doing so." In point of fact, they saw little "inconvenience" in giving up the rights and privileges of communities with long-standing charters to specific territories and jurisdictions, many of which extended beyond the watershed. As a result, the boundary line was as impractical as it was abstract. For in the end, the Caro–Ornano

99. N. G. Pounds, "France and 'les limites naturelles' from the seventeenth to the twentieth centuries," in *Annals of the Association of American Geographers* 44 (1954), 51–62; FAMRE Limites, vol. 439, no. 91: "Premier mémoire sur la frontière du Roussillon," September 24, 1777, fol. 2, reinterpreting the provisions of the 1659 Treaty.

100. FAMRE Limites, vol. 459, nos. 91–92 (correspondence between Spanish ambassador and French foreign minister, August 1761); FAMRE Limites, vol. 463: "Commissions de Limites, Espagne, 1784–1825; FMG AAT MR 1084, no. 75: "Reflexions sommaires sur la fixation des limites des Pirinées . . . en Navarre"; the text of the accord may be found in C. Koch, *Table des traités entre la France et les puissances étrangères dépuis la paix de Westphalie jusqu'à nos jours*, 2 vols. (Basel, 1802), 2: 477 et seq.

commissioners were drawn to a natural frontier shorn of history—not simply the ancient history of pre-Roman peoples but the historical claims of local valley communities.[101] It is worth quoting at length a perceptive but anonymous memoir of May 1780:

There have always been disputes among French and Spanish neighbors in the Pyrenees concerning fishing rights, rights to wood, and rights to pasture in the mountains. Treaties signed between the two crowns did give some kind of boundary to the two great powers by stipulating provinces, districts, and even villages in each canton. But the plenipotentiaries active in the Treaty of Vervins and the Treaty of the Pyrenees neglected to name commissioners after the treaties to *fix the territorial jurisdiction of the two states*. . . . The terrain which separates the two states has until now been the subject of discussions and altercations between the frontier inhabitants; frequently these people have signed accords and conventions between themselves, always for a limited time, in order to be able to use this or that mountain, or this or that forest, during a fixed and limited time.[102]

In fact, the author was describing local accords called *lies et passeries* in French and *facerias* in Spanish which long antedated the two treaties between France and Spain. By these conventions, the valley communities in the central and western Pyrenees maintained usufruct rights in the pastures and mountains of others across the watershed.[103] Drawing a linear boundary to determine the "territorial jurisdiction" of each state involved ignoring these local accords. It is no wonder that the valley communities of the western Pyrenees refused to recognize the 1785 treaty between France and Spain.

Although the Caro–Ornano commissioners never reached the eastern Pyrenees, news of the impending delimitation of the boundary soon did. The reactions of the intendant of Roussillon, the bishop of Urgell, and the town council of Llívia summarize the evolution of the Old Regime frontier of France and Spain. Raymond de Saint Sauveur, the

101. J. Sermet, *La frontière des Pyrénées et les conditions de sa délimitation* (Lourdes, 1983), passim; on the parlement of Navarre's rejection of the accord and its conception of the frontier as both a political division and one founded on pastoral exchanges, see C. Desplat, "Le parlement de Navarre et la définition de la frontière franco–navarraise à l'extrême fin du XVIIIe siècle," in Nail et al., *Lies et passeries dans les Pyrénées*, 109–120.

102. FAIG 4.3.2.8: May 1, 1780 (emphasis added).

103. H. Cavailles, "Une fédération pyrénéenne sous l'Ancien Régime," *Revue historique* 55 (1910): 1–32, 241–274, reprinted in Nail, et al., *Lies et passeries*, 1–67, remains the classic study of the form and evolution of these conventions—sometimes called "letters of peace"—among valley communities throughout the western and central Pyrenees. They are largely absent in the eastern or Catalan Pyrenees where, except for Andorra, the corporate character of the valley communities was less developed.

Roussillon intendant, underscored the "long-recognized need to fix precisely the limits of the two Cerdanyas." The Llívia accord, he continued, had left properties "mixed, confused, and intertwined." Concerned about the monarchy's fiscal frontiers and its ability to tax properties in the borderland, the intendant proposed to return to the division of 1660, and

> without regard for the dependencies of the villages of the two kingdoms, simply establish lines and demarcations as the locale permits, without following [the boundaries of] private possessions on this or the other side of the lines. These possessions must follow the fate of their location.[104]

The intendant of Catalonia had made a similar proposal in 1765, opposing the practices that had emerged from the 1732 survey of Llívia, but his counterpart in Roussillon had opposed the plan. Now, twenty years later, both crowns were prepared to give a territorial definition to their fiscal jurisdiction over properties.

In 1785, the bishop of Urgell, "hearing that they are trying to regulate the frontiers of Navarre," sent a memoir to the Conde de Floridablanca, the Spanish prime minister. The bishop cited "the difficulties of governing well, and the occasions for disputes about etiquette, as well as other problems which occur at every step when the customs and systems of government of the two kings are very different." He went on to propose the exchange of his ecclesiastical jurisdiction over the French Cerdagne for jurisdiction over the Aran Valley, dependent on the French bishopric of Comminges and located on the French watershed. He thus proposed to give up the ecclesiastical frontier that his predecessors fought so hard to preserve in favor of a territorial—and natural—boundary of France and Spain.[105]

Perhaps the most revealing petition came from the town council of Llívia, which wrote to Floridablanca when rumors reached the town that it might be ceded to France. Llívia's memoir contained several different arguments about why it should remain Spanish.[106] The first two

104. FAMRE Limites, vol. 461, no. 3 bis (intendant to French foreign minister, June 16, 1784).

105. SAHN Estado, libro 673d, fols. 158–159 (bishop of Urgell to Spanish first minister, December 9, 1785). There was no small irony to this proposal. In the Figueres conferences of 1666, the French commissioners had upheld the example of the bishop of Comminges, who had named a vicar-general to administer the Spanish villages of his jurisdiction: ADPO C 1393. The Urgell bishop's proposal in the late eighteenth century suggests the extent to which France had succeeded in extending its control over the ecclesiastical affairs of the French Cerdagne.

106. SAHN Estado, libro 673d, fols. 190–197 (municipality of Llívia to first minister, January 7, 1785).

focused on its excellence as a site for a fortification, "reputed to be one of the best in Spain." Even if the Llívia accord of 1660 forbade the construction of fortifications, that accord "was neither approved, nor would it ever be approved, by Philip IV or any of his successors." Such arguments, which reminded the Spanish crown of the Cerdanya's role in the military frontier, would have found little resonance among the Madrid court. A half-century earlier, military memoranda had urged the government to rebuild the walls of Puigcerdà, but the central government instead chose to spend a small sum rebuilding the two feudal ruins of Bar and Aristot, perched high above the Segre River far from the boundary, controlling the passage between the Cerdanya and the Urgellet.[107] Since the construction of Mont-Louis, France had remained firmly in control of the military frontier.

The town of Llívia also evoked its economic interests, and those of the crown more generally. The town claimed to provide some 63,000 reales de vellon annually to the royal treasury, including 20,000 raised in customs dues. The loss of Llívia would result, moreover, in "the total commercial decay of Puigcerdà and the rest of the Spanish Cerdaña," since the French would establish their businesses and factories directly on the boundary and fraudulently introduce merchandise into Spain, and since it would no longer be possible to smuggle cattle into France, which greatly profited Barcelona and all of the principality. The memoir concluded that if Llívia were ceded to France, the inhabitants would be forced to abandon or sell their houses, "since they would expatriate and come back to live in the lands of Your Majesty, since they [would rather be] the vassals of so liberal, generous, and benign a monarch."[108]

Although there was much standard rhetoric in Llívia's claims, the 1785 petition also reveals the concerns of local society in the Spanish borderland, including its potential impoverishment by the development of French industry. But the petition went unheeded. The later eighteenth-century states were increasingly concerned with the territorial dimensions of sovereignty, but they remained unconcerned with local society. The problem of territory was posed, but it was not developed with respect to the concerns, struggles, and conflicts of those who lived on its boundaries, nor with the way in which the inhabitants of the two Cerdanyas thought about themselves in relation to the states.

107. SBCM Documentos (3–2–5–3), no. 26; and SBCM Aparici 15 (1–2–6–12): "Datos sacados de Simancas sobre plazas y organizaciones de fronteras," esp. the memoirs of June 1748 sent to the Spanish first minister, the Marques de Enseñada).
108. SAHN Estado, libro 673d, fols. 190–197.

The state's insistence on territory in the late eighteenth century was also several steps away from becoming the vehicle or excuse of international strife. "Territorial violations"—and the associated formulas of "rape," "pillage," "usurpation," and "theft"—properly belong to the Age of Nationalism. Between the "territory" of the late eighteenth century and that of the late nineteenth lay the world of the French Revolution and the early nineteenth century. Before a simple border incident could become a casus belli in the late nineteenth century, territory—and its boundaries—would become nationalized and politicized.

3
Resistance and Identity under the Old Regime

FRENCHMEN, SPANIARDS, AND CATALANS IN THE SEVENTEENTH CENTURY

In 1626, Doctor Lluis Baldo, "honored citizen" of Perpignan, wrote a pamphlet directed to Philip IV urging the creation of a new province composed of the two counties of Roussillon and Cerdanya, distinct from the Principality of Catalonia. In strengthening his claim, lest Madrid think the demand involved separatist intentions, he evoked the courage and loyalty of the Roussillonnais, and especially of the Cerdans, in their struggles against the French, underscoring their Spanish identity.

> The people are naturally more Spanish than those of the other provinces of Spain; they have such a notorious antipathy and natural hatred of the French, their neighbors, that it cannot be described in writing. Their feelings are so extreme that a son, born in the counties, abhors with a natural hatred his father, born in France.

But a marginal note in one copy of Baldo's pamphlet expressed the doubts of "a curious reader," for whom the loyalty of the Roussillonnais and Cerdans was indelibly marked by their origins. "The plain is half-populated by Frenchmen," he scribbled, "and the other half are sons of Frenchmen, so that there are no natives upon whom one can count."[1] Baldo and his anonymous reader concurred as to the impor-

---

1. Baldo and his reader are quoted in J. S. Pons, *La littérature catalane en Roussillon,* 1: 8–9; copies of Baldo's pamphlet and responses by Geronymo Margarit may be found in FBN F01 48. On resentment against the hegemony of the principality and move-

103

tance of the French presence in Roussillon and Cerdanya: their disagreement reflected and reproduced the ambivalence of local attitudes among the Cerdans toward France and Spain before the Treaty of the Pyrenees.

On the one hand, the early seventeenth-century inhabitants of the Cerdanya saw the French as their enemy: having long suffered from periodic attacks and pillages by soldiers in the pay of the French king, they had developed a strong anti-French sentiment. Their hatred of "them" was grounded in their love of their local patria, as described by an anonymous Cerdan in the early seventeenth century.

> The people are . . . so attached to their patria that they will put their lives at stake, not fleeing from the enemy even when they know him to be close at hand. . . . They have known how to defend themselves from the pestering and hateful assaults of the French Huguenots, and they have even staged ambushes, forcing the French to turn around with a loss of men and reputation.

The term "Huguenot," as Núria Sales has shown, was widely used to describe the bandit factions of the Catalan Pyrenees, and their noble leaders would sometimes cultivate such an identification.[2] But when the Cerdans expressed their resentment and antagonism toward the "French Huguenot" armies in defense of their local patria, they identified the foreigner by religious affiliation. Turning churches into stables and murdering priests, the French armies in the Cerdanya during the fifteenth-century occupation had behaved with equal heresy, but it was only when religious difference and political affiliation coincided—that is, after the midsixteenth century—that the French were called Huguenots. Not that the Catalans themselves were unmoved by heretical currents in the later Middle Ages; yet as Jean Sebastian Pons wrote in his study of popular Catalan literature in Roussillon during the seventeenth and eighteenth centuries, "the hatred of the heretic was the foundation of patriotism."[3] For the Cerdans, it was a patriotism defined by a local sense of place in opposition to the French.

---

ments of "separatism" of the two counties, especially during the 1620s, see Elliott, *Revolt*, 263–266, and R. García Cárcel, *Historia de Cataluña, siglos XVI-XVII*, 2 vols. (Barcelona, 1985), 1: 53–60.

2. BC MS 184, fols. 3 and 6: "Descripción del condado y tierra de Cerdaña," n.d., ca. 1610; N. Sales, "'Bandoliers espaignols' i guerres de religió françeses," *L'Avenç* 82 (1985): 46–55; see also the sources on Catalan banditry cited in the Introduction, n. 28.

3. Pons, *La littérature catalane*, 16; on the fifteenth-century occupation, see X. de Descallar, "Episodes de la domination française dans les comtés de Roussillon et de Cerdagne sous le regne de Louis XI," *RHAR* 3 (1902): 359–371; and S. Galceran i Vigué,

On the other hand, the familial metaphor used by Baldo was particularly appropriate given the presence of French immigrants in the two counties and the ways in which they were assimilated into Catalan society. "Roussillon was almost French by blood before it was by the flag," affirm Nadal and Giralt in their essential study of French immigration into Catalonia during the early modern period. In 1542, the French in Perpignan made up almost 20 percent of the population.[4] In Osseja, adjoining the Cerdanya plain, a third of all spouses married in the parish between 1566 and 1602 were born in France, most in the neighboring bishoprics of Alet and Pamiers.[5] The southern movement of French immigrants from the Pyrenees, Languedoc, and the Massif Central into Spain reached its apogee in the late sixteenth century and declined rapidly after the midseventeenth, but large numbers of French men and women continued to arrive in Catalonia throughout the Old Regime. Although they practiced a wide range of trades, most of the French immigrants—almost 84 percent in 1627—were nonspecialized workers in agriculture, stock raising, fishing, or transport who tended to settle in rural areas, while those practicing more specialized trades, including artisans and merchants, settled in the towns and cities of Catalonia.[6]

For the most part, French immigrants were assimilated with relative ease into Catalan society; through marriage and residence in the towns or countryside of the principality, they acquired the rights and responsibilities of Catalan "nationality" or "citizenship" (naturaleza). It is true that first-generation Frenchmen frequently did not gain the full status of Catalans, with rights to exercise public offices. The legal re-

La ocupación francesa de Cerdaña desde 1462 al 1493 y Nuestra Señora de Gracia (Ripoll, 1971), which describes the creation of the Cerdanya's patron saint, "Our Lady of Grace," as a reaction to the sacrilegious dimensions of the French occupation. On French heretics in Catalonia during the late middle ages, see J. Ventura Subirats, Els heretges catalans (Barcelona, 1963); and E. Le Roy Ladurie, Montaillou: The Promised Land of Error, trans. B. Bray (New York, 1979).

4. J. Nadal and E. Giralt, La population catalane de 1553 à 1717: L'immigration française (Paris, 1960), 90–91.

5. ADPO 7J (Parish registers, Osseja [Sant Pere]).

6. Nadal and Giralt, Population catalane, 130–153; on French immigration to Spain more generally, see J. P. Poussou, "Les movements migratoires en France et à partir de la France (fin XVe-début XIXe siècle): Approche pour une synthèse," in Annales de démographie historique (Paris, 1970): esp. 23–51. On the status and activities of Frenchmen in early modern Spain, see A. Gutierrez, La France et les français dans la littérature espagnole: Un aspect de la xénophobie en Espagne, 1598–1665 (Saint-Etienne, 1977), . 94–121; and A. Girard, Le commerce français à Séville et Cadix au temps des Hapsbourg: Contribution à l'étude du commerce étranger au XVIe et XVIIe siècles (Paris, 1932). On French immigration to Catalonia during the eighteenth century, see E. Moreu Rey, Els immigrants françesos a Barcelona (Barcelona, 1959), 35–50.

quirements specified by the Catalan *constitucions* of 1422 and 1481 defined Catalans as those born in Catalonia or as the sons or grandsons of Catalans who established permanent residence in the principality, although individuals could also gain the status of Catalans by "letters of naturalization" given by the Corts and decreed by the Diputació. Frequently, first-generation Frenchmen, even if married to Catalans, were explicitly prohibited from holding public office, as for example by the legislation of 1553, yet their sons, born and residing in Catalonia, gained the full status of Catalan citizenship.[7]

The first-generation immigrants from France could thus be excluded and singled out for attack, often as a direct result of international tensions and rivalries between the French and Spanish states. At moments when the French–Spanish opposition became more marked in political life—at the turn of the century, during the early 1620s, or in the years surrounding the outbreak of war in 1635—the social boundaries of exclusion hardened. In the political and cultural centers of France and Spain, pamphlets and poetry, theater and ballet made reference to the "antipathy" of the Spanish and French nations. In Catalonia, literary and polemical attacks on the French tended to coincide with actual ones in moments of political and economic crisis, during which French subjects—and especially French merchants—frequently served as scapegoats. In Spain more generally, literary xenophobia at moments of crisis took shape as an unrelenting attack on Frenchmen, or *gavachos*.[8]

The origin of the term gavacho is uncertain. It can be linked etymologically to the Latin name of the Gévaudan region in France, from where many poorer immigrants to Spain emigrated. But the term was also used by the people of Languedoc as a derogatory description of neighboring Gascons—generally poor and in search of subsistence—who settled among them. In general, gavachos were outsiders, foreigners who did not speak the language, men and women who were less civilized. In early modern Spain, gavachos were Frenchmen. In the literary culture of the Golden Age, Frenchmen and Spaniards

7. On the definition of Catalan naturaleza, see N. Sales i Bohigas, "Naturals i aliénígenes: Un cop d'ull a algunes naturalitzacions dels segles XV a XVIII," in *Studie in honorem Prof. M. de Riquer* (Barcelona, 1987), 1: 675–705; V. Ferro, *El dret públic català*, 319–322; and García Cárcel, *Historia de Cataluña*, 1: 132–136.

8. Gutierrez, *La France et les français,* passim; on Catalonia, see Carrera i Pujal, *Historia economica de Catalunya*, 4 vols. (Barcelona, 1946), 1: 376 et seq., which summarizes the pamphlet attacks and polemics of Dalmau, Soler, and Peralta concerning Frenchmen.

formed a characteristic opposition—"two bellicose nations . . . in natural contestation," wrote Lope de Vega. As more generally in early modern Europe, national identities were frequently defined by counteridentities.[9]

Seventeenth-century Catalans demonstrated more ambivalence toward the *gavatxos* among them. In 1684, for example, the confraternity of Sant Miguel in Puigcerdà protested when a Frenchman proposed to open a store in the town. Certain royal privileges, moreover, prohibited the sons of French natives from holding municipal offices in the Cerdanya. In Barcelona and other towns, popular xenophobia helped perpetuate the French confraternities and French domination of membership in certain guilds.[10] But in the countryside, French immigrants, less differentiated by trade or customs, were excluded with more difficulty. Although native Catalans could express resentment against the gavatxos who came to work the harvests, those who married and established themselves in the villages became functioning members of the community. They became "neighbors," enjoying the full prerogatives of local citizenship, such as rights of pasturing and firewood, but also sharing in the responsibilities, including defending the village territory—and the valley—against the marauding bands of "Huguenots."[11]

If the attitudes of Cerdans toward the French were ambivalent, their identities as Spaniards were even more problematic. Not only did loyalty to Spain differ among distinct social groups—as was also the case of French loyalties and identities in the two centuries after the annexation—but national identity was evoked contextually and often in the service of local interests.

The municipal council of Puigcerdà, for example, frequently evoked

9. For possible etymologies, see Nadal and Giralt, *Population catalane,* 113–114; Gutierrez, *La France,* 30–31; and Moreu-Rey, *Els immigrants françesos,* 13. Lope de Vega is quoted in Gutierrez, *La France,* 459. For other examples of complementary oppositions in the definition of national identities, see O. Ranum, "Counter-Identities of Western European Nations in the Early-Modern Period: Definitions and Points of Departure," in P. Boerner, ed., *Concepts of National Identity: An Interdisciplinary Dialogue* (Baden-Baden, 1986), 63–78.

10. Puigcerdà, see AHP MA 1683–1689, fol. 39: September 13, 1684; and ibid. 1707–1715, fol. 136 (March 24, 1705); on Barcelona, see Moreu Rey, *Els immigrants,* 15–17; and Sales, "Naturals i aliénígenes," 700–701.

11. Nadal and Giralt, *Population catalane,* 155–160; Gutierrez, *La France et les français,* 30–32; see also AHP Correspondencia (letters of 1749 between the governor of Puigcerdà and the viguier of the French Cerdagne concerning the conflicts between the French agricultural workers and the native population). The institutions of local citizenship in the Cerdanya communities are discussed below, chap. 4 sec. 2.

its history as a "Spanish" town: founded by royal charter, a frontier stronghold on the confines of France, Puigcerdà had long been favored by the Aragonese kings with a large number of privileges and exemptions. But throughout the seventeenth century, the defense of "Spanishness," distinct from loyalty to the Aragonese or Castilian monarchs, pervaded the discourse of both municipal elites and families of the rural nobility which had long petitioned the crown for favors (*mercedes*).[12]

The use of national identities in petitions for additional privileges, titles, or tax remissions, was a rhetoric of loyalty that was easily unmasked in the midseventeenth century. But during the revolt of the Catalans between 1640 and 1652, the Cerdans were forced to choose their loyalties and identities. The initial reaction of most of the peasants of the district, as all across Catalonia, was to oppose and chase out the royal troops of Castile from the land. Royal officials concurred that the Cerdanya was a rebellious place. "There is no more obedience here than heretics have for the Pope," wrote a military official in the fall of 1639. "Holland is no more rebellious than the Cerdanya," echoed Don Diego Perez Villa, "only preachers are needed to make them lose faith with obedience."[13] But among the privileged groups, the choice was more complex, for resisting Castile meant joining the party of the French alliance.

Pau Claris, canon in the town of Urgell near Puigcerdà, and *diputat* in Barcelona, negotiated and signed the "alliance" of Catalonia and France in 1640. A few Cerdanya families of the older nobility were linked to the Claris' "party," but anti-French sentiment was more widespread among most of the titled bourgeois and lesser nobles of the valley.[14] The town council of Puigcerdà stubbornly opposed French

12. S. Galceran i Vigué, "Els privilegis reials de la Vila de Puigcerdà," in *Primer congrés internacional d'història de Puigcerdà* (Puigcerdà, 1983), 124–133; on the privileges pertaining specifically to the town's role as a military frontier stronghold, see SBCM Documentos (3–2–13–20): "Algunos datos sobre los antiguous fortificaciones de Puigcerdà, 1366–1678"; examples of petitions from the municipality underscoring its Spanish identity and loyalty include BCM Documentos (2–2–1–46): "Representación de Puigcerdà al Emperor Carlos V pidiendo se la fortifique"; and ACA CA leg. 210 (petition of consuls of Puigcerdà, July 6, 1667); examples of petitions from individuals include ACA CA leg. 208 (Juan de Mir, January 21, 1690); and ACA CA leg. 339 (Don Francisco Bages, April 5, 1691; and Don Jacinto Descatllar, November 8, 1690).
13. The quotations are from Sanabre, *La acción de Francia*, 45–46.
14. On intersections of levels of politics—Pau Claris' support of the French against the bishop of Urgell's support of Castile—see Elliott, *Revolt*, esp. 342–344 and 480–483; and J. Sarrète, "La Cerdagne pendant la révolution catalane et jusqu'au Traité des Pyrénées," *Guide-annuaire du Roussillonnais*, 1910: 175–213. Claris' supporters included Don Francesc de Sans (one of nine Catalan deputies sent to Philip IV in 1640), Don Llorens de Barutell (who took the title of archdeacon of Cerdanya in 1656), and Don

demands for manpower and arms, which over the next decade added much to social distress during years of plague, flood, and drought. The councillors offered gestures of loyalty—oaths and petitions to their new king Louis XIV in 1643, for example—as they stalled and renegotiated every tax and demand for soldiers imposed on the district.[15] But only in the very last moments of the Catalan revolt, as Barcelona was about to fall to Philip IV, did the hidden resistance become open. In May 1652 more than 150 inhabitants of Puigcerdà chased out the French troops and sympathizers: led by the first consul Jaume Morer, they shouted *Viva España*—surely in Catalan, although Morer's petition is in Castilian—as they returned the town to His Catholic Majesty. The same cry in Catalan—*Visca Espanya*—was recorded the next year, shouted by the peasants of the Cerdanya defending their villages against French attempts to reconquer the valley.[16]

In the choice of France or Spain—and such was often the option on the frontier during the early modern period—most of the Cerdans chose Spain, not so much because they were attracted to Spain as because they were repelled by France. The strength of this opposition was such that as an object of political identity and loyalty, Catalonia was rarely evoked. It was structurally possible for Catalonia to stand in opposition to both France and Spain, as suggested in an early seventeenth-century text from the borderland. Entitled "Comparatió de Espanya amb Fransa," the text compared the "national characters" of Spain and France, although the sympathies of the anonymous author lay decidedly with Spain.

> The temperament of the Spaniards is hotter and more humid, and their color darker; that of the French is colder and less humid; their skin is softer and

Rafael Ferran (whose father had been inquisitor-general in Catalonia and whose brother Hyacinthe was named governor of Roussillon and Cerdanya in 1649). On the social foundations of political alliances during the revolt, see J. Vidal Pla, *Guerra dels segadors i crisis social: Els exiliats filipistes (1640–1652)* (Barcelona, 1984).

15. See the town's register of municipal deliberations, AHP MA, during these years; and on the extreme misery in the valley, in which floods and plague added to the general insecurity of the local population, see J. Sarrete, "La Cerdagne de 1642 à 1652," *RHAR* 4 (1903): 183–190.

16. AHP MA 1650–1653, fols. 174–183v; see also the petitions for *mercedes* by Morer and Manuel Maranges, in ACA leg. 241, no. 112, and leg. 226, no. 36; and Sanabre, *La acción de Francia*, 520–521. For a description of the 1654 campaign from a French perspective, see *Les mémoires de Messire Roger de Rabutin*, 1: 504–553; from the Spanish side, see SBN MSS 2384, fols. 81–83: "Relacion de la campaña del año 1654;" and ibid., fols. 270–278: "Successos de la guerra en los condados," 1653, 1654, and 1655; and from a Catalan perspective, ADPO 7J 105: "Relatio de la entrada que feren los françesos en la terra de Cerdanya als 19 de juliol 1654 diu de diumenj," anon., s.d.

whiter. French women are more fertile than Spaniards. . . . The French speak with more ease, the Spaniards keep quiet and know how to dissimulate better.

The content of the unsigned manuscript resembles a well-known literary genre of the seventeenth century—the comparison of French and Spanish national characters—and it may even be a copy of one such description.[17] But its significance lies in the choice of language: the Catalans are invisible in the text, but Catalan served as the language *of* the text, a neutral set of signs used to describe the opposition of Frenchmen and Spaniards. In a parallel manner, it was possible for Catalans to evoke an identity as Catalans in opposition to Castile, or to both Spain and France. Joan Guardia, a peasant from Corco, decried the brutality of both Spanish and French soldiers in a diary entry of 1655, claiming that "it was enough to make any man who is a Catalan cry."[18] But such a conscious self-identification with Catalonia was rare in the seventeenth-century borderland and, when faced with a choice between France and Spain, the Cerdans tended to side with Spain.

The complex and ambivalent attitudes of the Cerdans toward France and Spain suggest the need to rethink what has become an accepted model of identity and loyalty of European rural society—the image of encompassing concentric circles. This model of "social belonging" has deep Indo-European roots and is frequently used to describe the parochialism of village life (*esprit de clocher* in French, *sociocentrismo* in Spanish).[19] In the concentric circle model, a sense of identity decreases

17. BC MS 184: "Comparatió de Espanya amb Fransa," fols. 37–40. When describing languages, however, the author distinguished the Castilians as one "people" among many that were "Spanish." Thus "the Spanish language is more severe, that of the French smoother. The Castilians, who are many [que son milissimes pobles], use language the most elegantly among all the Spaniards. In France, one hardly knows what language the French speak, since the tongue is spoken more among the nobility and courtiers than in any humble place." Cf. Carlos Garcia, *La oposición y conjunción de los dos grandes luminares de la tierra, o la antipatía de franceses y españoles* (Madrid, 1617; reprint Edmunton, 1979); Guez de Balzac, *Le Prince* (Paris, 1631), esp. chap. 19; and La Mothe le Vayer, *Discours de la contrariété d'humeurs qui se trouve entre certaines nations, et singulièrement entre la française et l'espagnole* (Paris, 1636); on these and others, see Gutierrez, *La France*, passim.

18. Cited in García Cárcel, *Historia de Cataluña*, 1: 176.

19. E. Benveniste, *Le vocabulaire des institutions indo-européennes*, 1: 293–319; for the Spanish case, see J. Caro Baroja, "El sociocentrismo de los pueblos españoles," in his *Razas, pueblos, y linajes* (Madrid, 1957), 263–292; for southern France, see H. Balfet, C. Bromberger, and G. Pavis-Giordani, "De la maison aux lointains: Pour une étude des cercles de référence et d'appartenance social en méditerranée nord-occidentale," in *Pratiques et représentations de l'espace dans les communautés méditerranéennes* (Paris, 1976), 27–75.

in relation to geographical and spatial distance from a social "ego"—a village or parish. Thus, in generic terms, the strongest sense of self-identification occurs in relation to a village community, followed by the sense of belonging to a group of villages, then a valley, then a region or province, and finally a nation. In the historical case considered here, an early seventeenth-century Cerdan would identify first with his or her parish or village, then in succession with his quarter or district within the valley or County of Cerdanya, with the "Two Counties" of Roussillon and Catalonia, with the Principality of Catalonia, and finally, with Spain itself (fig. 1). The development of national identity, in this model, implies an eventual reversal of the poles of loyalty: the relative decreasing sense of loyalty to the locality would be matched by the increasing attachment to the nation.

But the concentric circle model of identity fails to take account of the oppositional character of identities and loyalties, particularly visible in the Catalan borderland and found more generally throughout European society. The social and political expression of loyalties and affiliations is also an expression of difference and distinction. But differences are always relative, potentially fused in a higher and more generalized opposition. Anthropologists, writing of lineage-based societies, have called this a model of segmentary organization. But the basic principles of segmentation, complementary opposition, and struc-

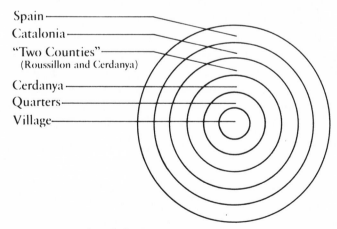

Figure 1. Circles of Identity

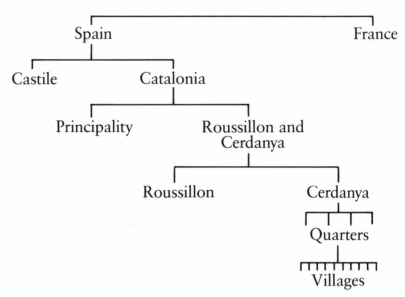

Figure 2. Counter-Identities

tural relativity may be usefully applied to the study of peasants in complex societies.[20]

Identities and loyalties in the Cerdanya were not fixed in a permanent hierarchical order but remained constantly shifting among levels—and this despite the constant flow of personnel from France into Catalonia (fig. 2). In defense of their local patria, in defense of the "two counties," in defense of Catalonia, and in defense of Spain, the Cerdans maintained their inherited antipathy toward the French. At the same time, individual villages could oppose each other; "quarters" (districts of villages within the valley) saw their counterparts as rivals; the Cerdanya stood opposed to the Roussillon; the two counties contraposed themselves to the principality; and Catalonia opposed Castile at the

20. Classical formulations of the segmentary system include E. E. Evans-Pritchard, *The Nuer* (Oxford, 1940), who also used a concentric circle model in his discussion of concepts of space and time (114–116); M. Fortes and E. E. Evans-Pritchard, *African Political Systems* (London, 1940); M. Sahlins, "The Segmentary Lineage: An Organization of Predatory Expansion," in R. Cohen and J. Middleton, eds., *Comparative Political Systems* (Austin, 1967), 89–115; and more recently, J. G. Galaty, "Models and Metaphors: On the Semiotic Explanation of Segmentary Systems," in L. Holy and M. Stichlik, eds., *The Structure of Folk Models* (New York, 1981), 63–92.

most generalized level of segmentation. When the Treaty of the Pyrenees divided the valley into two parts, it created the conditions for a new distinction—the French and Spanish Cerdanyas. The actualization of this further segmentation constituted the historical development of national identities in the borderland.

STATE AND NATION IN THE
SEVENTEENTH CENTURY

Writing in 1777, David Hume suggested one of the secrets of successful state building under absolutist governments: the leveling of differences between old and new subjects.

> When a monarch extends his dominion by conquest, he soon learns to consider his old and new subjects on the same footing; because, in reality, all his subjects are to him the same. . . . The provinces of absolute monarchies are always better treated than those of free states. Compare the *pais conquis* of France with Ireland, and you will be convinced of this.[21]

In late seventeenth-century France, the subjects of the "conquered provinces" automatically became regnicoles, subjects of the French king, indistinguishable in political status from other French subjects. The political framework of citizenship in prerevolutionary France was founded on a juridical distinction in use since the end of the thirteenth century, between regnicoles and *étrangers* (foreigners).[22] In Old Regime France, a body of customary rules and a growing (if not always concordant) jurisprudence defined a French nationality (*naturalité*) by birth, descent, or letters of naturalization. In this, as in many respects, the French experience was distinctive. There was no legal concept of a "Spanish" nationality in Spain during the early modern period. There were "subjects" or "vassals" of the Spanish king (technically, of the Count of Barcelona, or King of Castile, or the Two Sicilies, and so forth). But the different political entities which made up Spain were composed of different "nationalities": Navarese, Aragonese, Castilians, Catalans, Portuguese. In France, where the crown subsumed and defined nationality, the provinces were rarely identified as nations; in

---

21. D. Hume, "That Politics May Be Reduced to a Science," 1777, in H. D. Aiken, ed., *Moral and Political Philosophy* (Oxford, 1963), 298–299.

22. B. Guenée, "Etat et nation en France au Moyen Age," *Revue historique* 237 (1967): 25–26; for the early modern period, see M. Vanel, "Evolution historique de la notion de français d'origine du XVIe siècle au code civil" (Thèse de droit, Université de Paris, 1945), esp. 10–12.

Spain, by contrast, there was no national monarchy, and peripheral provinces were juridically distinct nations.[23]

Catalonia had preserved its own political institutions and administration in the constitutional ordering of the Spanish empire, and the Catalans retained their own "nationality" (naturaleza), different from their political status as subjects of the King of Spain. Even Louis XIII had recognized this nationality by granting letters of Catalan naturalization during the years before the Treaty of the Pyrenees.[24] During the late 1640s and early 1650s, when French troops occupied Puigcerdà and the Cerdanya, French officials ordered the town to allow Frenchmen to enter the municipal scrutinies, thus breaking down the statutory and institutional definitions of Catalan nationality.[25] But it was the formal annexation of Roussillon and the Cerdanya which brought an end to the Catalan nationality of their inhabitants, who were treated as the juridical equals of French subjects beyond the province. In 1663, an ordinance of Louis XIV, registered by the Sovereign Council of Roussillon, revoked all local statutes prohibiting Frenchmen and their descendants from holding political office in the Counties of Roussillon, Conflent, and Cerdanya. The following year, a royal ordinance, modifying article 56 of the Treaty of the Pyrenees, declared that those Catalans from the principality who came to live in Roussillon would also be treated as "our true and natural subjects," but that they could only pass on their goods to other regnicoles, subjects of the French king. With these two acts, the crown established a formal equality between its new subjects and its old, drawing a clear and unambiguous boundary between its subjects and foreigners. In Spain, Catalans of the principality preserved the privileges of their own nationality until the Bourbon reforms of Nueva Planta in 1714.[26]

23. For France, see F. Brunot, *Histoire de la langue française*, 12 vols. (Paris, 1905–1953), 6: 133–143; W. F. Church, "France," in O. Ranum, ed., *National Consciousness, History, and Political Culture in Early Modern Europe* (Baltimore, 1975), 43–66; G. Dupont-Ferrier, "De quelques synonymes du terme 'province' dans le langage administratif de l'ancienne France," *Revue historique* 161 (1929): 202–203; and V. L. Tapié, "Comment les français voyaient la patrie," *XVIIe siècle* 25–26 (1955): 37–58. For Spain, see H. Koenigsberger, "Spain," in Ranum, ed., *National Consciousness*, 144–172, and J. M. Jover Zamora, "Monarquia y nación en la España del XVII," *Cuadernos de historia de España* 13 (1950): 101–150.

24. Examples of Catalan naturalization from the 1640s may be found in ADPO B 401 and B 394; see Sales, "Naturals i alienigenes," esp. 680–690; and R. Gilbert, "La condición de los extranjeros en el antiguo derecho español," in *L'étranger* (Brussels, 1958), 151.

25. AHP MA 1648–1650, fol. 214v (September 17, 1650); and ibid. 1650–1653, fol. 39 (March 4, 1651).

26. ADPO 2B 29 (ordinances of 1663 and 1664); Sales, "Naturals i aliénígenes," 689–696.

The legal redefinition of the juridical status of Catalans in France mattered relatively little in practice: the Catalans of Roussillon and the Cerdanya were hardly "foreigners" with respect to the Catalans of the principality. In the first years after the annexation, a large number of nobles and peasants—estimated at over two thousand—left Roussillon and the Cerdanya to settle in the principality, continuing a long-standing tradition of southward movement in the region.[27] Many were the families that maintained kinship relations, not to mention property holdings, on one side or the other of the political boundary. Linked by a common language and a shared set of cultural conventions, the Roussillonnais and Cerdans were integral members of the Catalan ethnolinguistic community.

Paris ministers, provincial officials, and military officials in the late seventeenth and the early eighteenth century were concerned with what they sometimes identified as a "Catalan spirit," synonymous with rebellion and disloyalty, among the inhabitants of the Roussillon. "The people of the district are Catalans," wrote an anonymous official in the late 1690s, "and they never waste a single occasion to revolt, given their great inclination for the Spanish nation."[28] But their attachment was less to Spain than to Catalan culture, at least according to the Roussillon Intendant Barrillon in 1710:

> The [Catalan] nation is generally speaking war-like and has to an eminent degree the qualities with which to make good soldiers: the people have much self-esteem, pride, and courage, and are accustomed to a hard life, with a sobriety unknown in our lands. But these good dispositions are made useless by their natural ferocity and the savage, close-minded education that the youth receives, by their boundless hatred for order and subordination, by their specific aversion to French domination, and by their attachment to Catalan mores.[29]

Given such attitudes, harsh military repression of "rebellions" was not unknown. Following the events of 1674, for example, the provincial authorities executed the leaders of the Vilafranca conspiracy.[30] But if this violence was not altogether exceptional, the crown's policies,

---

27. On the resettlement of the population in the years before and after the Treaty of the Pyrenees, see Nadal and Giralt, *La population catalane*, 184–186, and Sanabre, *La acción de Francia*, 598–600.

28. France, Bibliothèque de l'Arsenal (hereafter FBA) MS 6935: "Mémoire pour la seurté [sic] des peuples du peis [sic] de Roussillon, Conflans, et Cerdagne française," n.d., but ca. 1695; the orthography suggests the official was originally from the Languedoc.

29. Comte de Boulainvilliers, *Etat de France: Extrait des mémoires dressés par les intendants du royaume par ordre du roi*, 5 vols. (London, 1737), 5: 263–295.

30. Marcet, "Les conspirations de 1674 en Roussillon," 275–296.

articulated by the provincial intendants, focused less on physical repression than on breaking the social and cultural ties that bound the Roussillonnais and the Catalans of the principality.

The French state encouraged Catalans loyal to France to acquire lands in Roussillon.[31] More severely, in 1678, Intendant Camus de Beaulieu, acting on the instructions of War Minister Louvois, went so far as to prohibit marriages between the Catalan "subjects of the Catholic King" and the Roussillonnais. The policy was directed toward "certain persons of quality," and there is no evidence that such prohibitions were ever respected. Instead, according to the anonymous official of the 1690s, the inhabitants of Roussillon continued to marry other Catalans, and "do not wish to make alliances with Frenchmen":

> There are a few young women who married Frenchmen frivolously, and without the approval of their parents, but there is not a single example of a man from Roussillon marrying a woman from France.[32]

But there was more to French policy than breaking the ties of family and property that bound the Roussillonnais and the Catalans of the principality, for the crown also introduced a number of measures designed to propagate the French language and culture—and simultaneously to create a new provincial elite, part native and part French, loyal to the French king. The policies coincided with the very period that saw the consolidation of France's military frontier in Roussillon and Cerdagne, thus establishing a link between military imperialism and cultural expansion which was to endure through the nineteenth century.[33]

In 1682, the same year that the French war minister Louvois created the province's military administration, the Sovereign Council of Roussillon registered an edict admitting to the liberal professions and its own administrative offices only those Roussillonnais who could speak French. Military and cultural consolidation went together—though ul-

31. FAMRE MD FR MS 1746, fol. 31 (intendant to foreign minister, November 1, 1664). As late as 1787, the Chevalier de la Grave, in a printed *Essai historique et militaire sur la province du Roussillon* (London, 1787), wrote that "Individuals in Roussillon still have goods in Catalonia, which maintains the Catalan spirit of the inhabitants of the province, and they should rid themselves of these goods, since one must have that [spirit] of the Prince under whose domination one lives."

32. FBA MS 6935: "Mémoire pour la seurté [sic] des peuples." On marriage prohibitions, see FMG AAT A(1) 613, no. 132 (intendant to war minister, September 24, 1678), and no. 163 (August 28, 1678).

33. Cf. G. Livet, "Louis XIV et les provinces conquises," *XVIIe siècle* 22 (1952): 504, who denies the links between "military imperialism" and "cultural expansion." For an overview of "gallicization" policies in late seventeenth-century Roussillon, see Sanabre, *La resistència del Rosselló*, passim.

timately the latter was more important. The process of administrative gallicization culminated in the edict of 1700 by which Louis XIV decreed that the persistent use of Catalan in official acts—including wills and testaments—was "contrary to our authority, to the honor of the French nation, and even to the inclination of the local inhabitants"— even if the "local inhabitants" (such as the consuls of Perpignan) opposed the measure without success.[34]

Introducing French as the language of law and administration—of public and political life—in its annexed provinces during the seventeenth century was a standard practice of the French monarchy.[35] So was encouraging French nationals to settle in the new provinces, though the absolutist state in France undertook nothing on the scale of English resettlement of Ireland. The French crown encouraged traders to establish themselves in Perpignan and populated the new fortress towns such as Mont-Louis with merchants and artisans—"good French citizens"— from neighboring Languedoc.[36] But the heart of Louis XIV's policy of state building and national unification did not lie so much in encouraging immigrants to settle in the newly annexed provinces. As he explained in his memoirs,

> In order to strengthen my conquests by closer union with my existing territories, and as I did not see myself practicing what the Greeks and Romans did, namely sending colonies of their own subjects into freshly conquered territories, I tried to establish French customs in them.[37]

Instead of importing old subjects, the crown and its officials recognized the need to form a loyal indigenous elite and focused its efforts on the linguistic and pedagogic assimilation of the Roussillonnais into Frenchmen.

The intention of the French crown was to demonstrate to the inhabitants of its newly conquered provinces both the honor and the utility of adopting French language and customs. In 1661, for example, when Cardinal Mazarin established in his will the "College of the

---

34. Sanabre, *La resistència*, 177–178; see also A. Brun, *L'introduction de la langue française en Béarn et en Roussillon* (Paris, 1923), 68–70.

35. Marquis de Roux, *Louis XIV et les provinces conquises*, 251–270; H. Peyre, *La royauté et les language provinciales* (Paris, 1933), 107–168; and Brunot, *Histoire de la langue française*, vol. 5, chap. 11. Making French the official language was an extension to the conquered provinces of the Villers-Cotterêts ordinance (1539); on the latter, see D. Trudeau, "L'Ordonnance de Villers-Cotterêts et la langue française: Histoire ou interprétation," *Bibliothèque d'humanisme et rennaissance* 45 (1983): 461–472.

36. ADPO 2J 57 (anonymous memoir, April 29, 1718).

37. Quoted in M. M. Martin, *The Making of France: The Origins and Development of the Idea of National Unity*, trans. B. and R. North (London, 1951), 149.

Four Nations" in Paris for the education of the privileged children of France's new provinces, he stressed that their Parisian education would show them "the advantages of being the subjects of so great a king."[38] In Roussillon, the crown's policy was less than hegemonic, since it tolerated the persistence of instruction in the Catalan language while patiently introducing French.

In August 1661, Louis XIV established a Jesuit college for the sons of privileged families in Perpignan, "for it is not only proper but advantageous in the future for the youth to learn according to the practice in France." That the Jesuit fathers had to learn Catalan in order to teach in Latin is significant, for it suggests that the crown did not yet believe in the exclusive use of French. As the Intendant Carlier proclaimed in 1672,

> Since there is nothing which upholds the union and love among people of different nations more than a common language, by which they understand each other in useful conversations and facilitate their interactions, His Majesty has resolved to use this means to unite all the Frenchmen who live in this province with the native inhabitants.

He thus prescribed the establishment of primary schools in Perpignan

> where children of both sexes can be taught the letters of the alphabet, syllables, dictions, and speech in the French language and in the local language, and even to write in the two languages, in order that they slowly become more and more alike.[39]

Ten years later, the Sovereign Council of Roussillon envisaged a policy of linguistic assimilation in which French (and Latin) would be taught in royal schools established in *all* the towns and villages of the provinces. The project, which depended exclusively on local funding, remained a dead letter, especially in the Cerdanya.[40]

In sum, the French state's policies of gallicization remained limited to administrative and political contexts. Certain French officials envisioned a more dramatic linguistic and cultural integration with an emphasis on a tolerant and gradual assimilation. Other officials perceived

---

38. Quoted in Marquis de Roux, *Louis XIV*, 265–266; see also Martin, *Making of France*, 149–150.

39. Quoted in Marquis de Roux, *Louis XIV*, 263; on French linguistic and educational policies in Roussillon, see P. Torreilles, *La diffusion du français à Perpignan, 1660–1700* (Perpignan, 1910); and P. Torreilles and E. Desplanques, "L'enseignement élémentaire en Roussillon dépuis ses origines jusqu'au commencement du XIXe siècle," *BSASLPO* 50 (1895): 145–398, which, unfortunately, excludes material from the Cerdagne.

40. Sanabre, *La resistència*, 175–176.

that the effects of such policies would inevitably remain limited so long as the local inhabitants felt no need to assimilate themselves. The anonymous official of the 1690s described how the local inhabitants refused to learn French:

> The evil stems from the fact that this people has never been in a situation where they need to learn French, and to enter into relations with the other peoples of the kingdom, for they preach, contract, sue, and render justice in Catalan; they have not a single item of business outside their district necessary to their lives.

Fifteen years later, the Intendant Barrillon proposed to introduce French "mores and customs" by placing the province under the jurisdiction of the Toulouse *parlement,* "thus forcing the Roussillonnais to settle their affairs in France," but no such reforms were ever seriously considered.[41]

Although the French Cerdagne remained untouched by the limited educational policies of the late seventeenth-century state, there were nonetheless a few local privileged families who, at this early stage, betrayed an interest in serving the French crown and identifying themselves with France. In Perpignan, Catalan refugees from the principality who had sided with France filled the highest posts in the new French administration, families like Segarra, Fontanella, and Trobat. In the Cerdanya, loyalties were also forged at the moment of annexation. Generations of the same families were to appear again and again under the Old Regime, at least until the appearance of new actors during the revolutionary decades in Spain and France. French loyalists in the Cerdagne were led by the Sicart family, which had settled in the Cerdanya from Foix during the thirteenth century. Having sided with the French in 1640, a dynasty was born: all but one viguier of the French Cerdagne were Sicarts, servants of the crown, and tied to France through the patronage of the quasi-hereditary governorship of Roussillon—a position monopolized by the Ducs de Noailles.[42] The sole royal notary of the French Cerdagne—a position less prestigious than that of viguier but more important structurally as a mediator of state and society—was held by the Gallard family, passed from father to son until the French Revolution. There were other nobles who served France, men

41. FBA MS 6395; Barrillon is quoted in Sanabre, *La resistència,* 136–137.
42. G. Miquel de Riu, "Extrait des souvenirs de M. François Sicart d'Alougny," *BSASLPO* 33 (1897): 167–178; and H. Aragon, "Les Ducs de Noailles," *Ruscino* 10 (1919): 49–68; 11 (1919): 137–145; 12 (1920): 5–24.

such as Don Francesc de Travy of Palau, who belonged to the Royal-Roussillon infantry for over twenty-five years, even though his father had been knighted by the Spanish king Philip IV in 1640.[43]

Other than this handful of privileged families, there were few, if any, signs of loyalty for the French among the mass of Cerdans. The nearly continuous occupation of the Cerdanya valley by Spanish and French troops until 1714 was hardly conducive to voluntary assimilation, especially since villages suffered greatly under the weight of military occupations, supplying men and foodstuffs to foreign troops.[44] Population statistics between 1642 and 1694 for villages of the Cerdanya, although hardly reliable indices, suggest that those of the "annexed district" suffered greater losses of population than those of the Spanish Cerdaña which were closer to Catalonia. In the center of the plain, the population of the Carol Valley fell from 410 to 299 inhabitants, while that of Càldegues dropped from 69 to 57 inhabitants. Such losses seem due to out-migration across the boundary into Catalonia, a movement already visible during the French reconquest of the valley in 1653 when some 50 Cerdans petitioned the Council of Aragon for foodstuffs.[45] On the other side of the boundary—which, in any case, had little significance to either military commanders or peasants—the Spanish Cerdaña also suffered greatly from the nearly continuous military occupation. Life was especially harsh during the War of the Spanish Succession, when the population of the Spanish villages declined almost 10 percent, from 5972 in 1696 to 5539 in 1718. Most of the villages near the border experienced more significant drops in population, and Puigcerdà itself was reduced to some 1100 inhabitants—leaving the decimated town not much larger than a prosperous village of a century later.

Those Cerdans who stayed in the valley—especially the inhabit-

---

43. Biographical information on these and other privileged families may be found in C. Lazerme, *Noblesa catalana,* 3 vols. (Perpignan, 1975); A. Puyol Safont, *Los hijos illustres de Cerdaña* (Barcelona, 1926); and J. Capeille, *Dictionnaire de biographies roussillonnaises* (Perpignan, 1914). See also the notes and documents collected by Albert Salsas, in ADPO 7J. On the Royal Roussillon, see Marquis de Roux, *Louis XIV,* 273–275.

44. A. Pladevall and A. Simon, *Guerra i vida pagesa a la Catalunya del segle XVII* (Barcelona, 1986). In fact, the *muncipality* of Puigcerdà grew wealthy—although most of the inhabitants of the town itself did not—through contracts to provision troops, and the town undertook a systematic policy of buying back its constituted rents (censals) between 1680 and 1702, as evidenced by the ten acts listed between 1680 and 1702 in S. Galceran i Vigué, ed., *Dietari de la fidelíssima vila de Puigcerdà* (Barcelona, 1977), 130–142; see also A. Alcoberro, "Entre segadors i vigatans: L'ocupació francesa de 1694–1698," *L'Avenç* 109 (1987): 45.

45. ACA CA leg. 208 (consulta of June, 1655); see the population statistics in app. A.

ants of the town of Puigcerdà—continued to express their anti-French sentiment, defending themselves by force of arms, often framed by religious metaphors and practices. For example, when the Duc de Noailles entered the plain in April 1678, the council of Puigcerdà ordered the town's convents to light votive candles and to assist the municipality in preparations for the siege. Puigcerdà's own chronicle, the *Dietari de la fidelíssima vila de Puigcerdà,* noted that "the loyalty of all those of the town could not have been greater; their enthusiasm was so great that this siege cost the French more than any in Flanders, as the French themselves admit." But after 1678, as the governor of Mont-Louis consolidated French strength in the borderland, the towns and villages of the district found themselves all too often forced to comply.[46]

French policy could be harsh and uncompromising. In 1691 Intendant Trobat wrote to Louvois, "It is possible that no provisions are to be had this year, even when we threaten to burn their houses, which are made of stone and earth, and as easy to reconstruct as they are to raze." But it could also recognize the limitations of forcible repression. "The peasants retain their liberty and privileges," wrote Trobat about the Carol Valley, "which force us not to push them too hard, and which we should only alter with care and moderation."[47] Indeed, concerning the hated salt monopoly (gabelle), originally imposed in 1661, the intendancy consistently negotiated reductions of salt-purchasing agreements for the Cerdagne in order to "conserve the peace" and to avoid that "the inhabitants go elsewhere" for salt—since the Cerdans provisioned themselves principally from the mines at Cardona, in Catalonia.[48]

Local resistance to the harsh demands of the French state took a range of forms, but "rebellion" was less common than flight, delay-

46. AHP MA 1673–1678 (April 29, 1678); *Dietari de la fidelíssima vila de Puigcerdà,* 130.

47. FMG AAT A(1) 1106, no. 159 (intendant to war minister, June 30, 1691); and A. de Boislisle, ed., *Correspondence des Contrôleurs Généraux des Finances avec les Intendants de Province,* 3 vols. (Paris, 1874–1897), 1: 415–416.

48. On the administration of the salt monopoly in the French Cerdagne, see for example ADPO MS 32, vol. 8, fols. 37–37v (extract of an ordinance of the Sovereign Council concerning a seizure of salt, August 13, 1683); FAN G(7) 508 (memoir of the intendant concerning the "treaty with the inhabitants of the French Cerdagne for the salt which they are accustomed to get in Roussillon," January 29, 1709). The quotation itself is from FAN G(7) 507, a letter from the intendant, April 27, 1703, following a disputed seizure of salt near Sant Pere de Cedret, out of which also emerged the first attempt to define cartographically the limits of the two Cerdanyas; see illustration 4.

ing tactics, and a simple lack of enthusiasm for celebrating their conquerers. When Puigcerdà was retaken by the French in April 1691, the governor of Mont-Louis ordered the town to celebrate "galas for the conquests of the Catholic King" in Mons, which the town did, but reluctantly. "As the fireworks were lit with little demonstration of joy," the governor ordered them celebrated a second time.[49]

Nor were attitudes much transformed during the alliance of French and Spanish Bourbons in the Spanish War of Succession. Neither the privileged groups nor the peasants of the Cerdanya willingly supported the Bourbon cause. Indeed, the only Cerdan to seek refuge in the fortress of Mont-Louis during the conquest of the valley by the armies of the Austrian Habsburg Pretender in October 1705 was the viguier of the French Cerdagne, Sicart. The rest of the Cerdans, repulsed by their "natural aversion for the French" and attracted by the promises of new privileges, accepted the sovereignty of the Archduke Charles.[50] But in September 1707, when the Bourbon armies reconquered the valley, the town council of Puigcerdà asked the bishop of Elna for a portrait of Philip V of Spain to hang in the place of the Archduke Charles, and for the next seven years the town and both Cerdanyas lived quite faithfully under Bourbon rule. Several older noble families with estates in the Spanish Cerdaña, such as the Maury and Bages, saw their lands confiscated as they continued to support the Austrian heir, as did a number of wealthy peasants in the French Cerdagne and much of the clergy throughout the Catalan Pyrenees. But most of the privileged classes remained nominally loyal to the Bourbons, and many found employment in the new administration in Catalonia.[51] The political parties that divided Catalonia thus divided the inhabitants of the Cerdanya; yet not, it would seem, along national or territorial lines. If the inhabit-

49. AHP MA 1689–1692, fols. 100–102 (April 29–May 3, 1691).

50. FMG AAT A(1) 1887, nos. 124, 210; ibid. 1888, no. 6 (letters of the commander of troops at Mont-Louis to the chancellor, October–November 1705); and ibid. 1891, nos. 219 and 302 (intendant to war minister, September–October 1705). Don Juan de Miguel, commander of the Archduke Charles' troops, wrote to Sicart, offering him titles of nobility if he would deliver the fortress, but without success: see FMG AAT A(1) 1888, no. 8 (letter and translation, November 1, 1705).

51. On the French reconquest of Puigcerdà and the Cerdanya, see FMG AAT A(1) 2053, nos. 179 et seq. (correspondence of the governor of Roussillon and the war minister, September–October 1707); Puigcerdà's request for a portrait of Philip V is recorded in ibid., no. 251 (bishop of Elna to war minister, n.d., ca. 1707); documentation of Puigcerdà's "compliance" under the Bourbon occupation is dispersed throughout this series, as well as in the town's municipal record: AHP MA 1707–1715. For a further discussion of the political and social divisions in the Cerdanya, see Sahlins, "Between France and Spain," 579–589; and more generally in Catalonia, see N. Sales, "Els Botiflers," in her *Senyors bandolers, miquelets i botiflers*, 141–219.

ants of the Cerdanya initially opposed the joint Bourbon armies of France and Spain, most of them later demonstrated a tacit acquiesence during the eight-year occupation, often dissimulating but never shedding their "notorious antipathy" for the French.

STATE AND LOCAL CULTURE IN THE EIGHTEENTH CENTURY

With the end of the military frontier after the War of the Spanish Succession, French policies of cultural assimilation, which had only been directed toward provincial elites, disappeared altogether. In Spain, while the reforms of Philip V abolished the political institutions and public law of Catalonia, the eighteenth-century Spanish state developed no consistent programs designed to assimilate culturally Catalonia into Spain. Enlightened absolutism was benign neglect—although perhaps even more in France than in Spain. As Pierre Poeydavant, an official employed in the Roussillon intendancy during the 1780s, wrote:

> It is not by being violent toward the inhabitants of a country, far removed from the center of the kingdom, who hold on to the mores, customs, and prejudices transmitted by their fathers that we will succeed in having them adopt the tastes, customs, and methods which can be accepted in the rest of France. Each nation has, as it were, its own nuances of character. We should wait for the benefits of time and the force of example to perpetuate the spirit of imitation; progress will certainly correct local imperfections and slowly install advantageous changes.[52]

Both his tolerance of local customs and his optimism in the long-term attraction of French civilization were shared by many of the eighteenth-century intendants, by government officials in Paris, and even by historians of this century.[53]

The possible exception to such policies of neglect in the Roussillon was the administration of religious affairs, where the eighteenth-century battle over gallicization was fought. Just as religion in Catalonia often provided the metaphors and language of resistance to the French, so too were ecclesiastical policies a central element in the French crown's policies of cultural hegemony. During the later seventeenth century, the crown made an effort to fill the convents of the

---

52. Poeydavant, "Mémoire sur la province de Roussillon," *BSASLPO* 51 (1910): 130.

53. For example, L. Febvre, "Politique royale ou civilisation française: Remarques sur un problème d'histoire linguistique," *Revue de synthèse historique* 38 (1924): 37–53.

province with regnicoles who were, in the ideological context of absolutism, supporters of the Gallican church. In 1735, the Sovereign Council extended the language edict of 1700 to include the acts of the parish clergy, requiring that all baptisms, marriages, and deaths be recorded in French and registered with the local courts—an edict reiterated in 1754.[54] And by the 1730s, as we have seen, the French administration had gained control of ecclesiastical administration in the French Cerdagne; a decade later, the intendant required parish priests to be able "to speak and understand French adequately."

Yet such policies failed to touch upon the practice of local religion and amounted to little more than an added measure of administrative gallicization. Except in the case of the cathedral church of Perpignan—concerning which, in 1676, Louis XIV expressed his joy that "preaching has started in French"—the crown was never concerned with the practice of religious life.[55] As Jean Sebastian Pons has documented, religious sensibility found its expression in Catalan throughout the eighteenth century: liturgical and devotional books continued to be printed in Catalan, as did religious drama, lives of martyrs and saints, and the popular *goigs* or hymns. In the Cerdanya, the continuing strength of Marian cults in the eighteenth century, the annual pilgrimages to chapels and shrines devoted to patron saints, the religious confraternities dedicated to the Rosary, and the female religious societies (*pabordessas*) helped perpetuate what Pons has called the "traditional spirit," a spirit (it hardly need be noted) expressed in the Catalan language.[56] Inasmuch as the continuing strength of Catalan cults and confraternities expressed a popular religious sentiment tied to a distinctive local culture, the French crown failed to reshape those indigenous beliefs.

The fate of the Belloch Chapel during the eighteenth century confirms the inability of the French state to restructure religious practices,

54. Sanabre, *La resistència,* 153–185.

55. Ibid., 174. In 1688, the intendant formally forbid the preaching of Catalan in the cathedral but not elsewhere; even in 1835, French masses were said in only two central parishes of Perpignan and nowhere else in the diocese: see D.M.G. Henry, *Histoire du Roussillon,* 2 vols. (Perpignan, 1835), 2: 522.

56. Pons, *La littérature catalane,* vol. 1, passim; on religious institutions in the Cerdanya, see M. Delcor, *Les vierges romanes de Cerdagne et Conflent* (Barcelona, 1970); D. J. Marti Sanjaume, *Las virgenes de Cerdaña* (Lérida, 1927); J. Sarrète, *La confrérie du Rosaire en Cerdagne* (Perpignan, 1920). For a list of the religious institutions of the Old Regime Spanish Cerdaña, see SAHN Consejos, leg. 7106: "Estado de las Cofradias de la Villa de Puigcerdà y Pueblos de su Partido," May 12, 1778. The very existence of this document suggests a form of policing of religious affairs in rural society under Charles III which never developed to the same extent under Louis XV.

even in a rare moment when it sought to do so. Situated on a small rise within the territory of Dorres, the chapel—affiliated with the Order of Servites—dated from at least the fifteenth century and was under the protection and patronage of the town of Puigcerdà. In 1663, Louis XIV forbade the three or four clerics to take their orders from Barcelona and commanded the ecclesiastical province of Marseilles to take over the jurisdiction of Belloch. In 1742, the French crown mandated the reduction of the order's holdings and Belloch was disbanded. Both Puigcerdà and the villages of the French Cerdagne protested loudly, claiming that Our Lady of Belloch was the "patron and protectress of both Cerdanyas," to whom constant appeals were made, especially during times of drought. Under the pressure, the state rescinded the order, and Belloch continued to thrive until the Revolution.[57]

If the French government made few inroads into Catalan religious culture, it made even less of an impact on other aspects of local life. The veneer of administrative gallicization was thinly spread and seemed to become thinner with altitude and distance from Perpignan—and Paris. The rural communities of the mountain districts were slow to comply to the royal edicts of 1700, 1735, and 1758, requiring the sole use of French. From 1700 onward all official correspondence between the communities and the administration was in French; yet the parish registers of Hix were kept in Catalan until 1748, while those of Osseja began to use French in 1749 but returned to Catalan from 1759 to 1764.[58] More importantly, Catalan remained not only the spoken language of local society but its written language as well. For example, surviving peasant account books, including the records of peasants who served as revenue collectors for local nobles and clergy, were kept in Catalan throughout the eighteenth and nineteenth centuries.[59]

On the Spanish side of the border, Catalan survived to an even greater extent as a written language. The legislation of the Nueva Planta made Castilian the language of public law and administration,

57. ADPO C 1335 (esp. Louis XIV to Belloch clerics, April 19, 1663, as well as the memoir and petition of the syndic of the French Cerdagne sent to the intendant and the provincial governor, June 1742); and M. Gouges, "Le monastère de Belloch," *Tramontana*, 465–466 (1963): 55–58. Not unexpectedly, the jurisdiction of Belloch was subject to dispute from the Spanish side: see AHP Correspondencia B, fol. 413 (provincial head of the Servites of Spain to consuls of Puigcerdà, November 28, 1705).

58. Sanabre, *La resistència*, 169–185; J. G. Gigot, "Situation à ce jour de l'état civil anterieur à 1792," *CERCA* 29 (1965): 165–195.

59. ADPO 2J 91 (account books of Joan Sicart of Dorres, 1722–1750, and Josep Margall of Dorres, 1775–1791); and the sources quoted in J. Sarrète, "La paroisse d'Hix," *BSASLPO* 46 (1905): 313–345.

just as it restructured Catalonia's political institutions on Castilian models. Following the municipal reforms of 1717, the registers of the municipality (*ayuntamiento*) of Puigcerdà switched abruptly from Catalan to Castilian in 1720, in which language they remained.[60] Although all official correspondence in Bourbon Spain was in Castilian, the so-called "literary decadence" of written Catalan in eighteenth-century Catalonia is a misnomer, as Núria Sales has recently argued. Medical works, technical manuals, educational texts, not to mention religious tracts, notarial registers, and private contracts, all continued to be written and published in Catalan.[61]

The Bourbon reforms in Catalonia imposed Castilian public and administrative law, although the Nueva Planta recognized the continuing validity of Catalan private or family law. Such legislation was more ambitious but ultimately less effective than the policies of the French Bourbons. At the moment of annexation, Louis XIV recognized the continuing validity of the Catalan *constitucions* in the province, and though he reserved the right "to abolish them and make new laws and ordinances," there was no systematic attempt to do so. As a result, Catalan written and customary law remained the principal body of legislation in French Catalonia under the Old Regime and even—in certain cases—until the late nineteenth century.[62] The only concern of the Sovereign Council during the eighteenth century was to assure a French *jurisdiction* over civil law and property disputes in the borderland, where properties were intertwined. The court forbade acts of sale before "foreign notaries," prohibitions reiterated in 1680 and 1713, and it later ruled that "foreigners" could be judged by French courts for their lands in the French Cerdagne.[63]

In the eighteenth century, the French intendancy was unconcerned with diminishing the Catalan "spirit" through educational institutions. Although the reform-minded Barrillon saw that "the [Catalan] nation must be obliged to stop seeing itself as totally separate from French

60. AHP MA 1720–1730; the first use of Castilian in the town of Puigcerdà's own ceremonial book—the *Dietari de la fidelíssima vila de Puigcerdà*—came only in 1732, and Catalan was used consistently until the 1760s, and sporadically until 1829. On the Bourbon reforms of the Catalan municipalities, see J. Torres i Ribes, *Els municipis catalans de l'antic règimen, 1453–1808* (Barcelona, 1983), 143–220.

61. N. Sales, "Els Botiflers," in *Senyors bandolers*, esp. 207–219; see also S. Sole i Cot, "La llengua dels documents notarials catalans en el periode de la Decadència," *Recerques* 12 (1982): 39–56.

62. L. Assier-Andrieu, *Le peuple et la loi: Anthropologie historique des droits paysans en Catalogne française* (Paris, 1987), esp. 69–87.

63. Cases summarized in ADPO MS 32, vol. 1, fols. 296–297: see "étrangers."

subjects," neither he nor his successors focused on schools as a means of assimilation. The few schools in the French Cerdagne went largely unregulated by the intendency, which seemed to have its eye on everything else; run by local priests, these schools appear to have taught in Catalan.[64] Their prestige was less than those of Puigcerdà, since most of the privileged—and socially aspiring—families of the French Cerdagne, despite a royal edict of 1686 prohibiting children from studying in Spain, sent their sons to school across the boundary. Although Puigcerdà boasted a school in 1603, the Piarist school established in 1728 became the principal local institution "for the instruction of reading, writing, counting, grammar, and the Christian doctrine." The Piarists taught Latin and Castilian, not because the Spanish government forbade the use of Catalan in grammar schools—only in 1768 and 1781 were modest proposals made along those lines—but because these were the languages of a successful public life throughout Catalonia during the eighteenth century.[65]

In sum, neither France nor Spain demonstrated any particular concern for nation building under the Old Regime. Uniformity of political administration was their goal—though less in France than in Spain— but uniformity of culture, language, law, and religious practices did not form part of an agenda imposed from above. The states were concerned to extract revenue and to maintain a minimum of civil order and political loyalty. Beyond that, although the diffusion of national cultures was desirable, state policy remained disengaged from the ideal.

RESISTANCE TO STATE BUILDING IN THE BORDERLAND

From the perspective of the Cerdanya peasants, the Old Regime state remained a distant entity. Yet the status of the Cerdanya as a frontier district made its presence more visible than in other peasant communities. It was also more of a burden and an unwelcome presence

64. ADPO C 1355 (mentions of schools in Err and Osseja villages in the biographical information on the parish priests in the French Cerdagne); J. L. Blanchon, "En Cerdagne: Les écoles de la Belle Epoque," *Font de Segre* 7 (1981): 10–11; J. Sarrète, "La condition des non-privilégiés sous l'Ancien Régime à Osseja," *RHAR* 6 (1905): 108–110. On the absence of policies of linguistic uniformity in Old Regime France, see J. Monfrin, "Les parlers en France," in François, *La France et les français,* 745–775.

65. Sources on Puigcerdà's schools include BC MS 184, fols. 32–33: "Descripción de Puigcerdà"; AHP MA 1656–1660, fol. 43 (includes a description of the curriculum); AHP Correspondencia B, fols. 426, 440–441, and 492. On Charles III's linguistic policies, see Sales, *Senyors bandolers,* 207.

in the valley, as evidenced in the French communities' cahiers de doléances drawn up in the spring of 1789. Although many of their complaints are couched in the political and rhetorical style of the Old Regime—full of standard evocations of the poverty of the district and of the loyalty of local inhabitants, for example—the cahiers also reveal a moment when the authority of the state was breaking down and when the communities began to raise their voices in opposition.[66]

On the eve of the Revolution, the Cerdans revealed what they had long felt silently and without commentary. Their complaints and claims failed to evoke their identity as Catalans, nor did they refer to any specific privileges or rights that, as Catalans, they might have held.[67] Rather, the grievances reveal peasant opposition to the state based on what James Scott calls "forms of everyday resistance," daily and piece-meal confrontations with the tax collectors, customs guards, and political authorities.[68] Thus the communities complained about the recently raised salt prices, about the taxes on their cattle, the wood that they were required to bring to Mont-Louis, the oppressive tickets and the crooked customs guards who gave them out, and so on. It was less the weight of the state's presence which the French communities opposed than the arbitrariness by which taxes were allotted or regulations enforced: in particular, the Viguier Sicart was often singled out for his favoritism in dispensing tax exemptions.

On the whole, the village communities managed to maintain their autonomy and to keep the Old Regime state at arm's length. But when life became too difficult, the Cerdans either resisted or fled. Because of the proximity to the boundary, the choice of "voice" or "exit"—to borrow Albert Hirschman's distinction—was weighted in favor of the latter.[69] It was not as if overt resistance were absent. The Cerdans did not participate in the seventeenth-century salt revolts largely because they were provisioned directly from the Cardona salt mines in Cata-

66. Cahiers de Doléances, ed. E. Frenay, passim.

67. Only the Provincial Assembly of Roussillon, meeting in Perpignan in April 1789, explicitly asked "that the Treaty of Péronne of 1641 be forever upheld as the safeguard of our privileges, of our laws and our specific customs sanctioned by the oath of Louis XIII and by the Treaty of the Pyrenees, and since confirmed by several royal ordinances": cited in Sanabre, La resistència, 150–151. On the perceived "contractual" dimension of the province's relation to the crown see M. Brunet, Le Roussillon: Une société contre l'État, 1780–1820 (Toulouse, 1986), 17–20.

68. J. Scott, Weapons of the Weak: Everyday Forms of Peasant Resistance (New Haven, Conn., 1985).

69. A. Hirschman, Exit, Voice, Loyalty: Responses to Decline in Firms, Organizations, and States (Cambridge, Mass., 1970).

lonia. But criminal statistics for 1757 note several "rebellions" against tax-collectors in the Cerdanya. The armed struggles of dozens of smugglers against customs and salt-farm guards, already mentioned, were more impressive still. The prosaic and everyday struggles of peasants against soldiers and customs guards fill the archives of the Old Regime intendancy, underscoring the extent to which contraband trade formed a basic subsistence activity in the last decades of the eighteenth century. Contraband trade was not just an activity, it was a morality, a way of life. "Every soldier is seen in the Cerdagne as a barrier established to prevent fraud," wrote a military official in 1836, although he might just as well have been writing a century earlier, "and Catalan pride never allows them to forget that this country used to have the privilege of defending itself."[70]

It is hardly surprising, then, that the Cerdans demonstrated a notable unwillingness to enlist in the French army. After posting notices for voluntary enrollments in 1761, the viguier Sicart noted that "I would be pleased if the success responded to my own zeal, but I fear I will be no happier this year than last, since the inhabitants of this district are little inclined to serve in regular troops"—a sentiment constantly echoed among Old Regime military and civilian officials.[71] The French state enjoyed a slightly greater success in enlisting French Cerdans in the irregular troops of *miquelets,* the Catalan auxiliaries used to prevent desertion, impede contraband trade, and assist the royal police (*maréchaussée*) in the maintenance of public order. But even these modest enrollments failed to match the number of French Cerdans who enrolled as miquelets on the other side of the boundary, or who joined the Walloon Guards and other regular troops in Spain, according to Intendant Le Bon.[72]

70. FMG AAT MR 1222, nos. 112–113: "Mémoire militaire sur la Cerdagne française," March 1836; material on smuggling in the eighteenth century is dispersed in ADPO C 1270–1272; on the content and social organization of smuggling operations, see Brunet, *Le Roussillon,* 73–171; and below, chap. 4.

71. ADPO C 433 (viguier to intendant, January 11, 1761).

72. Ibid. (intendant to war minister, March 17, 1764, referring to the failures in fulfilling Perpignan's share of the thirty-two provincial regiments created by the Royal Ordinance of September 6, 1763). On the irregular troops of miquelets (or *fusillers de montagne,*) which traditionally served in defense of the province but were used by the French state against the Protestants of the Cévennes, see N. Sales, "Miquelets catalans i protestans del Llenguadoc," in her *Senyors bandolers,* 105–137; and see the drawings of uniforms and baggage in FAMRE MD 1948, fols. 170–172: "Etat militaire, écclésiastique, et politique du Roussillon," 1758. The French crown also tried to organize and control the *somatens,* local militias built on avowed self-interest, raised only in moments of imminent danger in defense of the district itself. But the royal ordinance of November

More generally, the French Cerdans remained drawn to Catalonia, toward which they tended to "exit" rather than to voice their opposition to the French state. "They live on the frontiers of Spain," reads a military memoir of 1717, "and they take refuge there when they break the law."[73] But neither flight from criminal prosecution nor an abstract loyalty to Spain can account for the movement of French Cerdans across the boundary. As the Roussillon intendant complained in 1764, the Cerdans were "drawn to Spain by the ironworks of Catalonia and employment in the fortress of Figueres." It was the relative prosperity of Catalonia in the second half of the eighteenth century—the economic "take-off" of the principality, the subject of Pierre Vilar's magisterial study—which attracted the French Cerdans and which greatly preoccupied the provincial administration.[74] For example, an enlightened employee of the intendancy elaborated upon the need of the French crown to police this current of emigration toward Catalonia:

> Because the Roussillon is on the frontier of Catalonia, the richest, most flourishing district of Spain, a continued surveillance is necessary to prevent such easy emigrations; in fact, these can only be prevented by making the people of Roussillon happier at home, with a lenient government, moderate taxes and duties, and an honest prosperity.[75]

Catalonia remained a pole of attraction for temporary workers and permanent migrants from the French Cerdagne until the twentieth century; many were those who went to work in the urban centers of the littoral in trades typically associated with their pastoral products (butter, sausages, and other meats).[76]

But it was not simply the economic growth of the principality which drew the French Cerdans and Roussillonnais across the boundary. Catalonia remained their homeland, "where ancient habits call them" wrote the Provincial Assembly of Roussillon in 1788, and the Cer-

---

10, 1733, was unsuccessful in linking the defense of the local patria with that of the nation: see FAN K 1221, no. 6: "Memoire sur les milices et soumettans du Roussillon," n.d.; and FMG AAT MR 1084, no. 25: "Memoire sur les comtés du Roussillon, Conflent, Cerdagne et les milices desdites pays," by the *Inspecteur Général des Milices,* March 1760.

73. FMG AAT MR 1083, no. 28.

74. Vilar, *La Catalogne dans l'Espagne moderne,* passim; ADPO C 433 (intendant to foreign minister, March 17, 1764); see also the officials cited in Sanabre, *La resistència,* 28–29.

75. FAN K 1221, no. 8: "Notices sur l'Intendance de Roussillon."

76. BC MS 1535: P. Cot Verdaguer, "Notes folkloriques pirinenques," 1928; Vila, *La Cerdanya,* 202.

dans retained a profound identification with those who spoke the same language (with minor differences of dialect) and who practiced the same customs.[77] The failure of the boundary to transform the identity of Catalans, in cultural or ethnic terms, meant that the Roussillonnais and Cerdans were able to melt into Catalan society. Throughout the eighteenth century, immigrant Cerdans and Roussillonnais in Catalonia largely escaped their status as French nationals; crossing the boundary without passports, failing to register at the French consular offices, the Cerdans and Roussillonnais moved south freely, assimilating themselves into Catalan society across the boundary.[78]

This secular movement of the population from the two counties to the principality has already been noted; what began to change in the course of the eighteenth century was the direction of the movement. The Bourbon reforms of Nueva Planta and the military occupations during the War of the Pyrenees drove a certain number of Catalans to settle, permanently it would seem, in the French Cerdagne. Between 1719 and 1721, over thirty inhabitants of Puigcerdà, for the most part agricultural workers, went to live in villages of the French Cerdagne. More generally, just as French Cerdans fled south out of economic necessity, poverty could drive Spanish Cerdans into France. In 1777, a list of forty-four peasants from Llívia unable to pay the catastro "absented themselves from Puigcerdà and went to live in France."[79]

Soundings in the parish registers of five French Cerdan villages suggest that such movements were not isolated. More marriages were contracted between French Cerdans on the one hand, and Spanish Cerdans and Catalans on the other, than a century earlier. In Odello, the percentage of spouses born in villages of the Spanish Cerdaña more than tripled, reaching 15 percent, and nearly doubled in La Tor de Carol (to 39 percent) and in Santa Llocaya (to 31 percent), both near the boundary (see app. B and table B.4). In Osseja, almost a quarter of the marriages celebrated between 1774 and 1783 involved Catalans from the principality, although the number of spouses from the Spanish

77. The Provincial Assembly, which noted that "entire families [leave to] seek their means of subsistence in the neighboring kingdom," is cited in J. Sagnes, ed., Le Pays Catalan (Pau, 1983), 572–573.

78. See Brunet, Le Roussillon, 238–240, although he fails to distinguish between legal status and actual practices of "binationality" in the borderland.

79. In May 1719, a lieutenant general wrote to the military governor of Catalonia that "each day there are many people who to escape their debts take refuge in France": AHP Correspondencia, May 14, 1719; AHP Catastro, 1719–1722; and AMLL Catastro, 1777.

Cerdaña remained what it had been a century earlier. In sum, despite counterexamples such as the village of Càldegues, the parish registers suggest that significant numbers of Spanish Cerdans and Catalans were settling in the French Cerdagne, even while increasing numbers of French Cerdans emigrated to the principality. Rather than being divided by the boundary, the two Cerdanyas were becoming more closely related.

Thus it might be argued that the boundary, essentially permeable, was increasingly ignored by the inhabitants of the Cerdanya. The persistence of the Catalan language and culture across the boundary occurred in the context of local opposition to the state and activities of everyday resistance. Yet if the two Cerdanyas were becoming more and more connected, this process grew out of the perceptions of differences introduced by the division of the valley and the states' emerging definition of a linear boundary. The boundary reshaped population movements, but also the social relations of class and community in the borderland. And it provided opportunities for the deliberate and strategic deployment of national languages and identities in the service of local interests under the Old Regime.

4

Community, Class, and Nation in the Eighteenth-Century Borderland

In 1778, a group of notables of Puigcerdà—lesser nobles, clergy, merchants, and other "distinguished persons"—founded, under the tutelage of the district administration, the "Sociedad Económica de Puigcerdà." This Economic Society, one of the earliest in the Spanish-speaking world, was intended to promote "agriculture, industry, and the arts" in the district. In memoirs sent to Don Francisco de Zamora, the Puigcerdà elite looked back to a golden age of commerce and manufacturing in the thirteenth and fourteenth centuries, when the town had flourished as a center of cloth production and trade. Don Thomas Bresson, one of the founding members, offered a sophisticated analysis of the town's subsequent decline.[1] But the Economic Society's founders also looked across the boundary to the French Cerdagne. Not only had the division of the valley resulted in restrictive regulations that stifled commerce and manufacturing, but the French Cerdagne seemed to have grown wealthier than the Spanish Cerdaña, and at the latter's expense. The perceived prosperity of the French Cerdagne—its agriculture, stockraising, and commerce—formed the principle motive for the establishment of the Puigcerdà Economic Society.[2]

1. Biblioteca del Real Palacio (hereafter, BRP) MS 2436, fols. 166–194: "Noticia de las sucesas que han causado el extermino de la antiguas fabricas de la villa de Puigcerdà," Don Thomas Bresson, April 27, 1779.

2. Documentation on the Puigcerdà Economic Society is dispersed in ACA RP Procesos de Bailia Moderno 1779.6.0; *Real sociedad económica matritense de amigos del*

The reality is more difficult to gauge. Less than any actual difference in economic development of the two Cerdanyas, what is significant is the *perception* of the differences introduced by the political boundary. The memoirs and commentary of the Puigcerdà Economic Society evoke the existence of two Cerdanyas, French and Spanish, the distinctive identities of which were defined by a set of (potential) national differences. The eighteenth-century history of the borderland reveals, moreover, how the propertied elite, the village communities, and ordinary peasants attempted to appropriate these differences in the service of their local economic interests.

AGRICULTURE, COMMERCE, AND INDUSTRY IN THE BORDERLAND

There is little evidence to support the Puigcerdà elite's perception of the Spanish Cerdaña's agricultural underdevelopment, since all signs point to the relative stagnation, if not immobility, of agricultural and pastoral production in the French Cerdagne until the middle decades of the nineteenth century. There is no evidence of an "agricultural revolution" in the French Cerdagne before then, if by this is meant the introduction of new techniques, the development of artificial meadows and forage crops, the abandonment of biennial rotation, or the use of artificial fertilizers. In 1741, the viguier of the French Cerdagne calculated that each jornal of moderately good land yielded three and a half to four cargas of rye, not including deductions for reseeding (one-half carga of grain), taxes, and tithes. (The yield in revenue to the owner amounted to one carga.) In 1805, the communities of Osseja and Santa Llocaya estimated identical yields and deductions for reseeding. "The new knowledge has not penetrated into the Cerdagne," wrote Bonaventure de Pera at that time. A lawyer born in Palau, his opinion was confirmed in numerous descriptions of both Cerdanyas during the Old Regime. "The old methods of agriculture have stayed the same," wrote a French army captain as late as 1836, "not a sign of improvement can be found." Even the introduction of the potato in the late eighteenth century had met with great resistance.[3]

país, leg. 22, no. 7; and AHP MA 1778–1781, fol. 24 et seq. The memoirs and questionnaires sent by members of the society to Francisco de Zamora may be found in BRP MS 2436. For an overview of economic societies in Catalonia see E. Lluch, *El pensament econòmic a Catalunya, 1760–1840* (Barcelona, 1973); and more generally, R. J. Schaffer, *The Economic Societies in the Spanish World, 1763–1821* (Syracuse, 1958).

3. ADPO C 884: "Etat des journeaux de terre, champ, et pred, et revenues par es-

In describing the wealth and prosperity of the French Cerdagne, members of the Puigcerdà Economic Society were referring less to an actual state of affairs than to a potential difference of political economy that resulted from the division of the valley: a shift in the legal framework of the agro-pastoral system of production on the French side. They referred to the 1769 edict abolishing in the Roussillon the collectivist agrarian tradition of usufruct rights to meadows and arable lands. The edict permitted landowners to enclose their meadows and thus to gain control of a second harvest, and the landowners of Puigcerdà—many with properties in the French Cerdagne—saw in the edict the cause of their dependency on the better-quality livestock across the boundary in France.

Meadows formed the critical link between agriculture and stockraising in the Cerdanya. "The meadows have no utility except that of nourishing the animals, and these have none except to fertilize the lands, which without manure, would not provide a yield" reads an anonymous memoir of 1794. Although owned as "private property," meadows were subject to collective grazing rights, one form of the collective usufruct rights known as *empriu*. "It is the custom for farm owners to pasture their animals in all the fields and meadows of the other farms of the same village territory," providing that the harvest had been gathered, wrote the viguier in 1768. By speaking of the "owners of meadows composing the common prairies," he revealed the ambivalent notions of "private" and "communal" property among the mountain peasantry, in Catalonia as throughout Old Regime Europe.[4]

Inspired by physiocratic theories of agricultural wealth, the French monarchy and provincial parlements attacked these collective grazing rights, especially in the late eighteenth century. The 1769 enclosure edict in the Roussillon came from the pen of Raymond de Bertin, who had been intendant of the Roussillon in the 1750s and was later

---

timation," 1741; ADPO O Osseja and Santa Llocaya: "Statistique Agricole, an XIII"; ADPO 7J 105: "Mémoire à l'occasion de fixer les limites de la Cerdagne," Bonaventure Pera, year XI (1803). On the "persistence of routine" in late eighteenth-century Spanish Cerdaña, see BRP MS 2346, fol. 170v; and for the French Cerdagne in 1836, FMG AAT MR 1222, nos. 112–113: "Mémoire militaire sur la Cerdagne française," Captain Puvis, March 1836. For a similar assessment of the state of agriculture in the department as a whole during the first half of the nineteenth century, see G. Gavignaud, *Propriétaires-viticulteurs en Roussillon,* 2 vols. (Paris, 1983), 1: 291–324.

4. ADPO L 393: "Observations sur les preds et bestiaux," n.d., ca. 1794; ADPO C 1501: memoir of the Viguier Sicart, November 24, 1768; on the second hay harvest more generally, see M. Bloch, *French Rural History,* trans. J. Sondenheimer (Berkeley and Los Angeles, 1966), 210.

comptroller-general and secretary of state in Paris.[5] But the movement toward "agrarian individualism"—to use Marc Bloch's famous phrase—had begun in the Cerdanya long before. The Viguier Travy reported in 1768 that many proprietors in the French Cerdagne had been trying to enclose their meadows since the early 1700s, especially in the territories of Palau, Sallagosa, and Estavar. In the Spanish Cerdaña, too, wealthy landowners had undertaken efforts to enclose meadows and reap the second harvest. A member of the Economic Society wrote that

> [F]or some time now, eyes have been opened in this district, and many meadows have been closed, from which the owners themselves enjoy the fruits; but it has been costly and very difficult, and innumerable law suits about the matter are currently pending in almost all the courts.[6]

Looking across the boundary, the Puigcerdà Economic Society saw in the enclosure ordinances of Roussillon a device for increasing the total wealth of their community. The members urged their own government to pass a similar ordinance, since enclosing meadows would allow their owners to increase their livestock and, in turn, produce more manure that would increase the yields of rye. Others stressed how the "natural liberty to dispose of the fruits of one's meadows during the whole year" would result in greater total wealth; and with more money in circulation "there would be an increase in salaries, in consumption, and all this would contribute to the establishment of factories and manufacturers."[7]

In point of fact, none of these improvements took place in the French Cerdagne despite the enforcement of the 1769 ordinance. Neither the overall numbers nor the composition of livestock in the French Cerdagne experienced any significant changes before the late nineteenth century—especially since the increased herds among meadow owners would have been offset by losses among the poorer peasants.[8] If the

5. Bloch, *French Rural History*, 209–213; on Bertin, see Gavignaud, *Propriétaires-viticulteurs*, 1: 119. Compare the promotion of agrarian individualism in Burgundy, where the intendancy helped the village communities resist the reforming activities of the crown: Root, *Peasants and King*, 105–154. On the intendancy and "enlightened administration" in the late eighteenth century, see the still-useful P. Ardasheff, *Les intendants de province sous Louis XVI*, trans. L. Jousserandot (Paris, 1909), esp. chaps. 3–5.

6. BRP MS 2346, fols. 176v–179v; ADPO C 1501 (memoir of 1768).

7. BRP MS 2346, fol. 169; see also the "Discurso sobre la naturaleza de la Cerdaña, parte considerable del corregimiento de la villa de Puigcerdà su capital," in the *Correo Général de España*, 1770, 313–340 (copy in FBA H 2172 [2]).

8. That the edict was enforced is evidenced by its incorporation into the village ordinances (e.g., ADPO Bnc no. 439, Carol Valley, 1786); and the numerous disputes that

quality of horse breeding in the French Cerdagne was clearly superior to that on the Spanish side—prices of horses and mules bred in the French Cerdagne reached three times those paid for animals in Spain—it was because of the Royal Stud Farms of Roussillon. Originally created in 1718 and reorganized in 1751, the farms of *haras* consisted in the placement of stallions in the mountain communities at the expense of the inhabitants. If the Cerdagne communities complained bitterly about the added fiscal burden in their cahiers de doléances, they also drew a great profit from the presence of quality horses, crossing them to produce mules sold in Catalonia. The memoirs of the Puigcerdà Economic Society failed to mention the French stud farms in their memoirs, so intent were they on convincing the Spanish crown that enclosures lay at the basis of the wealth and prosperity of the French Cerdagne.[9]

But it was not only the possibility of enclosures in the French Cerdagne which led the Puigcerdà elite to decry their own misery and the general prosperity across the boundary. In a memoir of 1790, the Economic Society elaborated upon their commercial dependency on the French Cerdagne and, in particular, on the border settlement of les Guinguettes:

> At this point, the [Spanish] Cerdaña is in debt to les Guinguettes, which is made up of only several houses in France a short distance from Puigcerdà. Many of the names of Puigcerdà's own inhabitants are inscribed in the account books of les Guinguettes. The [Spanish] Cerdans owe these storeowners for their countless purchases of cloth. . . . To pay for these goods, they sell to the French merchants the fruits of their lands while these are still in the ground. Both in the sale of their goods as in the means of payment, the [French] merchants gain an excessive profit.

arose over rights of passage (e.g., ADPO C 2078, Osseja 1777; ADPO C 2055, Dorres 1776). On the composition of livestock in the eighteenth century, see J. Balouet, "L'élévage en Roussillon au XVIIIe siècle," *CERCA* 4 (1959): 181. Between 1730 and 1866, changes in the numerical quantity and species ratio of livestock in the French Cerdagne were statistically small: from 8.8:1 sheep to cows in 1730 to 7.3:1 in 1866, although the ratio dropped to 4.2:1 in 1888. Statistics for these calculations were gathered from official tabulations in ADPO C 2045 (1730) and ADPO Mnc 2620 (1866); see also J. de Pastor, "La Cerdagne française (Étude de géographie humaine)," *Bulletin de la société languedocienne de géographie* 3 (1933): 140.

9. BRP MS 2346, fol. 144, on the "abundance of livestock much greater than in Spain, especially the horses." On the Royal Stud Farms, see Poeydavant, "Mémoire sur la province de Roussillon," *BSASLPO* 37 (1896): 397–407. On crossbreeding horses to produce mules, see J. A. Brutails, "Notes sur l'économie rurale du Roussillon à la fin de l'Ancien Régime," in *BSALSPO* 30 (1889): 300–303; and on the importance of mules in the economy of eighteenth-century Catalonia, see N. Sales, "Ramblers, traginers i mules (s.xviii-xix)," *Recerques* 13 (1984): 64–81.

In short, the Economic Society accused the merchants of les Guinguettes of illegal trade, just as Francisco de Zamora had described the stores in 1787 as a "harmful settlement, the origin of contraband trade, and the very cause of the destruction of the famous town of Puigcerdà."[10]

The growth and development of les Guinguettes in the course of the eighteenth century is dramatic testimony to the importance of such trade, and more generally to the role of the boundary in the differentiation of commercial activities in the Cerdanya. Traditionally, Puigcerdà was *la vila*, "the town," and with its weekly markets, it brought together peasants from all over the valley to buy, sell, and trade. The Roussillon intendancy's attempts to create markets in Ur and Estavar (1758) and in Sallagosa (1787) never succeeded in siphoning significant commercial activity away from Puigcerdà.[11] Only the merchants and artisans of Mont-Louis, whose retail sales derived in large part from the garrison of soldiers, provided a significant alternative to the markets of Puigcerdà—at least until the development of les Guinguettes (Catalan: la Guingueta), an offshoot of the village of Hix. Located at the center of the plain, just across the boundary formed by the Raour River, the village of Hix had once been the capital of the early medieval County of Cerdanya but, with the founding of Puigcerdà (1178), was reduced to a simple parish until well after the Treaty of the Pyrenees. In the 1690s, Andreu Giraut of Ur bought property next to the Raour, and in 1717 received royal permission to run an "inn, cabaret, breadshop, salt and meat dispensary." Thus was born the "Taverns of Hix."[12]

It was more than a half-century before the new settlement flourished; by 1767 there were two large stores, one owned by Joseph Ramonatxo, "manufacturer of local cloth and clothing merchant." By 1773, however, les Guinguettes had drawn the wrath of the merchants of Mont-Louis, who requested the enforcement of a royal ordinance of 1772 "prohibiting all warehouses of merchandise within four leagues

10. BRP MS 2346, fol. 150v: "Memorias topográficas del condado de Cerdaña"; and F. de Zamora, *Diario de los viajes hechos en Cataluña*, ed. R. Boixareu (Barcelona, 1973), 92.

11. ADPO C 2047 (establishment of markets of Ur and Estavar); and ADPO C 2049 (creation of a market at Sallagosa [following a denial of a petition of the *syndics* of the French Cerdagne to have a market established at Hix, June 1786]).

12. B. Alart, *Notices historiques sur les communes du Roussillon*, 2 vols. (Perpignan, 1878), vol. 2, chap. 2; J. Sarrete, "La paroisse d'Hix," *BSASLPO* 46 (1905): 313–346.

of the frontier." The subsequent debates between Ramonatxo and the merchants of Mont-Louis reveal that the latter had indeed built a substantial business that went well beyond the "local consumption" that Ramonatxo claimed to supply. Although the Mont-Louis merchants clearly exaggerated when they declared Ramonatxo's inventory worth 300,000 livres, Ramonatxo himself was less than honest in declaring that only cloth and material originating in France were for sale. The comptroller-general of the Royal Farms in Roussillon, who sided with the merchants of Mont-Louis, found "olive oil, string, tin, shoes, cotton hats, and silk handkerchiefs" from Catalonia, of which the import was forbidden.[13]

The General Council of the French Cerdagne publicly supported the merchants of Hix, as did the viguier himself, arguing that such sales were advantageous to foreign exchange and that if the depots were suppressed, "the merchants of Puigcerdà would draw off all the money." Such arguments prevailed, and les Guinguettes remained in place. Although the intendant refused to establish a weekly market in Hix in 1786, just as the prefect would refuse in 1819—both arguing that by its proximity to the border such a market would only encourage contraband trade—the growth in size and status of les Guinguettes was such that the Duc d'Angoulême, upon his return to France after the "100 Days" in 1815, gave it the name of Bourg-Madame in honor of his wife.[14]

Les Guinguettes was not the only commercial center in the late eighteenth-century French Cerdagne. Two other border villages—La Tor de Carol and Osseja—also gained a reputation for contraband trade, just as the populations of these villages increased dramatically in the course of the eighteenth century. It was not due to the biological growth of the population that the Valley of Carol went from 574 to 1122 inhabitants, or that Osseja grew from 283 to 904. Both communities were located on the boundary, and both lay across important passes in and out of the Cerdanya, along a single passage that went from Toulouse to Barcelona. Like all the other settlements bordering Spain, according

13. J. L. Blanchon, "Hix: Les Guinguettes d'Hix, Bourg-Madame," *Conflent* 45 (1968): 114; ADPO C 2048 (petition of Mont-Louis to the intendant, July 7, 1773; memoir of comptroller-general of the Royal Farms, n.d., ca. 1774).

14. ADPO C 2048: Petition of the General Council of the French Cerdagne, "meeting on the basis of a privilege given by Philip II, November 30, 1585," November 7, 1774; memoir of the viguier, October 21, 1773; ADPO O Bourg-Madame, no. 2 (petition of municipality, 1819); and on the renaming of Les Guinguettes, Sarrete, "La paroisse d'Hix," 345–346.

to an anonymous memoir of 1790, "they trade imports and exports despite the surveillance of the customs guards of both nations."[15]

Smuggling was not only an important means of livelihood throughout the Cerdanya, it was also one of the most important productive activities in the district. Great fortunes were built on the profits of smuggling. The merchant François Durand, whose grandfather settled in Perpignan from Montpellier in 1720 and who, with his brother Raymond, became the principal supplier of Napoleon's troops in Catalonia, built a vast fortune on the profits of illegal trade.[16] In the Cerdanya, most of the larger merchants who organized contraband trade were from the French side of the border. Some built their fortunes on the sale of mules and horses in Catalonia; others, men such as Joseph Ramonatxo of Hix and François Garreta of La Tor de Carol, maintained trade contacts in Paris, Flanders, Lyon, and as far as Odessa; importing bullion from Spain, they exported cloth and merchandise brought to their warehouses on the frontier. At his death, Garreta's estate was worth over a million francs.[17]

For most of those who did the actual smuggling, participation in the illegal trade was less a way to make a fortune than to survive. Poorer peasants, agricultural workers, and lesser craftsmen employed by the merchants as packeteers, earned much of their livelihood from smuggling. When caught, their most common defense was that "they had to do it in order to live, and preferred to be smugglers than ordinary bandits." The local judicial system remained indulgent, and the provisions of the 1786 convention between France and Spain, requiring the mutual restitution of convicted smugglers, were little enforced. Although most of the smuggling bosses were French, the packeteers were both French and Spanish, drawn from throughout the Catalan Pyrenees. They manipulated the juridical uncertainties of nationality in the borderland just as they used the territorial uncertainties of an ill-defined boundary. In 1823, for example, the customs director of Perpignan was informed that "Spanish mule drivers who cannot send wine across the border turn themselves into smugglers, after taking French names, in order to get passes as far as the frontier in Cerdagne."[18]

15. ADPO 7J 14: "Statistique de la Cerdagne," n.d., ca. 1790.
16. Brunet, Le Roussillon, 103–111.
17. H. Ramonatxo, Un Cerdan témoin de l'histoire: François Garreta (Luzech, 1978).
18. Quoted in J. Sacquer, "La frontière et la contrabande avec le Catalogne dans l'histoire et l'économie du département des Pyrénées-Orientales de 1814 à 1850" (Diplôme d'études supérieures, Historie économique, Université de Toulouse, 1967), 84; on judicial

Smuggled goods moved in both directions across the boundary. The lesser, local contraband generally went from Spain to France, including salt from the Cardona mine; Mediterranean foodstuffs such as oil, garlic, and oranges (which only rarely found themselves on the tables of poorer peasants); and items of the traditional Catalan costume. The large-scale contraband trade, with the exception of Spanish currency, tended to originate in France; it included manufactured goods, cloth, silk, leather, as well as the grain and cattle which often provisioned the city of Barcelona and the towns of the Catalan littoral.[19] This imbalance, combined with the fact that much of the trade was organized by French merchants who flaunted their fortunes by building impressive houses, led the Puigcerdà Economic Society to stress that the French Cerdagne was growing wealthier than the Spanish side of the valley.

Yet in comparing the growing agricultural and commercial prosperity of the French Cerdagne against the underdevelopment and dependency of the Spanish side of the valley, the memoirs of the Economic Society failed to reveal the way in which industrial development in the borderland actually favored the towns of the Spanish Cerdaña. The proto-industrial, mechanized production of woolen and cotton hosiery in the Spanish Cerdaña, and the persistence of "putting out" routines on the French side, reproduced the regional differentiation—running counter to a more generalized set of national differences—of the industrial "take-off" in northern Spain which anticipated and overran that of southern France.[20]

That the French Cerdagne was *not* growing wealthier in the late eighteenth century is evidenced by the increased reliance on the production of woolen stockings among the poorer peasant classes. Although such a "cottage industry" had probably existed for centuries, it was only after 1750 that it took on an important role in the subsistence economy.[21] The household manufacture and distribution of woolens

testimony of smugglers, see M. Llati, "La contrabande en Roussillonais au début du XIXe siècle" (Diplôme d'études supérieures, Université de Montpellier, 1955), 114; on the indulgence of the judicial system, see Brunet, *Le Roussillon*, 159–160; and on the enforcement of the 1786 convention, see the dossier in ADPO Mnc 1944.

19. Brunet, *Le Roussillon*, 73–92.

20. E. Badosa i Coll, "La indústria rural a Catalunya a finals del s. XVIII," in *Primer congrés d'història moderna*, 1: 345–349.

21. On the importance of spinning and weaving in the rural Pyrenean communities, see J. F. Soulet, *La vie quotidienne dans les Pyrénées sous l'Ancien Régime du XVIe au XVIIIe siècle* (Paris, 1974), 145–172; and more generally in Old Regime France, O. Hufton, *The Poor in Eighteenth Century France* (Oxford, 1974), chap. 3; and G. Gullickson, *Spinners and Weavers of Auffay: Rural Industry and the Sexual Division of Labor in a French Village, 1750–1850* (Cambridge, 1986).

was intimately linked to a sexual division of labor and seasonal patterns of out-migration: women combed, wove, and knitted during the winter months, while men left the valley to peddle last year's finished product in France. Whereas in 1758, the dozen self-declared "stocking merchants" furnished the primary material and paid a reduced price for each finished product, the next several decades witnessed a "democratization" in the French Cerdagne of the manufacture and distribution of stockings. According to the prefect in 1808, the canton of Sallagosa had an annual emigration of some 150 peddlers.

> They leave their homes around the beginning of October, after the winter crop is sown, and head off toward Lyons, Grenoble, and Nice, where they have already sent the woolen stockings . . . knitted the preceding winter. . . . From there, they disperse into the Cévennes, Alps, Piedmont, Switzerland, and in all the northern and eastern departments of France. They return toward the end of June, bringing back cloth, stockings, and cotton bonnets which they resell in their district or in Spain.[22]

Rarely was the manufacture of woolen stockings organized in factories or mechanized beyond the domestically made looms: only Joseph Ramonatxo had in the 1760s established a shop employing some two dozen workers, but it did not last, and no evidence of factories or "industries" of any kind can be found at the end of the eighteenth century. Not only did the provincial government fear that factories built near the border would be used as outlets for contraband trade, but municipal officials opposed such factories, which, according to one, "would cause great damage to the inhabitants who weave the wool themselves."[23]

On the Spanish side of the border, however, proto-industrialization was much in evidence during the 1770s. To be sure, the peasant families of the district continued to spin and weave woolen stockings themselves: in 1778, according to the governor of Puigcerdà, more than 2400 women were so occupied in their homes.[24] But in the towns of Puigcerdà and Llívia, the manufacture of woolen and linen stockings, along with other goods, became a more centralized, industrial opera-

22. FAN F(20) 435 (prefect's report, June 27, 1808).
23. ADPO C 2086 (director of the Royal Farms at Narbonne to interior minister, August 11, 1787); ADPO O La Tor de Carol, no. 3 (letter of mayor of La Tor de Carol, October 13, 1835); ADPO 1J 160: "Statistique de la Cerdagne"; and ADPO Mnc 872/1 (industrial statistics of 1814).
24. SAHN Estado leg. 2934 (letter to First Minister Floridablanca, February 26, 1779).

tion using mechanical looms and relying exclusively on salaried wage earners, most of them men.

Puigcerdà thus recovered some of its medieval glory as a manufacturing center—though in the late eighteenth century, mechanized spinning and weaving developed at the margin of the guild organization, which itself was in decline.[25] In 1772, Pere Bruguère established a factory (*fabrica*) of 45 looms for weaving wool, yarn, and cotton stockings in quantities "never before seen in the district"; the following year, Pere and Isidre Calvet, from the village of Guils, opened another factory that employed 57 operators on 24 looms, producing almost 2,500 dozen pairs of men's and over 3,000 dozen pairs of women's stockings annually, in addition to two dozen hats a week. By 1787, Calvet and Company employed 120 workers, out of the more than 500 workers in the town of Puigcerdà. Production peaked around the turn of the century, as the town boasted some 200 industrial looms employing 700 men—and women—in combing cotton, making linen, and weaving silk and wool stockings.[26]

The town of Puigcerdà grew wealthy on the sale of its products to France; and many were the Frenchmen who, because the French government refused to allow factories near the frontier, came to Puigcerdà and the Spanish Cerdaña to open industrial establishments.[27] By the end of the eighteenth century, this mechanized production had completely outstripped the unmechanized spinning and weaving of the French Cerdagne's cottage industry, undercutting stocking production in France. It was a pattern that remained in place throughout the nineteenth century, and which became even more evident during Puigcerdà's second industrial renaissance in the 1860s and 1870s.[28]

Eighty years earlier, when the effects of mechanization were already visible, the Puigcerdà Economic Society had failed to mention the industrial activity of the Spanish Cerdaña, insisting on the growing prosperity of the French side. Though it is true that few of the members of

25. On manufacturing in medieval Puigcerdà, see S. Galceran i Vigué, *La indústria,* passim; and S. Bosom i Isern, *Homes i oficis de Puigcerdà al segle XIV* (Puigcerda, 1982). On the industrial growth of Catalonia in the eighteenth century in relation to the guilds, see M. Ardit et al., *Història de los països catalans de 1714 a 1975* (Barcelona, 1980), 74–76.

26. Galceran i Vigué, *La indústria,* 153–156; AHP: "Plet sobre mitges."

27. ADPO C 2086 (director of the Royal Farms at Narbonne, August 11, 1787); and ADPO O La Tor de Carol, no. 3 (Victor Berges to prefect, August 6, 1835).

28. BCM Documentos (3–1–1–29): D. Joaquim Barraquer y Puig, "Memoria descriptiva . . . de Puigcerdà," 1875.

the society were actual investors in the industrial plants, their silence on the subject is nonetheless significant. Petitioning their government for assistance and reforms, they used the French Cerdagne to highlight a series of contrasts. The realities of the economic differentiation in the two Cerdanyas at the end of the Old Regime are more difficult to gauge. Enclosures in the French Cerdagne do not seem to have produced higher agricultural yields overall; and although great fortunes in contraband trade were made in the French Cerdagne, proto-industrial activities were firmly located on the Spanish side. What matters more than the actual differentiation were the perceived differences occasioned by the political boundary. Such perceptions extended to the identities of the Cerdans as well.

PROPERTY RELATIONS IN THE BORDERLAND

"Because the territory of Ur borders on Spain," wrote the community of inhabitants in their cahiers of 1789, "many Spaniards have bought up and continue to buy a considerable number of estates and properties." Verdrinyans, across the valley floor, saw its inhabitants heavily burdened by seigneurial dues, the owners of which were "all Spaniards, living in Spain." Similar complaints against the Spaniards, voiced by the communities of Llo, Dorres, Càldegues, the Carol Valley, and others, focused on the exemption of Spanish landowners from the capitation and their successful removal from the local tax rolls, which increased the fiscal burden of the communities. In opposition to the Spaniards, the communities even affirmed their loyalty to the French king and their identity as his subjects. As the municipality of Santa Llocaya wrote, "full of love for their king,"

> Far from being jealous of this exemption which these foreigners solicit so passionately, the inhabitants see in paying the capitation the proof of a most precious title for them, that of a Frenchman.[29]

---

29. *Cahiers de Doléances de Roussillon,* 288–334; more generally, see B. F. Hyslop, *French Nationalism in 1789 According to the General Cahiers* (New York, 1934). Hyslop finds "an essentially national spirit" produced in France on the eve of the Revolution (p. viii); she uses a concentric circle model centering on Paris (p. 228), and finds that "the *généralités* along the frontier lagged in their concept of French nationality and mingled appeals to provincial and class status with such statements as they made," (p. 62) although a marked hostility to foreigners existed in certain frontier provinces (e.g., Lorraine and Alsace, p. 35).

If such claims form part of the stylized dialogue of peasant communities and crown officials, they also describe the experience of social hierarchies in the Cerdanya. Just as many of the privileged families making up a majority of the large landowners in the French Cerdagne chose to identify themselves as Spanish in order to gain exemptions from personal taxes, so too did the village communities affirm their opposition to these Spaniards in national terms. More than mere rhetoric, the discourse of both peasants and elites suggests how the division of the Cerdanya into two parts by the Treaty of the Pyrenees had transformed, 130 years later, the way in which the inhabitants of the Cerdanya thought about themselves. The boundary created a potential opposition of nationalities that dovetailed with a geographic configuration of social class in the valley; in pursuing their local economic interests, the Cerdans identified themselves with their respective nations.

The village communities of the Cerdagne resembled tens of thousands of others across Old Regime Europe. The traditional "agricultural community" had a marked corporate character: arable land was held as "private property," though subject to collective usufruct rights, while woodlands, pastures, and waste remained common lands, although these were often converted into "private property." In the French Cerdagne, the eighteenth-century village communities collectively exploited common pasture lands and woods, regulated the grazing of stubble on private meadows, and generally enforced an institutionalized moral presence on the constituent households.[30] The village communities were represented politically by elected syndics, chosen from among the "principal inhabitants," whose purpose was to defend the corporate identity of the community against encroachments by other communities or supra-local powers. (Before the Treaty of the Pyrenees, the Cedanya valley maintained a "General Council of Syndics," representing the four "quarters" of the valley, although the political character of the valley community was not as well developed in

30. On the European village community, see J. Blum, "The European Village as Community: Origins and Functions," *Agricultural History* 45 (1971): 157–178; L. Assier Andrieu, "La communauté villageoise: Objet historique, enjeu théorique," *Ethnologie française* 17 (1987): 351–360. On the rural communities of the Cerdanya, see P. Sahlins, "The Nation in the Village: State-building and Communal Struggles in the Catalan Borderland during the Eighteenth and Nineteenth Centuries," *Journal of Modern History* 60 (1988): 242–243; G. Gavignaud, "L'organisation économique traditionelle communautaire dans les hauts pays catalans," in *Conflent, Vallespir, et Montagnes Catalanes,* Fédération historique du Languedoc méditerranéen et du Roussillon (Montpellier, 1980), 201–215; and L. Assier-Andrieu, *Coutumes et rapports sociaux: Etude anthropologique des communautés paysannes du Capcir* (Paris, 1981).

the Cerdanya as in the central or western Pyrenees. After the Treaty of the Pyrenees, the French Cerdagne still elected two syndics to represent the collective interests of the remaining two "quarters" of the valley to the crown and provincial administration, while in the Spanish Cerdaña the corporate interests of the villages tended increasingly to be represented by the municipality of Puigcerdà.)[31]

The village communities, in the Cerdanya as elsewhere, defined their corporate identity through local ordinances. These *crides* (in Catalan), renewed periodically, established the rules and penalties for exploiting common resources, limiting their enjoyment to the constituent households. The corporate identity of the village community defined, in strict terms, the boundaries of insiders and outsiders. The former were "neighbors" (veïns), household heads who had lived in the village at least a year and who thus enjoyed the full rights of local citizenship. Outsiders or foreigners (forasters) were excluded from the full enjoyment of communal resources. Owners of fields and meadows within a communal territory who did not reside in the village were considered "outsiders," and their use of local pastures, woodland, and waters was more restricted, increasingly so in the course of the eighteenth century.[32]

The corporate organization of the agricultural community, however, was no "primitive communism," despite Marx's own romanticization of the village community as a "unique stronghold of popular liberty and popular life."[33] For the village was also a miniature theater of class

---

31. On village assemblies and political representation in village Communities, see J. Blum, "The Internal Structure and Polity of the European Village Community from the Fifteenth to the Nineteenth Century," *Journal of Modern History* 43 (1971): 541–576; and for France, see J. P. Gutton, *La sociabilité villageoise dans l'Ancienne France* (Paris, 1979), 69–154; and Root, *Peasants and King*, 66–104; On the "General Council of Syndics" in the Cerdanya, see Introduction, n. 22; and on the political assemblies of Pyrenean valley communities, see J. Poumarède, "Les syndicats de vallée dans les Pyrénées française," in *Les communautés rurales,* Receuils de la société Jean Bodin pour l'histoire comparative des institutions, vol. 43 (Paris, 1984): 385–409.

32. ADPO Bnc 439: "Criées et ordonnances de la Cerdagne française, 1684–1789." On the institutions of village citizenship in Catalonia, see Assier Andrieu, *Le peuple et la loi*, 147–168; and for the Aran Valley, including evidence of a similar restriction of admission during the late eighteenth century, see M. Angels Sanllehy i Sabi, "L'afillament a les communitats araneses (segles XVII–XIX)," *L'Avenç* 115 (1988): 32–37. More generally, see Gutton, *La sociabilité villageoise,* 31–40; P. Ourliac, "La condition des étrangers dans la région Toulousain au moyen age," in *L'étranger,* Receuils de la société Jean Bodin (Brussels, 1958) vol. 10: 101–108; and P. Toulgouat, *Voisinage et solidarités dans l'Europe du moyen age* (Paris, 1981).

33. Marx to Zasulich, March 8, 1881, in E. J. Hobsbawm, ed., *Pre-Capitalist Economic Formations,* (New York, 1964), 144–145. Marxist scholars of the French peasantry have tended to play down the importance of internal stratification under the Old Regime: see, for example, the classic statement by A. Soboul, "The French Rural

warfare, opposing the richer landowners of the "great houses" to small proprietors and a growing rural proletariat. The hierarchical character of property ownership within the Old Regime French Cerdagne communities was particularly marked, since there were few middling peasants between the larger landowners and those who barely managed to subsist on their small plots of land and meager livestock.

Although there was a notable absence of "very large" properties in the Cerdanya—there were no great noble estates to speak of—the disparity between "large" and "very small" landowners was striking (see app. B and tables B.5 and B.6). In Angostrina, to take the most dramatic example, almost two-thirds of the landowners owned less than five jornals, far below the amount necessary to support a family, and together this group held only a fifth of the arable land—hardly any of it meadows, and most of it of poorer quality. Whereas there were no "large" landowners at all in Angostrina, the community of Santa Llocaya boasted a majority. There, in the center of the fertile plain, neither the number of proprietors nor the number of houses changed much between 1694 and 1775. In 1775, the richest 40 percent of the landowners owned 90 percent of the arable land in properties greater than fifty-six jornals (quite large by Cerdan standards), while the poorest 40 percent owned about 2 percent of the land, made up in parcels of less than ten jornals each. In Estavar and Bajanda, "medium" proprietors, 20 percent of the landowners in 1694, had shrunk to less than 3 percent in 1775, thus recapitulating a general trend toward impoverishment. Unable to support themselves from the land alone, using their meager flocks for fertilizer, the poorer landowners depended on the collective resources of the community, including grazing rights on private meadows. They also came to depend on seasonal employment on the larger estates, rural spinning and weaving, and, increasingly, seasonal and then permanent migration from the valley.[34]

All evidence suggests that the class of rural proletarians in the Cerdanya, as throughout the Pyrenees, increased during the eighteenth

Community in the Eighteenth and Nineteenth Centuries," Past and Present 10 (1956): 83–85.

34. ADPO C 2050 (tax rolls of the vingtième, 1775); and Gavignaud, Propriétaires-viticulteurs, 1: 254–258. For a fuller discussion of property hierarchies in selected villages of the French Cerdagne from the late-seventeenth to the midnineteenth century, see Sahlins, "Between France and Spain," 475–482, and app. 2. On seasonal migration in the Pyrenees, see L. Goron, "Les migrations saisonnières dans les départements pyrénéens au début du XIX siècle," RGPSO 4 (1933): 230–272; and for France more generally, A. Poitrineau, Remues d'hommes: Les migrations montagnardes en France aux XVIIe et XVIIIe siècles (Paris, 1983).

century, due in part to the increasing population pressure on limited ecological resources.[35] The population of the French Cerdagne grew from 4054 inhabitants in 1730 to 6987 in 1787, a 72 percent increase in less than fifty years, with the population density reaching 15.5 inhabitants per square kilometer. In general, villages near the valley floor had lesser rates of growth than those perched on the mountain side, although border villages in the plain grew more dramatically due to their role as centers of contraband trade. The population of the Spanish Cerdaña grew less dramatically, although the 59 percent increase between 1718 and 1787—from a population of 6646 to 10,580—placed additional strains on already limited resources.

Although there does not seem to have been much conversion of pasture to arable land in the Spanish Cerdaña, statistics and descriptions for the French side suggest extensive clearings (défrichements), encouraged by the royal edicts of 1761, 1764, and 1766.[36] Whereas the French viguier wrote in 1730 that "all the land which might be turned to cultivation is already cultivated," his successor in 1768 noted an increase, which he admittedly underestimated, of almost 1500 jornals, and calculations suggest a 16-percent increase in arable land between 1741 and 1775.[37] The increase in land under cultivation, however, hardly matched the population growth, and later eighteenth-century descriptions of the Cerdanya emphasize its overwhelming poverty. "They are reduced to eating soups made with rotten rye bread and salted lard, often rancid," wrote J. B. Carrere in 1788, referring to the growing class of rural proletarians. "Indigence" was a serious problem at the end of the Old Regime: of all the French Cerdagne villages, only Santa Llocaya reported no indigents in 1790, while Enveig claimed that almost a third of the peasants had no means of subsistence.[38]

35. Cf. A. Zink, Azereix, la vie d'une communauté rurale à la fin du XVIIIe siècle (Paris, 1969).

36. Gavignaud, Propriétaires–viticulteurs, 1: 133–137. According to Pierre Vilar, the Spanish Cerdaña was the only district of Catalonia which failed to show a progression of the surface area under cultivation: La Catalogne, 2: 197.

37. ADPO C 2045: "Etat de 1730"; ADPO C 2048, no. 46 (report of 1768); see also C. Mach, "Production agricole et population du Roussillon au XVIIIe siècle" (Mémoire de maîtrise, La Sorbonne, 1968), 18–19. The increase in land under cultivation shown in the study of property hierarchies (app. B and table B.6) suggests even greater increases: 22 percent between 1694 and 1775 in Angostrina, then another 100 percent increase between 1775 and 1830; nearly 40 percent in Santa Llocaya between 1694 and 1775, and another 36 percent before 1830. However, these figures tell more about the changing nature of the land tax, and forms of its administration, than they do about statistically significant measures of the increase in arable land.

38. J. B. Carrere, Voyage pittoresque de la France (Paris, 1788), 128–129; ADPO L 190–191 (reports of number of "indigents," by village, 1790).

At the other end of the property hierarchy were the large land-owners, who seemed to grow richer in the course of the eighteenth century. In sheer percentage, property held by noble and privileged groups was relatively scarce in the French Cerdagne: more than 95 percent of all arable land was held by families without exempt status. But measuring land ownership through the status categories of Old Regime society says little about property hierarchies themselves. In Estavar, for example, the richest 5 percent of landowners had estates over fifty-six jornals, which together totaled almost half of the arable land and meadows in the community.[39] Nor did all of their wealth come from land ownership. Strictly feudal dues were vestigial, and more than half of the village communities in the Cerdagne had no complaints against seigneurial burdens or "feudalism" in 1789. But many of the richer families often owned the tithes (delmes) as well as whatever "feudal" dues remained. More onerously, they held large numbers of constituted rents or annuities which guaranteed a stable income from peasant indebtedness.[40] In addition, most owned large herds of cattle, up to several thousand sheep, and several hundred cows, trafficking in these animals as well as in the products of their estates.[41] Their vast stables and granges harbored these livestock, classed by species and age, alongside the hay, straw, and rye which their lands and meadows produced. Staffed by servants, agricultural workers, and other hired help, their estates were magnificent by mountain standards. According to a report of 1810, which described the losses suffered by the French Cerdans during an invasion of Spanish "insurgents":

> In this isolated district, the principal proprietors cannot display their wealth either in luxury clothes, or in the tables they lay; and they devote great efforts to maintaining their houses well furnished, especially those of Saint Leocadie, where M. Sicart, previously a very rich proprietor and viguier of the French Cerdagne, had a house which was the source of much emulation.

In 1730, the worth of the Sicart family—including its ownership of the Carlit Mountain, acquired from the town of Puigcerdà in 1715—was estimated at almost 137,000 livres, making it one of the wealthiest

39. M. Guibeau, "Enquête économique du Roussillon," BSASLPO 43 (1902): 317–324; and below, app. B.

40. On the "feudal" system in Roussillon at the end of the Old Regime, see Brunet, Le Roussillon, 21–28; on seigneurial dues and constituted rents owed in the French Cerdagne, see ADPO C 2044 (cens, censals, and delmes paid in the French Cerdagne villages, 1725).

41. ADPO 2C 991: "Enregistrement, Bureau de Saillagouse," declarations of herds pasturing on the Carlit Mountain; ADPO C 2045: "Etat de 1730."

families in the district. In 1810 the Sicart house—twenty-five meters long, half as wide, four stories high, and constructed "in great taste"— was worth, with a grange measuring fifty by nine meters, 55,000 francs.[42]

Sicart was not alone, since the report of 1810 described nine other equally grand houses in the French Cerdagne burnt to the ground by the Spanish rebels. But the case of Sicart was in many ways exceptional. The family was distinguished by their status as viguiers of the district, and by the fact that unlike other wealthy families in the French Cerdagne, the Sicart owed little to contraband trade. More importantly, most of the larger landowners in the French Cerdagne in the eighteenth century were or claimed to be residents of Puigcerdà and Llívia, and thus were or could claim to be Spaniards.

The larger landowners in the eighteenth-century Cerdanya constituted a class drawn from distinctive groups and social strata. As a class, they were dominated by families of ancient nobility, who had estates on both sides of the boundary and throughout the Catalan Pyrenees.[43] Some of these nobles were seigneurs of villages in the French and Spanish Cerdanyas: Pastors was seigneur of Enveig, Mora acquired the title of Marquis de Llo in 1754, Codol was seigneur of Ur. But the larger landowners also included members of an urban nobility (including several "honored citizens" of Barcelona and three "bourgeois" of Puigcerdà) who served as notaries, lawyers, and doctors in the town of Puigcerdà. Finally, the class of larger landowners included non-privileged members, many of them rich peasants whose wealth might rival that of the higher nobility. The Carbonell of Gorguja, for example, were peasants who held estates and property among the largest in the entire valley.[44]

What distinguished and united these thirty or so families—besides their extensive property holdings and, among the privileged groups,

42. ADPO Mnc 831/1: "Rapport sur les pertes de l'invasion du 29 septembre 1810"; ADPO C 863: "Etat des biens en fonds de Sicart," 1730.

43. On the nobility in Roussillon and the Catalan Pyrenees, see N. Sales i Bohigas, "Classes ascendents i classes descendents a la Catalunya francesa d'antic régimen: La noblesa rossellonesa, arruïnada i disminuïda?" in R. Garrabou, ed., Terra, treball i propietat: Classes agraries i règim senyorial als Països Catalans (Barcelona, 1986), 24–41.

44. Documentation of these families has been gathered from a variety of sources: in addition to those cited in chap. 3, n. 43, see the eighteenth-century catastros of Puigcerdà and Llívia; and the widely dispersed tax rolls for the French Cerdagne, including ADPO C 793–794 (tax rolls from the capitation); ADPO C 883 (rolls of "privileged" families, 1741); and ADPO C (tax rolls from the vingtième for the French Cerdagne communities, classed by communities).

their extensive ties of intermarriage—was the fact that most of them resided in either Puigcerdà or Llívia. It is true that several noble families, including the Mir of Bajanda, the Montellà of Saint Llocaya, the Pont of Osseja, and the Travy of Palau, had their estates and ancestral homes in villages of the French Cerdagne. In fact, several of these families were descendants of younger sons of families in Puigcerdà or Llívia who had settled in villages in the French Cerdagne. And there were other large proprietors who lived in the French Cerdagne, without apparent ties to the towns of Puigcerdà or Llívia—families such as the Girves of Llo. Yet beginning in the sixteenth century, most of the local propertied elite had left their lands in the villages to be leased or farmed out, and had moved into the nearby towns to live off their rents—at least according to an eighteenth-century description of the concentration of landowners in the two towns.[45]

The "domination of the town over the countryside," a common feature of Mediterranean society, predated the Treaty of the Pyrenees in the Cerdanya.[46] But after 1660, it took on a distinctive quality, since the "towns" remained Spanish while part of the "country" became French. Thus in 1665, many more Spanish inhabitants held lands in France than vice versa: declarations to the French and Spanish commissioners at Figueres reveal that Spaniards owned 51 estates of 122 separate properties totaling more than 400 jornals while Frenchmen owned 25 estates of 41 properties amounting to 109 jornals.[47] (Not only are these figures underestimated, but they also include the property held by French and Spanish peasants located in the soon-to-be-disputed borderland of Llívia: see table 1.) The distinction of French and Spanish landowners reproduced, overlay, and redefined in national terms the direction of the economic domination of the town over the countryside.

This distinction of nationalities became especially relevant in the exemptions that Spaniards gained from the capitation tax. In 1713, the French comptroller-general decreed that "foreigners . . . who presently have their usual home in France are subject to the capitation, and especially those who reside there at least six months during the year."[48] In order to gain exemption from the capitation tax, the large landowners

45. BRP MS 2346, fols. 125–126 and 149–149v: "Memorias topográficas del condado de Cerdaña"; see also Vilar, La Catalogne, 2: 515.
46. Gavignaud, Propriétaires–viticulteurs, 1: 263–266.
47. ADPO C 1360 (lists of January 1665).
48. ADPO C 793 (decree of March 19, 1713).

manipulated their residences so that they would conform. In 1726, Don Juan de Barutell could claim at least three domiciles: Concellabre (a large farm [*masia*] dependent on Osseja), a farm house in Enveig, and his house in Puigcerdà. He claimed the latter and was exempted. Even if landowners were born in France—as was the case of François de Mir, who petitioned for an exemption in 1744—they might be exempted for residing "outside the kingdom," as long as their presence in the French Cerdagne was only "occasional," for the purposes of overseeing their estates. By the mideighteenth century, few nobles or bourgeois— "privileged" families—were left in the French Cerdagne: according to the Viguier Sicart, they had "leased out their lands and gone to live in Spain."[49]

How much land did Spaniards own in the villages of the French Cerdagne? In 1744, exemptions from the capitation rolls show that Spaniards owned roughly 30 percent of the fields and meadows in Osseja; 20 percent in Enveig, Sallagosa, and Hix; 15 percent in Ro, Santa Llocaya, and Càldegues; and 10 percent in Carol, Ur, Angostrina, and Eyna. These were, moreover, some of the larger estates in the valley: by Cerdan standards, the 130 jornals owned in Santa Llocaya by a notary of Puigcerdà, Domingo Aldran, was a lot of property, and most Spanish properties in the French Cerdagne averaged around 50 jornals, still large by local standards.[50] Attempts to measure accurately these holdings, however, run into considerable discrepancies in the data (see table 2).

The 1665 declarations to the Figueres commissioners, showing a total of 400.5 jornals of fields and meadows, were admittedly underestimated. François Sicart's 1774 estimation of land held by Spaniards was plainly exaggerated. As its critics (the Spanish proprietors) pointed out, the tabulation included lands originally held by Spaniards which had passed into French hands either by sale or during the many confiscations and seizures of Spanish property during the late seventeenth century, and which were only recuperated after the introduction of the capitation tax. The 1794 report by the district administration, based on properties confiscated during the revolutionary war (1793–1795), is perhaps the most accurate assessment, although it can only be read

---

49. ADPO C 863–864 (capitation rolls and lists of Spanish exemptions for the French Cerdagne, 1729–1789); and Sahlins, "An Aspect of Property Relations in the Borderland," 415.

50. ADPO C 864: "Etat des charges qui composent chaque communauté de la Cerdagne française pour servir à la répartition de la capitation," 1744.

TABLE TWO   SPANISH PROPERTIES IN THE FRENCH CERDAGNE
(1665, 1774, 1794)
(measured in jornals)

	1665[1]	1774[2]			1794[3]
		Properties Held before Capitation	Properties Acquired Since	Total	
Angostrina	9	89	—	89	61
Càldegues	66.5	182.75	122	304.75	123
Dorres	—	—	—	—	56
Enveig	80	191.5	451.25	642.75	354
Err	28.5	19.5	59	78.5	16
Estavar	43.5	41	3	44	8.5
Bajanda	—	95	5	100	98.5
Hix	9	82.25	321.25	403.5	473.5
Nahuja	—	—	149	149	95
Osseja	13.5	338.5	300.5	639	327
Odello	—	—	25	25	30
Via	—	—	16.25	16.25	52
Palau	48	31.75	15.25	47.25	194.25
Sta. Llocaya	4	60.5	192	252.5	479
St. Pere	—	—	119	119	55
Sallagosa	—	—	88	88	110
Vedrinyans	7.5	1	—	1	—
Ro	—	—	—	—	66
Targasona	5	—	5	5	—
Ur & Flori	55.5	—	169	169	108
Vilanova	—	—	28	28	16
Carol Valley	7	150	23.5	173.5	135.5
TOTAL	400.5	1411.75	2093.25	3505	2858.25

[1] ADPO C 1360: Declarations by municipal councillors "del pahis adjacent de Cerdanya" (French Cerdagne) of properties and incomes held by Spanish proprietors, March 15, 1665.
[2] ADPO C 863: Report of François Sicart, viguier of the French Cerdagne, 1774.
[3] ADPO Q 123: "Etat des biens situés en France posédés par des Espagnols," pluviôse an II (February 1794).

as an approximate measure: many Spanish proprietors simply slipped through the cracks, since kinsmen in the French Cerdagne claimed to own their properties, just as many French Cerdans were sheltered as émigrés by their families on the Spanish side of the boundary.[51]

During the eighteenth-century debates over exemptions from the capitation tax, the large proprietors found themselves united as Spaniards; and as such, in 1777, they joined together to protest the measures taken by Sicart against their exemptions. They contested Sicart's statistical tabulation; they stressed the nature of the tax as a *personal* levy; and they offered the general argument that it was not in their interest to acquire lands across the boundary, preferring to own them in their native kingdom. Ignaci de Pera, whose case created a general policy in France, had himself used a similar argument in 1775:

> It is more natural to sell than to acquire goods in another kingdom, unless one plans on establishing and transferring one's domicile there. Each is naturally driven to gather his goods into his own country.[52]

For some large landowners, this may well have been true; surely it was in line with seventeenth-century attempts of the intendancy to break the link between the Roussillonais and the Catalans of the principality. Political concerns—if not loyalty—may well have been at the basis of an exchange of constituted rents (censals) between the Viguier Sicart and the noble Don Antoni de Manegat of Puigcerdà in 1705, who had held them on opposing sides of the boundary.[53] But for most of the ruling elite, the expressed desire to hold one's properties "in his own country" was more likely to have been a rhetorical claim than an economic interest.

For one, property situated in one kingdom or another was subject to a distinctive fiscal regime. By the mid eighteenth century, the Spanish catastro, which had initially weighed so heavily on the rural population of Catalonia, had become a substantially lighter burden. Before the introduction of the vingtième in the French Cerdagne in 1752, how-

---

51. See below, chap. 5. In fact, many Spanish proprietors managed to profit from having their properties confiscated, as they collected inflated revenues with interests from the departmental administration once these properties were returned after July 1795; see ADPO L 1411 bis: "Réclamations par espagnols reconnus dans le cas d'obtenir la restitution des fruits perçus sur leurs biens," n.d., ca. 1795; ADPO Q 148 (petition of the attorney of the Duc d'Hijar); and ADPO Q 369 (petition of Don Jacintho Destcallar, October 16, 1796).

52. ADPO C 864: "Mémoire pour Don Ignace de Pera" to comptroller-general, 1775.

53. AHP Protocols: Françesc Domech, January 4, 1705.

ever, land taxes were lighter still, averaging less than 3.8 livres per inhabitant, compared to the 5.6 (French) livres in the villages of the Spanish Cerdaña. With the doubling of the vingtième in the second half of the eighteenth century, and the further decrease in levies of the Spanish catastro, land taxes became marginally heavier on the French side, averaging more than 7 livres per inhabitant.[54] But the capitation made up as much as a third of this levy, so that the acquisition of land in the French Cerdagne became, for the proprietors of Puigcerdà and Llívia, a financially interesting proposition.

It was interesting as well because of the enclosures permitted in the French Cerdagne, as more than one landowner, also a member of the Puigcerdà Economic Society, was to note. For the larger landowners were also the innovators in the new agricultural techniques of estate management, and only in the French Cerdagne were they permitted to enclose their lands and meadows.

The Spaniards bought up land, for which, as foreigners, they gained exemption from the capitation; the village communities protested, identifying the richer landowners as Spaniards and themselves as loyal subjects of the French king. Within this national division lay a class struggle. Each group, out of its local economic interests, evoked its own national identity and its opposition to the other. Thus was a national identity as Frenchmen or Spaniards advanced in the service of local interests, collectively defined in terms of class and community. Such a nationalization of interest was even more visible among village communities divided by the frontier.

COMMUNAL STRUGGLES IN THE BORDERLAND

During the first half of the eighteenth century, competition among local communities divided by frontier gave shape to the territorial boundary of Spain and France and to their own identities, French and

54. On the catastro in Catalonia, see Ardit, *Història,* 19–25. Between 1720 and 1748, for example, the sum generated by the Puigcerdà cadastre fell from 3443 to 2079 French livres: AHP Catastro. Unfortunately, few of the catastros for the villages of the Spanish Cerdaña have survived. Calculations for the towns of Puigcerdà and Llívia, while possible, would be of little comparative value. The sums given here are based on the 1747 catastro of Das and Tartera (AHP Catastro), and on a few general tabulations from the end of the eighteenth century (AHP Catastro and Revolució françesa). Sources for the French Cerdagne include ADPO C 843–846 (tax rolls for the communities of the viguerie of French Cerdagne); ADPO C 883 (dixième of the French Cerdagne, 1734–1756); and ADPO C 1007: "Relévé général des impositions dues en Cerdagne en 1789," including a list of seigneurial jurisdictions.

Spanish. In these local disputes concerning the defense of pastureland, waters, and the extension of communal lands in the borderland, village communities appealed to their respective state administrations, invoking the national identities of their "enemy–neighbors" in the process. The result was a nationalization of interest, the first step in the construction of national identity.

The corporate character of the Old Regime village communities was most in evidence in their elaborate sense of village patriotism.[55] This esprit de clocher was manifest in a number of contexts. Compared with endogamy rates of villages in western and southern France, for example, those of the French Cerdagne were unusually high. The French Cerdagne communities became, moreover, more endogamous in the eighteenth century, drawing a greater percentage of spouses from within the village itself. Two-thirds of the marriages celebrated in Odello between 1777 and 1789 involved inhabitants of the community, more than double that of a century before. In Càldegues, none of the fifteen marriages that took place between 1770 and 1785 involved any "outsiders."[56]

In Roussillon, as elsewhere among European and Mediterranean peasantries, the defense of village parochialism was often the domain of young men. Responsible for the ritual and symbolic initiation of "outsiders" who settled into the community, they also defended the rights of their villages against neighboring communities, especially during festivals and fairs.[57] Village communities skirmished over a variety of issues. Although disputes could break out over such immaterial things as claims to preeminence of pilgrimage sites, most village conflicts had ecological interests at their base—rights to pastures, waters, or firewood. Municipal parochialism was most evident in the lengths to which communities went in defining and defending their local territories and boundaries.

The sense of local identity was grounded in a specific if disputed

---

55. Brunet, Le Roussillon, 50–68.

56. In the later eighteenth century, endogamy rates averaged about 63 percent in the French Cerdagne (below, app. B and table B.3). In the Vendée during the same period, Tilly found a comparable rate (61 percent), but within a density of settlement roughly five times that of the Cerdagne: The Vendée (Cambridge, Mass., 1974), 88–90. In Provence, Sheppard noted a similar rate, but within a parish that was nearly eight times as large as those of the French Cerdagne: Lourmarin in the Eighteenth Century (Baltimore, 1971), 33–35.

57. For Spain, see J. Caro Baroja, "El sociocentrismo de los pueblos españoles," 263–293; for France, see A. Varagnac, Civilization traditionnelle et genres de vie (Paris, 1948), 138–139 and 247–272.

local territory and its boundaries—disputed, in part, because the idea of a village territory was not yet completely formed in the earlier eighteenth century.[58] The idea of an enclosed territory—encompassing individually owned pastures and mountains which formed a contiguous, delimited space—was modified by the notion and practice of usufruct rights (*empriu,* in Catalan). The elaborate "webs of usufruct" that entwined the villages of the Catalan Pyrenees had spatial and territorial extensions beyond the contiguous space surrounding the nucleated village. Communities of inhabitants had collective rights of empriu to the pastures and forests in the possession of seigneurs and the crown, a right affirmed by the "*stratae* statute" of Catalan customary law.[59] Moreover, these collective usufruct rights—a form of "jurisdictional sovereignty" at the local level—had specific territorial boundaries. In 1674, an accord between the town of Puigcerdà and the College of Priests of Santa María, seigneurs of Bolquera, named definite boundaries and border stones demarcating those areas where the peasants of Bolquera could pasture their herds; the limits named were distinct from the boundaries of Bolquera's territory itself. And in 1705, the communities of Guils and La Tor de Carol signed an agreement that named the boundaries of their reciprocal grazing rights but not the boundaries of their territories.[60] The notion of an enclosed territory was only emerging slowly.

Yet the territorialization of the community was taking shape in the fifteenth and sixteenth centuries, as evidenced by the increasing attention paid by village communities and by the state to their local boundaries. Royal privileges and court decisions reveal growing disputes

58. On the emergence of the village community in the medieval period—which did not, however, imply the notion of village territory as a contiguous and enclosed space—see J. M. Font i Rius, "Orígenes del régimen municipal de Catalunya," and "Poblats i municipes a la Cerdanya medieval," in *Estudis sobre els drets i institucións locals en la Catalunya medieval* (Barcelona, 1985), 261–560 and 723–734; J. A. Brutails, *Etude sur la condition des populations rurales au moyen age* (Perpignan, 1891); and, for France more generally, Bloch, *French Rural History,* 167–189; Gutton, *La sociabilité villageoise,* 21–31; and P. C. Timbal, "De la communauté médiévale à la commune moderne en France," in *Les communautés rurales,* 337–348.

59. Assier-Andrieu, *Le peuple et la loi,* 91–146; on usufruct rights and traditions of collectivism more generally see, for Spain, R. Behar, *Santa María del Monte: The Presence of the Past in a Spanish Village* (Princeton, 1986), 189–264, and J. Costa y Martinez, *Colectivismo agrario en España* (Madrid, 1915); and for France, Bloch, *French Rural History,* 167–189 and 198–234.

60. AHP Protocols, Rafael Ferran: "Acte de Concordia entre la Universitat de Puigcerdà i Col.legi de Preveres . . . acerca la montanya dels Pasquers," September 6, 1674; SAHN Estado, libro 673d, fols. 96–97 (provisional treaty of Guils and Carol, September 18, 1736, discussing the earlier accord); and the dossier in ADPO C 2083.

among communities over the extension of their local territories. Although Angostrina and Llívia had disputed the priority to a shared irrigation canal as early as 1307 (a dispute revived in 1410 and again in 1754), questions concerning their territorial extensions were only raised in 1540, when Llívia erected boundary markers in the terrain called "El Nirvol," and Angostrina protested to the courts of the Royal Patrimony. Although Llívia lost the law suit, its appeals to Charles V resulted in royal "letters of maintenance" that described the boundaries in its favor. The same territory and boundary markers were contested in 1659 and after that, periodically until the later nineteenth century.[61]

Villages developed long-standing claims to lands and boundaries that overlapped; these boundaries also overlapped with parish and seigneurial jurisdictions. Local seigneurs were not much interested in defining the boundaries of the domains, content as they were to collect their revenues and rents from individual parcels of land. But occasional disputes among seigneurs—such as that in 1756 between the College of Priests of Puigcerdà and the Royal Jesuit College of Perpignan, over the collection of the tithe between Guils and Sant Pere de Cedret—reveal boundary stones distinct from those defining the village territories of their vassals.[62]

Disputes among communities over questions of boundaries and usufruct rights did not always involve overt conflict, since village representatives could settle differences amicably and locally according to customary and unwritten procedures. A mid-nineteenth century description of a boundary stone offers a rare glimpse of the rituals of dispute settlements—one that accords well with the boundary as a point of articulation of territories:

> In the meadow situated next to the wall which follows the stream, there is a large rock, covered with brambles, on which five holes had been made. It was used for the reunions of the mayors of Guils, Saneja, Maranges, and La Tor de Carol. The middle hole, larger than the others, was a neutral place. In each of the other holes, the mayors placed their staffs of office while they discussed the difficulties which had arisen among themselves.[63]

61. These court decisions and royal privileges were discussed and reproduced verbatim in a series of memoirs that resulted from the renewal of the rivalry in 1760 (ADPO C 2050–2051), and again in the debates among the Bayonne commissioners in the 1860s: for France, FAMRE Callier 16; for Spain, SAMRE TN 221–222.

62. Periodic *cabreus* (comparable to the French *livres terriers*) beginning in the fourteenth century contain lists of declarations of specific dues and rents paid on parcels of land but do not describe the boundaries of segneurial jurisdictions or village territories. Examples of such cabreus may be found in ACA RP BGC, and in AHP; on the 1754 dispute, see ADPO C 2084: "Transaction passée à Puigcerdà," n.d., ca. 1756.

63. ADPO Mnc 1924/1: "Rapport relatif à des usurpations," May 22, 1858; another

Yet such amicable settlements were consistently overridden by a marked propensity of the village community to appeal to higher authorities, to engage in costly and lengthy lawsuits in defense of its rights, privileges, and territorial integrity. Such lawsuits were often the result of contested seizures of cattle (*prenda, penyora,* and its related designation, *marca*). The customary right of seizing cattle was a ritual and juridical defense of the community's territorial identity, recognized and bureaucratically supervised by the state.[64] It involved sequestering an animal belonging to a "foreigner" who failed to comply with a community's ordinances. Seizures were most frequent during the annual transhumant cycles when herds, climbing or descending from their summer pastures on the mountain, crossed through village territories. If a fine was paid, the animal was released; if the seizure was contested, the result was a lawsuit that might drag on for decades.

Local struggles over territorial boundaries and usufruct rights thus long antedated the division of the Cerdanya, and they continued long after. Disputes among villages divided by the boundary, however, became vehicles for the development and expression of national identities in the Cerdanya. This process was not immediately visible, for, during the late seventeenth-century military invasions and occupations of the valley, the village communities were less divided among themselves than they were united in opposition to the occupying forces, be they French or Spanish. Registers of the lower courts in the district of the French Cerdagne reveal that although communal rivalries did not disappear during the later seventeenth century, such quarrels remained without appeal to the authority and powers of the state, and tended to be resolved or regulated by the communities themselves. There were a few exceptions, as when, in May 1665, taking advantage of their newly found jurisdiction as part of the "adjacent district," two representatives of Dorres went to Perpignan to introduce a lawsuit before the Sovereign Council of Roussillon against Puigcerdà over their rights to Matanova on the Carlit Mountain. No action was taken, however, and when the Spanish troops occupied the "adjacent district" in 1674, Puigcerdà took advantage to reassert its rights against Dorres and other challengers, pressuring them to sign "transactions" that specified re-

example, less well described, is reported by Pere Cot y Verdaguer in his "Notes folkloriques pirenenques" (1928), BC MS 1535.

64. E. de Hinojosa, *El elemento germánico en el derecho español* (Madrid, 1915), 79–106; P. C. Timbal, "Les lettres de marque dans le droit de la France médiévale," in *L'étranger,* 2: 110–138.

spective boundaries and rights.[65] Though such legally notarized agreements failed to settle these disputes definitively, local quarrels receded from view during the years of warfare in the later seventeenth century.

After the episode of the Spanish sanitary cordon (1720–1722) and the end of the military frontier, the local communities began to appeal to their separate administrations in defense of their rights and boundaries. Precisely in the period when the two states sought to consolidate their distinct jurisdictional frontiers, communities and landowners of the two Cerdanyas entered into contention over their boundaries and usufruct rights in the borderland, empowering themselves with the authority of their respective states while defining the territorial boundary line and their own nationalities in the process.

It is ironic that one of the earliest and most enduring local struggles in the Cerdanya after the Treaty of the Pyrenees—one that had *not* existed before 1660 but which was renewed periodically, most recently in 1984—occurred precisely where the two commissioners had established a natural frontier that "separated the two kingdoms." The Raour River, known as the Angostrina at its birth on the Carlit Mountain, rushed across the floor of the valley near the village of Hix. On the stretch between the Llívia bridge (which linked the so-called "neutral road" from Puigcerdà to the Spanish enclave) and its confluence with the Segre, the Raour River was disputed as the point of separation of the two kingdoms.

The Raour was a young and wandering river. As a memoir of 1750 was to note,

> The district is flat, and the river does not have a clear bed to contain even the smallest floods; it is a rapid river, which means that it drags along quantities of gravel and small stones; it fills up, lifts its bed, forms large piles of rock higher than the lands on its banks, and constantly forms new bends and sinuosities at each flood.

But it was not simply nature and geography that made the Raour a wandering frontier. For to protect themselves, "and even to enlarge their estates," riverfront proprietors in Hix built up dikes, barrages, and "defenses of various kinds" which threw the river onto the opposing bank.[66] From the early 1720s until the mid-1750s, competition be-

65. AHP MA 1662–1667, fol. 151v (May 27, 1665); fol. 170 (December 11, 1665); and MA 1673–1678, fols. 105–106 (September 29, 1674).

66. FAMRE Limites 450, no. 40: "Mémoire sur la partie des limites du côté des Deux Cerdagnes sur la rivière de Raour," n.d., ca. 1750.

tween riverfront proprietors escalated into a matter of state. In 1729, two town councillors of Puigcerdà signed an accord with the viguier of the French Cerdagne which fixed the width of the river and ordered the proprietors to keep the banks unencumbered, but the proprietors were not satisfied with the agreement. With the further construction of a farmhouse on the Spanish side by the convent of Sant Dominic of Puigcerdà during the early 1740s, and the seizures and counterseizures of sheep passing along the river banks, the conflict came to attract the attention of the provincial administration. The Viguier Sicart defended the inhabitants of Hix, claiming that "the Dominicans living in their farm on the edge of the Raour are as unrestrained as the Africans living on the Ocean shore where, seeing the shipwrecked vassals of France, they hurl endless abuses upon them."[67]

A severe flood in August 1750 forced the provincial governments to act. The Comte de Mailly, the reform-minded intendant of Roussillon—though he was not beyond lining his own pockets through the exercise of his power—wrote directly to the governor-general of Catalonia without consulting Paris. His counterpart, the Marqués de Mina, also acted without the approbation of Madrid. The two officials agreed to realign the Raour riverbed and, on November 23, 1750, two engineers—one French, one Spanish—planted border markers along its edges. Taken aback by the local initiative of a provincial intendant, the French foreign minister refused to countenance such a "treaty of limits" and named the accord "a simple convention." "Such small differences" as those settled by the convention, he argued, "cannot challenge the perfection of an arrangement between two such great powers as France and Spain."[68]

The struggles over the Raour had many contemporary parallels, in the Cerdanya as throughout the Pyrenees. Most of the local disputes in the 1730s and 1740s involved village communities appealing to political authorities, identifying their enemies as "foreigners" and "usurpers," and claiming to uphold the rights of their respective monarchies in the borderland. In the 1730s, for example, Guils and La Tor de

67. ADPO C 2051 (viguier's reports of July 1743 and July 1744); AHP MA 1720–1730, fol. 203; AHP Internacional (letter of Sicart to the consuls of Puigcerdà, in Catalan, May 21, 1750).

68. FAMRE Limites 459, no. 57 (Comte d'Argenson to intendant, December 13, 1750); ibid. no. 40 (intendant to foreign minister, September 9, 1750); ibid., no. 50 (accord, in Castilian and French, November 23, 1750); ibid, no. 59: "Mémoire sur le redressement du Raour," n.d.; and FAIG 4.3.2 (maps and memoirs of the 1750 rectification).

Carol began to struggle not for the extension of their usufruct rights—which had been settled by an agreement in 1706—but for the "incontrovertible division of their territories." The debate had begun with the establishment of the Spanish sanitary cordon in 1721 and was aggravated by a seizure of contraband goods along an already disputed tract of land in 1735.[69] But what is striking is the way in which the local communities gave significance to these political events.

In a lengthy petition of 1740, syndics from the Carol Valley argued that the boundaries of Guils and La Tor de Carol were to be a "dividing line along the watershed" between Guils and the hamlet of Sant Pere. In defense of their claims, they cited a tenth-century donation to a monastery, as well as article 42 of the Treaty of the Pyrenees—that "The Pyrenees Mountains, which anciently divided the Gauls from the Spains, shall henceforth separate the two kingdoms." Their argument was that article 42 evoked a "natural limit," which was nothing other than the watershed of Guils and Sant Pere. The syndics went on to denounce the "arbitrary actions of the Spaniards," who, "more enterprising than the petitioners, because of the great assistance they are given, have invaded a part of our territory, and aim to keep this usurpation."[70]

The community of La Tor de Carol argued that such "usurpations" had begun with the end of the military frontier, in 1722. Other communities struggling for their boundaries and rights during the first half of the eighteenth century evoked similar points of origin for their struggles: by their mention of the international competition of the military frontier, they sought to infuse their claims with a national significance, stressing the arbitrary exactions of a "national enemy" and calling on their own political authorities to defend their local rights. As the syndics of La Tor de Carol claimed in a petition of 1740,

> It would be sad if the inhabitants of La Tor de Carol and Guils had to resort to armed force over a measly pasture, yet that is precisely what there is to fear. For the inhabitants of La Tor de Carol are sensitive to the insults of the Spaniards which they have long experienced, and they could one day lose their patience. They do not fear their foreign neighbors, and the injustice inflicted upon them will one day force them to chase the Spaniards from the land. Despite the haughty and insulting manners of the Spaniards, the inhabitants have refrained from retaliating, if only because of the profound respect they have for His Majesty. But the inhabitants of La Tor de Carol, seeing themselves pushed to the limit, could easily retaliate. They hope that

69. SAHN Estado, libro 673d, fols. 96–97 (provisional treaty of September 18, 1726); ADPO C 2083 (procès-verbal of seizure, December 5, 1735).
70. ADPO C 2084: "Mémoire des Syndics de Carol," March 26, 1740.

His Majesty will have the good will to have the boundaries of this valley established clearly. . . . That is the only way to procure a peaceful enjoyment of their territory.[71]

Thus the rural communities sought to locate their quarrels, emerging from a local set of issues, in a political language and context of national significance. The first step in this "nationalization" of disputes consisted in naming their enemies by nationality. The identification of the inhabitants of Guils—fellow Catalans with whom they intermarried—as Spaniards did not attribute a specifically "French" character to themselves, other than their identity as subjects of the French king. Rather, it was their enemies whom they identified by a distinctive national identity, as Spaniards.

Such name calling was widespread throughout the village disputes of the eighteenth century. It was visible in the late 1720s, during quarrels between the Spanish town of Puigcerdà and the two French villages of La Tor de Carol and Enveig over the irrigation canal that supplied the town with its only water.[72] In the 1750s, the communities of Llívia and Angostrina renewed their quarrels over an irrigation canal and the respective limits of their territories, appealing to their provincial administration and evoking the nationality of their "enemies" in the process.[73] Nor were such appeals restricted to individual communities: in July 1777, a general council of the French Cerdagne meeting in Sallagosa requested that

[a]ll preference be given to the inhabitants of the French Cerdagne and the exclusion of the Spaniards to use and make use of the pastures of the *Pasquiers Royaux* in the Capcir. . . . The inhabitants offer His Majesty the same dues that the Spanish cattle pay annually, since it is natural that regnicoles have precedence over foreigners to the fruits which the territory offers.[74]

Those "Spaniards" were, of course, Catalans from beyond the Pyrenees—just as today, in the Roussillon, "Spaniard" means Catalan while other peninsular Spaniards are described by their provincial origins—Castilians, Andalusians, or Gallegos.[75] The sense of national difference emerged from the proximity of Frenchmen and Spaniards, even if the

71. Ibid.
72. AHP Asequia, passim; AHP MA 1720–1730, fols. 46, 51, 207–209, 249.
73. AMLL (unclassified documents on struggles with Angostrina, including a decree of the Sovereign Council of Roussillon concerning their mutual lawsuits, June 22, 1751); and the dossiers in ADPO C 2050–2051 on the renewed disputes of 1760.
74. ADPO C 2049: "Conseil Général de la Cerdagne française," July 5, 1777.
75. D. Bernardo and B. Rieu, "Conflit linguistique et revendications culturelles en Catalogne–Nord," *Les temps modernes* 324–326 (1973): 323, n. 44.

French Cerdans under the Old Regime defined themselves as regnicoles but not as Frenchmen.

Village communities struggling in the borderland thus appealed to their separate administrations, evoking the nationality of their enemies if not yet their own. These local conflicts provided an additional source of "territorial violations" which the two states were increasingly concerned with regulating. In attempting to "settle the question of limits," the governor of Puigcerdà and the viguier of the French Cerdagne signed an accord or "provisional treaty" in September 1736 which defined the limits of the two crowns—and of the territories of Sant Pere and Guils. Not only did the villagers of La Tor de Carol protest, but so too did the Spanish government in Madrid. Citing the antecedents of the Llívia convention, the Spanish court vehemently disowned the treaty.

> Care must be taken when treating incidents in the borderland, in order not to harm the rights of sovereignty; and although the governor of Puigcerdà innocently set out to confer with the viguier of the French Cerdagne, such a high and delicate matter might have caused much damage.

Disapproving the convention, and reserving such affairs for "the wills and accords of kings," the court urged its local political officials to "persuade the communities extrajudicially that these conventions are between themselves . . . a private transaction, like a contract."[76] As a result, many of the eighteenth-century disputes found an ephemeral resolution in such private accords. Puigcerdà signed conventions with La Tor de Carol and Enveig in 1731 and 1732, and Angostrina and Llívia came to terms in a convention of 1754. All of these were signed under the tutelage of the district administrations, but without their formal or official sanction.[77] These accords failed to quell local disputes, however, which erupted periodically until—and well after—the French Revolution.

CONCLUSION: THE USES OF LANGUAGE AND IDENTITY

The experience of the Cerdanya under the Old Regime suggests that the adoption of national identities did not necessarily displace local

76. SAHN Estado, libre 673d, fols. 111–119 (memoir of April 13, 1737).
77. ADPO Mnc 1859/3 (accords signed between Enveig and Puigcerdà [May 13, 1733]; and between Puigcerdà and La Tor de Carol [September 13, 1730]). See also the extensive documentation in AHP Asequia; and ADPO XIV S 111 (transaction between Angostrina and Llívia, July 8, 1754).

ones, and that the process of nation formation was not simply the imposition of politics, institutions, or cultural values from the top down and the center outward. Rather, the evocation of national identities by the propertied elite of Puigcerdà and by the village communities of the Cerdanya was grounded in local economic interests, and in a local sense of place.

Of course, it could be argued that national identity was only a strategic "mask" of essentially local concerns: the rhetoric of patriotism and appeals to national identity were either "ritual forms of civility" or "nationalist disguises" of personal or collective strategies defined by a local content, and in constant opposition to the "coercive authority of the state."[78] But such an interpretation falls short on two counts. First, the framing of local interests in national terms produced a transformation at both ends: a nationalizing of the local, and a localizing of the national. By voicing their local economic interests in national terms, both peasants and nobles brought the nation into the village, just as they placed themselves within the nation. That appeals were made to the authority of the nation at the end of the eighteenth century suggests a nationalizing of differences of the French and Spanish Cerdanyas—and of French and Spanish Cerdans—before any visible differentiation in landscape or culture in the borderland.

Second, the idea that appeals to political authorities evoking national identities were *only* strategic or instrumental—that nobody in the borderland really took their national identities seriously—disregards the varied and diverse ways in which peasants and rural propertied elites could define their identities. Ernst Gellner recently described the sense of identity in agrarian society as antithetical to that of modern nations. Among the peasants of Tibet, he believed,

> life-style, occupation, language, and ritual practice may fail to be congruent. A family's economic and political survival may hinge, precisely, on the adroit manipulation and maintenance of these ambiguities, on keeping options and connections open. Its members may not have the slightest interest in, or taste for, an unambiguous, categorical self-characterization such as is nowadays associated with a putative nation, aspiring to internal homogeneity and external authority.[79]

Gellner's idea that peasants might not need a "categorical self-characterization" could easily be reversed: neither, it should be argued, do members of a national society *necessarily* aspire to a clear and unam-

78. Brunet, *Le Roussillon,* 20, 58, and 178.
79. *Nations and Nationalism* (Ithaca, 1983), 12–13.

biguous status as national citizens at the expense of other ties of loyalty and identity. At the end of the Old Regime, the Cerdanya peasants resisted the "external authority" of the "putative nation," as I have suggested in chapter 3. But, they could also assert, along with and in opposition to propertied elites, their national identities. Grounded in collective interests, national identities were as much a legitimate expression and self-characterization as was the sense of identity tied to a particular place.

Consider the parallels with language. Just as Catalan remained the lingua franca of rural society in the borderland, so too could it coexist with the contextual adoption of national languages—French or Castilian—despite the failure of either state to pursue a policy of linguistic uniformity. But the use of French (or Spanish) did not make the Cerdans French (or Spanish). Just as "one could be French without knowing the language of the king," as Henri Peyre stated long ago, so too could one know the king's language without "being French."[80] In contrast to what the nineteenth century has led us to believe, there was no necessary relation between language and identity. The French Cerdans spoke or wrote, in specific contexts, French and Spanish, notwithstanding the persistence of spoken and written Catalan in others. And in the same way, the Cerdans could oppose the intrusions of the centralizing states as members of local society—though rarely as Catalans—while identifying themselves as French or Spanish in others.

In both Spain and France of the Old Regime, knowledge and use of "national" languages was essential to the self-definitions of elites. In Spain, the so-called "decadence" of the Catalan language among political and intellectual elites began in the sixteenth century, as Castilian became the language of prestige and power in the principality and Catalan was relegated to unofficial status and common use among the "lower orders."[81] In much the same way, proficiency in French in the Roussillon and French Cerdagne quickly became necessary for the assimilated elites.

But learning French was not only the province of elites. The use of French also signaled a participation in the world beyond the valley. Thus Francisco de Zamora, who traveled through the French village of Err on his way to Puigcerdà, noted that "even though they are Catalans

80. H. Peyre, La royauté et les langues provinciales, 21; see also L. Febvre, "Langue et nationalité en France au XVIIIe siècle," Revue de synthèse historique 42 (1926): 31.

81. J. Amelang, Honored Citizens of Barcelona: Patrician Culture and Class Relations, 1490–1714 (Princeton, 1986), 190–195.

as we are, almost all of those I heard spoke French, as hasn't happened with Castilian, even in Barcelona itself." Zamora's passage was a rapid one, however, and he probably ventured no further than the inn, "which was very good."[82] The inn was a place of contact with the world beyond—there, merchants, travelers, and soldiers came into contact with local society, and French was usually the language of these encounters. Moreover, families with commercial contacts in France inevitably spoke French. The business papers and account books of the stocking merchant François Jean Delcor (1733–1802) were in French, as was the cheaply printed copy of the *Aventures de Télémaque* which he brought back to Palau from a business trip. So too was the personal correspondence of the Garretas of La Tor de Carol, important livestock merchants who played the Paris *Bourse*. In the late eighteenth century, French was used in letters sent between close members of that family.[83] French was not just the language of the outside world, the world of authority and power and the state; nor was it simply useful in the world of profit and commercial interests—although it was certainly present in both these contexts.

At the end of the Old Regime, the boundaries of linguistic use and of identity remained fluid, just as the territorial boundary of France and Spain remained permeable. The French Revolution, as it was imposed from the center and experienced in the borderland, marked a turning point in the ideas of nationality and territory, creating a framework that has endured through the nineteenth century to the present day.

82. Zamora, *Diario de los viajes*, 92.
83. Delcor and Garreta family papers.

5
The French Revolution

THE REVOLUTIONARY REFORMS (1789–1792)

The revolutionary division of France into *départements* had precedents in late-eighteenth-century reforms that proposed to give the "province" a legal and juridical status. But reforms such as that of d'Argenson in 1764, the geographer Letrosne in 1779, or even the Assembly of Notables in 1787 never came to fruition and, at the end of the Old Regime, the "province" had no de jure existence. Instead, the 1789 Constitution Committee found that

> [T]he kingdom is divided into as many different divisions as there are kinds of regimes and of powers: in *diocèses,* as concerns ecclesiastical relations, in *gouvernements,* as concerns the military; in *généralités,* as concerns administrative relations; and in *bailliages,* as concerns the judiciary.

It was only by abolishing privilege as the basis of private and administrative law that the revolutionary government could institute a direct link between power and territory.[1] The creation of départements, which achieved their definitive shape under Napoleon, established a

---

1. Alliès, *L'invention du territoire,* 179; on the juridical status of provinces, see C. Berlet, *Les tendances unitaires et provincialistes en France à la fin du XVIIIe siècle* (Nancy, 1913), 24–43; A. Brette, *Les limites et les divisions territoriales de la France en 1789* (Paris, 1907), 57–81; and G. Dupont-Ferrier, "De quelques synonymes du terme 'province' dans le langage administratif de l'ancienne France," *Revue historique* 160 (1929): 241–267. The Constitution Committee is quoted in G. Dupont-Ferrier, "Sur l'emploi du mot 'province'": 262.

territorially defined administrative framework that served the different jurisdictional dimensions of state sovereignty—justice, taxation, ecclesiastical affairs, military governments, and policing of the interior.[2]

Yet many of the Old Regime provinces did have clearly defined social and political identities, and these proved decisive in the new division of France. During the winter of 1789, two competing schemes were put before the Constitution Committee. The Comte de Mirabeau proposed to divide France based on the identities of the provinces, whereas the lawyer Thouret's more abstract geometrical scheme claimed to divide France rationally into areas of equal dimensions. It was the former that won out at the end, and it was no small irony that the Constitution Committee gave a legal and administrative status to the "provinces" only to dismantle them into smaller and more manageable unities.[3] Such continuity was especially visible in the Roussillon, where the généralité, minus the Comté de Foix, and with the addition of some thirty-six (Occitan) villages from the neighboring district of Fenouillèdes in the Corbières, became the department of the Pyrénées-Orientales.[4]

Yet during the first years of the Revolution, the central government expressed little interest in the problem of France's national boundaries. Only in 1791 did the Constituent Assembly, shortly before dissolving itself, vote to urge the foreign ministry to "settle the questions of the limits with Spain."[5] The early revolutionaries insisted less on what divided states than on what united men—on the abstract universals of the Rights of Man and Citizen. Alexis de Tocqueville later gave expression to these ideals:

> The French Revolution had no territory of its own; indeed, its effect was to efface, in a way, all the older frontiers. It brought men together, or divided them, in spite of laws, traditions, character, and language, turning enemies sometimes into compatriots, and kinsmen into strangers; or rather, it formed, above all particular nationalities, an intellectual common country of which men of all nations might become citizens.[6]

2. J. Godechot, *Les institutions de la France sous la Révolution et l'Empire* (Paris, 1951), 91–112 and 586–599.

3. Berlet, *Les tendances*, chaps. 8 and 9; G. Mage, *La division de la France en départements* (Toulouse, 1924).

4. P. Vidal, *Histoire de la Révolution française dans le département des Pyrénées-Orientales, 1789–1800*, 3 vols. (Perpignan, 1885–1886), vol. 1, chap. 5.

5. FAMRE Limites, vol. 422, no. 90 (Buache de la Neuville to Montmorin, minister of foreign affairs, July 13, 1791); on the "expert" geographer Buache, heir of a dynasty of royal geographers, and his conception of the Pyrenees, see D. Nordman, "Buache de La Neuville et la 'frontière' des Pyrénées," in *Images de la montagne*, 105–110.

6. Quoted in R. Palmer, *The Age of Democratic Revolution: A Political History of Europe and America, 1760–1800*, 2 vols. (Princeton, 1964), vol. 1, facing title page.

The French revolutionaries threw their moral and political support to Patriot parties throughout Europe, while the government insisted on the principles of nonconquest, inscribed in the Constitution of 1791. Only later would territorial boundaries become important in terms of the opposition of France and its neighbors. Until the war years, the center had little concern for its periphery.

Nor was the periphery much moved by the tumultuous events at the center. For the French Cerdans in particular, the Revolution and its administrative redivision had relatively little impact on daily life. The territorial division of France did open up cracks in the unity of local society, as municipalities petitioned for administrative supremacy, while others sought independence from each other. The redrawing of tax rolls in 1790 was another occasion for struggles among neighboring communes, as each village took advantage of the breakdown of administrative order to assert its particular claims and specific boundaries.[7]

But the Revolution also opened cracks in the Old Regime façade of public order, into which cracks seeped a spirit—at first diffuse, and later quite open—of general resistance to the state. It was the revolutionary state's attempt to reorganize its different jurisdictions which catalyzed much of the local resistance in the Roussillon.

The creation of a national land register—the cadastre—and the abolition of the fictive distinction of real and personal taxes on landed property brought protests, not only from the French communities, but from Spanish proprietors as well. Those who had gained exemptions from the capitation tax protested loudly—but to no avail—when it was abolished.[8] More dramatically, the state's reconstruction of a customs frontier was the context of wide and open resistance. Indeed, when the first echoes of the Revolution reached the Roussillon, one of the first responses of the local inhabitants was to tear down the remaining cus-

7. ADPO L 1292: "Projets de réunions de communes," September 23, 1790; examples of struggles over boundaries include ADPO L 625 (Estavar and Bajanda); ADPO AC Dorres: "Registres des délibérations de la commune," fol. 57 (vs. Vilanova); and ADPO L 421 (mayor of Ur to prefect, January 21, 1791 [vs. Angostrina]). More generally, see M. Ozouf-Marignier, "De l'universalisme constituant aux intérêts locaux: Le débat sur la formation des départements en France (1789–1790)," Annales 41 (1986): 1193–1213; and J. N. Hood, "Revival and Mutation of Old Rivalries in Revolutionary France," Past and Present 82 (1979): 82–115.

8. ACSU leg. "Cerdanya françesa" (Girves y Tord, administrator of the Urgell bishop's lands to the bishop, September 1790); see also Vidal, Histoire, 1: 123. On the establishment of the cadastre, which in its final form combined maps and lists to form a land registry, see J. Konvitz, Cartography in France, 41–53, and below, chap. 7 sec. 3.

toms barriers along the Languedoc frontier, as well as those placed along the political boundary of Spain. In November 1790, the newly constituted customs administration asked the National Assembly for an ordinance "reestablishing the barriers at the extremities of the kingdom"; and although the law was immediately voted, the resistance of the Cerdagne municipalities delayed the erection of customs posts until July 1791.[9] For some time after that, the new administration had great difficulty in stopping what appears in the archives as a great increase in contraband activity—although this increase also reflects the new stringency and protectionism on the part of the state, as rigid in theory as it remained unenforceable in practice.[10]

The intrusion of the state during the first years of the Revolution, and the local resistance of the peasants, was nowhere more visible than the attempt to impose a new ecclesiastical boundary in the Cerdanya. The Revolution in Paris provided the opportunity in the Cerdanya for the provincial administration to gain de jure control of the ecclesiastical frontier, a recurrent goal of the intendancy since the annexation, and achieved de facto in the 1730s. The completion of this secular (and secularizing) policy formed part of the attack on clerical privileges in the first years of the Revolution. Out of its own fiscal needs, but affirming a long-standing goal of state control of church affairs, the National Assembly decreed the sequestration and sale of church lands in April 1790 and eventually adopted the Civil Constitution of the Clergy in July of that year. The Constitution made the clergy into a class of civil servants, requiring them to swear an oath to the Constitution and receive their pay from the state. Further, article 5 of the Civil Constitution abolished foreign ecclesiastical jurisdictions in French territory— and hence the bishop of Urgell's jurisdiction, with or without a general vicar, in the French Cerdagne, the protests of the bishop and the Span-

---

9. Brunet, Le Roussillon, 32 et seq.; Vidal, Histoire, 1: 175–182; ADPO L 883 (correspondence of July 1791); and ADPO L 409, no. 23 (municipality of Mont-Libre to Departmental Administration, September 9, 1971).

10. Although by 1793, there were some fifty-three customs guards in the border villages of the French Cerdagne (ADPO L 884), the customs administration, interrupted entirely during the revolutionary wars, only took on its definitive shape under the Empire. Until 1806, jurisdiction over the customs administration in the French Cerdagne remained based in Foix, much to the dismay of both local inhabitants and departmental officials: FAN F(20) 435. In 1810, Napoleon established a twenty-kilometer-wide zone between the political boundary of France and the interior, in which customs agents would exercise their jurisdiction: see Godechot, Les institutions, 656–683. On the mistreatment of customs guards by the local populations during the revolutionary period in Roussillon, see Brunet, Le Roussillon, 132–142.

ish ambassador notwithstanding.[11] The French Cerdagne was joined formally with the rest of the department to the bishopric of Carcasonne, which was moved back to Perpignan only in 1828.

Ironically, during the Revolution, the French state lost control over its newly acquired jurisdiction, since the bishop of Urgell refused to recognize the ecclesiastical annexation until Napoleon's Concordat of 1802. In the intervening decade, the bishop and Spanish clergy with lands in France retained authority over the hearts and minds of the French Cerdans, even if there were some dissenting voices. In some cases, as that of Ramon Carbonell of Gorguja (in Spain) and his lands in Ur (France), the tithe continued to be paid until 1802.[12] More generally, the bishop and Spanish clergy became the key actors of the veritable counterrevolution in the French Cerdagne.

In the years after France's passage of the Civil Constitution of the Clergy, the Spanish clergy directed a propaganda campaign in the French Cerdagne, sending out itinerant preachers and pastoral letters, and insisting that French parishioners cross the border to perform their religious duties. The Spanish clergy initially directed their campaign against those village priests in the French Cerdagne who had signed the oath of loyalty to the Constitution. There were six of them, out of a total of forty in the French Cerdagne.[13] On May 29, 1791, the gendarmérie of Sallagosa seized several pastoral letters that "circulated clandestinely" in the French Cerdagne, threatening the French clergy with excommunication. The parish priest Sarret of Sant Pere de Cedret wrote (in Catalan) to the departmental administration, recounting the threats he had received from the bishop and the "principal inhabitants of the village." By June, the six constitutional clerics, citing the extreme danger to their lives in accepting "the oath which ties them

11. For an overview of the Constituent Assembly's ecclesiastical policy, see Godechot, *Les institutions*, 255–266, and J. McManners, *The French Revolution and the Church* (New York, 1969), 24–46; on the protests of the Urgell bishop, see Vidal, *Histoire*, 1: 233–234.

12. Account books of Ramon Carbonell, Carbonell family papers. The persistence of the tithe in Spanish territory gave rise to numerous disputes between French and Spanish communities in the first decades of the nineteenth century: see Sahlins, "Between France and Spain," 354–355.

13. ADPO L 421, no. 155 (letter of departmental administration of Perpignan, March 19, 1791); P. Torreilles, *Histoire du clergé dans les Pyrénées-Orientales pendant la Révolution française* (Perpignan, 1890), 220; for the Pyrénées-Orientales department as a whole, 77 percent of the clergy signed the oath, compared to 70 percent in the neighboring departments of the Aude, and 66 percent in the Ariège: see G. Cholvy, "Réligion et politique en Languedoc méditerranéen et Roussillon à l'époque contemporaine," in *Droite et Gauche de 1789 à nos jours*. Actes du colloque de Montpellier, 9–10 juin 1973 (Montpellier, 1975), 41.

to the *patrie*" and claiming to be abandoned by the administration, resigned from their posts.[14] By the law of August 26, 1792, the "refractory clergy"—those who refused to sign the oath—were expelled from French territory, and 34 from the French Cerdagne (out of 375 from the department) left France, though many simply crossed the boundary and continued to minister to their parishioners, as we shall see.[15]

Given the extensive propaganda campaign of the Spanish clergy in the French Cerdagne, it is perhaps not surprising that sales of "national goods of first origins"—that is, church lands—were slow or nonexistent. In point of fact, most of the ecclesiastical property and incomes in the French Cerdagne were in the hands of the Spanish clergy from Puigcerdà and neighboring towns in Catalonia. Although the tithe was abolished on these lands, and although they were to be confiscated as "Spanish properties" during the war years (1793–1795), these lands were never sold. Thus the sale of church lands amounted to those belonging to the parish churches. Soundings in the registers of sales suggest that some of this property was acquired by private individuals. In Llo, three men bought eleven jornals of fields and meadows, worth 2,270 livres, while a merchant of Mont-Louis acquired two meadows worth 3,175 livres. Three residents of Via bought the lands of the parish church worth 2,470 livres, and other sales took place in Odello and Ur. And despite the protests of the town of Puigcerdà, which claimed to be patron of Belloch, this small chapel and convent of Servites was sold in 1792.[16]

There is reason, however, to suspect the validity of these sales. Decades later, a parish priest noted that François Ribalaigua of Dorres had bought Belloch on behalf of the community and that, in other villages such as Enveig, "good Christians" pooled their resources to acquire church lands and returned them after the revolutionary crisis. At the time, there was visible resistance to the private acquisition of these properties. The municipal council of Osseja refused to allow the sale of the parish house, claiming it was needed "to hold meetings" and "for a schoolteacher." More striking was the case of Joseph Ra-

14. ADPO L 358, nos. 12–21 (letters of priests withdrawing their oaths, May 1791); Vidal, *Histoire*, 1: 195–197.

15. ADPO L 1151 (list of priests and other clergy of the Pyrénées-Orientales who left French territory following the law of August 6, 1792); on emigration from the Cerdagne, see below this chap., sec. 3.

16. ADPO Q 117: "Biens nationaux, première origine, valeur approximative"; ADPO Q 138 and Q 239: "Registres de ventes"; on Belloch, see AHP MA 1789–1793, fol. 127; M. Gouges, "Le monastère de Belloch," *Tramontana*, 465–466 (1963): 55.

monatxo of Hix, who bought twelve jornals of fields and meadows from his parish church. Writing to the district attorney of Prades, he complained how "a superstitious respect seems to protect [the lands], and one could not encourage too much those [buyers] who have the force to contradict this public opinion." Those, like Ramonatxo, who did end up with church lands, were outsiders, fighting, as they well knew, the force of public opinion.[17]

Local struggles over the sales of national lands and the Civil Constitution of the Clergy suggest that the bishop of Urgell's influence over the French Cerdagne went far beyond the issues of a contested ecclesiastical frontier. The attempts to redefine that frontier provided the context of resistance to the secularizing policies of the revolutionary state. Religious issues gave a struggle and a language which helped to crystallize opposition to the Revolution. In their printed texts and sermons, the Spanish clergy thus identified France as a godless, heretical nation. Such religious opposition had long existed in the borderland, where from the thirteenth-century Cathars to the later sixteenth-century Protestants, the French were heretics. But the Revolution brought to this discourse a political opposition of unprecedented political and territorial implications: the events of the Revolution came to differentiate, not just the French and Spanish nations, but the territorial boundary of Spain and France. By giving to the emerging opposition of territories a cultural and symbolic significance, the Revolution instituted what Lucien Febvre has called "a ditch between nationalities . . . a moral frontier."[18]

It was along this moral frontier that the tensions of the French and Spanish governments increasingly found their expression. Already in August 1790, the Conde de Floridablanca, Spain's prime minister, complained to the French chargé d'affaires: "If I were to follow my own advice, I would put a cordon on the frontier, as for a plague." Floridablanca's "great fear," as Richard Herr has called it, disrupted the opening of Spanish society to French enlightened thought begun under Charles III. From then on, the Spanish government attempted to stem the tide of French revolutionary pamphlets, prohibiting all news from France from entering the country and seizing huge numbers of pam-

17. ADPO Q 138 (Llo); ADPO L 688 (petition of principal inhabitants of Osseja, an V [1796–1797]); ADPO L 1054 (response of the public attorney of Prades to Ramonatxo's petition of August 9, 1790); B. Cotxet, Noticia històrica de la imatge de Nostra Senyora d'Err (Perpignan, 1855), 29.
18. "Frontière: The Word and the Concept," 214. Examples of clerical propaganda may be found in ADPO L 421, no. 155.

phlets and books at the frontier. But without the excuse of a plague, all Floridablanca could do was to place troops along the frontier, for the ostensible reasons "of keeping troublemakers and vagabonds from disturbing the friendship of the two countries."[19] At the same time, despite formal declarations of peaceful intentions by diplomats, army officials and engineers on both sides of the Pyrenees produced a flood of military memoranda on strategies of both "defensive" and "offensive" warfare in the Pyrenean frontier, anticipating the conflicts that were soon to erupt.[20]

The presence of troops on the Spanish side of the boundary—evidence of the militarization of the French–Spanish opposition—changed the terms of local debate over the Revolution. The nearly universal opposition to France, under the influence of the Spanish clergy, gave way to fears of a Spanish invasion and conquest, and to some signs of support for France. Arguably, the concern of the French Cerdans was only for themselves. In July 1791, the municipalities of the French Cerdagne petitioned the departmental administration for weapons and for National Guardsmen "in order to keep the peace in this district and to restrain the brutality and fury of these enemies of civilization by means of a moat and a barrier." The petition did not necessarily reflect the desire of the Cerdans to fight in defense of the Constitution and France. When companies of National Guardsmen were organized in September 1792, the Cerdans showed little enthusiasm for enrolling. Pera of Palau, charged with drawing up the registration lists, was unable to provide the departmental administration with a list of enrollees—for there were next to none.[21]

The Cerdagne thus stood poised between France and Spain, waiting to see which way the winds would blow. The drama of events unfold-

19. R. Herr, *The Eighteenth-Century Revolution in Spain* (Princeton, 1958), Floridablanca is quoted on 253–254; on Spanish prohibitions, see also A. Matilla Tascon, *Catalogo de la colección de ordenes generales de rentas*, which lists the numerous prohibitions beginning in the fall of 1789; and M. Defourneaux, *Inquisición y censura de libros en España* (Madrid, 1963).

20. Spanish examples include SBCM Documentos 5–5–6–8: "Apuntes estadisticas y reconocimientos para la formación del plan de defensa de los Pirineos," n.d., ca. 1791; and ibid. 5–5–6–9: "Memoria sobre los medios ofensivos," 1792; French examples include FMG AAT MR 1083, no. 10: "Mémoire relative à une guerre défensive contre l'Espagne," 1792; and ibid. 1222, no. 10: "Mémoire sur les signaux à établir," n.d., ca. 1791.

21. ADPO L 1022 (petition of various municipalities of the Cerdagne requesting weapons and munitions, July 31, 1791); ADPO L 1024, no. 68: "Organisation de la garde nationale," September 23, 1792; see also J. F. Armangau, "Contribution à l'étude de la garde nationale dans les Pyrénées-Orientales pendant la période révolutionnaire, 1789–1799" (Mémoire de maîtrise d'histoire contemporaine, Université de Montpellier III, 1971), 34–66.

ing in Paris and the reactions of Madrid—the flight of Louis XVI and the campaign of Charles IV for an international alliance, the abolition of the monarchy, and eventually the regicide of January 1793—crystallized the opposition of the French Republic and the Spanish monarchy. In March 1793, citing the undeclared "state of war" on the French Republic, the Legislative Assembly declared war on Spain.

THE GREAT WAR (1793–1795)

Spain struck first, and most Cerdans welcomed the Spanish troops—their prior claims of loyalty to the French state notwithstanding. The Spanish military general Conde de Aranda, who had replaced Floridablanca in 1792, conceived the Spanish offensive, which was to penetrate into France along the Mediterranean frontier. Less interested in recovering the northern Catalan counties than in distracting France's war effort against the armies of the First Coalition, the Spanish offensive was felt closest to the frontier itself.[22] Almost without opposition, the Spanish armies crossed the boundary and occupied the border villages of the Roussillon in April and May 1793. In all but a few, the Spanish army was universally welcomed. Many municipal councillors, mayors, and other notables of the French Cerdagne welcomed the Spaniards as "liberators," their motives probably an admixture of compulsion, self-interest, and Spanish patriotism.[23]

The ambiguities of the reception of the Spanish are illustrated in the activities of Joseph Gallard, the sole royal notary of the French Cerdagne, whose story reveals the possibilities and constraints of the world of the French Revolution in the Cerdanya. Born in Sallagosa in 1736, Gallard was the heir to three generations of notaries who had faithfully served the French crown since the annexation. With a royal dispensation because of his age, the young Gallard, at twenty-four inherited his father's office,[24] and during the next four decades lived and carried out his duties in Sallagosa. As a large landowner, Gallard was certainly one of the "principal inhabitants"; but it was his status and authority as notary which made him a person of considerable power. As clerk of

22. A. Ossorio y Gallardo, *Historia del pensamiento político catalán durante la guerra de España con la República francesa* (Madrid, 1913), 22–23.

23. Brunet, *Le Roussillon*, 177–223; ADPO Q 38 (list of individuals designated mayors and concillors during the Spanish invasion, n.d.); the Spanish troops treated the French Cerdagne with surprising respect, distributing 200 *quarteras* of grain to the inhabitants: AHP Revolució françesa (letter to the municipality of Puigcerdà, June 8, 1793).

24. ADPO C 2060 Chancellor Lamoignon to intendant, March 12, 1761).

the local court, he decided which denunciations would be prosecuted. As notary, he was the essential mediator between the state and local society. Responsible for enforcing the edict of 1700 requiring the use of French, the notary registered all civil acts (sales, donations, wills, and contracts), interpreting and translating Catalan customary law into French—putting Catalan wine into French bottles.

Joseph Gallard had remained in the Cerdagne during the first years of the Revolution, but April of 1793 found him and his notarial papers in the town of Puigcerdà in Spain. Years later, in a petition written from a prison in Toulouse where he was serving a life sentence as a traitor, Gallard was to claim that the general commander of the Spanish troops had forced him to move, interning his papers and notarial writings. Gallard expressed in the petition his patriotism and loyalty to the French Republic, asking to be removed from the list of emigrants. But other testimony raised questions about his loyalty. Calvet and Cristofol, municipal officers of the revolutionary administration at Sallagosa, claimed that the notary had welcomed the Spanish troops with a white flag and had proved himself during their occupation to be a zealous persecutor of French patriots. (Since Gallard's loyalty was a borderline case, it is appropriate that he should have been arrested on the boundary—or rather, next to the "neutral road" that led from Llívia to Puigcerdà, near the Llívia bridge.)[25] Gallard was eventually released: his choice to take up a position in Puigcerdà, rather than to resettle in Sallagosa, had as much to do with the enemies he had made and the fact that the position of notary had been filled, than with his unambiguous loyalty to the Spanish nation.[26]

The Spanish occupation of the French Cerdagne during the late spring of 1793 was relatively brief, and the newly reorganized French armies, including several batallions recruited from the Pyrénées-Orientales, had little trouble reconquering the valley. On August 28, 1793, three thousand French soldiers under the command of General Dagobert routed the Spanish at the battle of the Perxa Pass near Mont-Louis,

25. FAN F(7) 5497 (dossier on Gallard, including his petition and local denunciations); see also AHP Internacional (petition of Antonia Gallard, September 9, 1799); and ADPO L 439, no. 113 (letter to the Departmental Administration, 13 thermidor an IV [July 31, 1796]).

26. The position had been filled by George Blanc, an immigrant from Catalonia, who married a French Cerdan in 1785 and was named notary ten years later—although he failed to register his own marriage act until 1799! See ADPO MS 34, vol. 3, fol. 220. Gallard continued to practice as a notary in Puigcerdà until May 1811 when, at the age of 75, he was arrested and taken to the Seu d'Urgell, accused of espionage for France: ADPO Mnc 833 (letter of May 4, 1811).

and the following day they passed with ease onto the valley floor and across the boundary. "At seven in the morning of the 29th, they took Puigcerdà," reads a local account of the occupation, "at 11:30 they took Bellver, and at 4:30 in the afternoon they came to Martinet, and after having taken possession of the town, they retreated to Bellver."[27] The French army thus took control of the entire valley, ignoring the political boundary and securing their position along an arc of villages on the valley's northern and southern flanks. For twenty-one months, until April 1795, the French remained in military control of the French and Spanish Cerdanyas.

During the French occupation, military authorities forming part of the war-time government of the Terror denounced the French Cerdans who had collaborated with the enemy. One revolutionary "mission representative," reporting in June 1794 to the Committee on Public Safety in Paris, identified no fewer than 126 men from the French Cerdagne as "enemies of the Republic." Many of the 18 mayors, 38 municipal officers, 15 public attorneys, 10 other notables, and dozens of merchants and others, all "infamous friends of slavery and despotism," managed to flee to Spain before they could be captured, although their goods were sequestered and their names placed on the lists of émigrés.[28]

Clearly, the list represented the vision of a Parisian revolutionary, for whom no degree of patriotism seemed satisfactory. The mission representative, Chaudron-Rousseau, even dismissed the municipal council of Mont-Libre (the renamed Mont-Louis), which had been in the early years of the Revolution the sole municipality to set up a "Popular Society" and to support the Civil Constitution of the Clergy.[29] Chaudron-Rousseau had omitted few names from his list, but local perspectives suggest certain divisions of the local population and as much compulsion as patriotism in collaborating with the Spanish armies. The Comité de Surveillance of the Cerdagne declared to the Parisian revolutionary that many of those on his list had held municipal office under the Spanish "against their will, and did not execute the orders of the Spanish commander." Others were there "because their ancestors were

27. AHP Catastro 1797: "Breve resumen de lo acaecido al corregto. de Puigcerdà en la invacion de los Franceses el 29 de Agosto de 1793"; and J. Fervel, Histoire des campagnes de la Révolution, 3 vols. (Paris, 1823), 1: 113–124.

28. FAN FII 134 (1035), no. 25 (4 messidor an II [June 22, 1794]).

29. ADPO L1022 (letter of town council of Mont-Libre expressing its revolutionary zeal, December 29, 1791); ADPO L 1070, nos. 62–63 ("resignation" of the municipal officers of Mont-Louis, 29 nivôse an III [January 18, 1795]).

noble," while some, such as Pierre Puig of Bajanda, were simply "imbeciles, incapable of harming the Republic or anyone else."[30]

The truth lies somewhere in between the vision of a Parisian revolutionary and the apologists for local society. Certainly there were some who supported the French: Calvet and Cristofol, who had denounced Gallard, had a stake in the new system, having acquired some of Gallard's lands when they were sold in 1794.[31] Other families even responded to the Republic's calls for soldiers in February and August 1793. Indeed, the cohorts of young men called up between March and October 1793 in the village of Enveig show that almost half had enrolled in the French armies, although the percentage was substantially less in the village of Osseja. Eight of fourteen enrolled from Ur, while Err—which was to become a center of resistance in the next several years—sent eight young men to defend the fatherland.[32]

It is impossible to give a socioeconomic breakdown of the household heads who did support the revolutionary government, but a study of the lists of local officials during and after the war years suggests a class of men relatively new to public life, often men whose descendants were to play important roles as mayors throughout the nineteenth century. The Patriot party, it is true, included several "ex-nobles." The son of Bonaventura Pera, an ennobled family and a lawyer at the Sovereign Council of Roussillon, became "justice of the peace"—an important position mediating between local society and the revolutionary state. François Montellà was a man "well-known for his civic spirit and confident belief in Republican principles."[33] But most of the active revolutionaries were men of common, if wealthy, birth. Laurent Delcasso, ex-Constitutional priest from Mont-Louis, was one of the earliest and most vocal of the Patriot party, while François Alart and François Marti of Palau emerged as premier spokesmen of the revolutionary cause in the Cerdagne during and after the French occupation. These were men who had not left their mark in politics before but who

30. Montellà family papers: "Tableau Général qui a été exigé par le Représentant du Peuple Chaudron Rousseau, du Comité de Surveillance séant à Estavar," 25 messidor an II (July 13, 1794).

31. ADPO Q 141 (acts of sale, 28 prairial an II [June 16, 1794]). Joseph Calvet was the largest of six purchasers of Gallard's lands: he acquired five parcels worth 2430 livres, almost a third of Gallard's estate.

32. ADPO L: "Communal Affairs," classified by commune; the lists date from late October and early November 1793.

33. For background on the Pera and Montellà families, see P. Lazerme, *Noblesa catalana*; the appraisal of Montellà is by the district administration at Prades: ADPO L 389, no. 117 (7 germinal an VII [March 28, 1799]).

emerged as local revolutionaries when the French took control of the Cerdanya.[34]

Yet over the course of two years of war, anti-French sentiment in the French Cerdagne came to outweigh by far local support for the Revolution. It was not just the number of Cerdans who fled across the boundary. Opposition—or at the very least, refusal to support the revolutionary cause—can be more directly measured by the failures of French military conscription in the Roussillon. Indeed, by 1795, there were few young Frenchmen left to enroll in the revolutionary armies.[35] By contrast, many French Cerdans enrolled in the popular militias raised in Catalonia to fight *against* France. Nearly half of the 245 volunteers raised in three companies of Catalan *miquelets* organized in nearby Berga in 1795 were young men from the French Cerdagne. Angostrina sent 17 young men, Sallagosa sent 13 (including Joseph Gallard, son of the notary), and Err sent 15. Except for Hix and Nahuja, every village was represented on the lists—an eloquent testimonial to the strength of anti-French sentiment.[36]

It was not as if the French troops actually treated the French Cerdagne as an enemy territory or an occupied country. Certainly the "house visits" and systematic searches in the French Cerdagne, with the goal of requisitioning all material necessary to the armies, alienated much of the population of the district.[37] Although the French Cerdagne was burdened by the presence of the revolutionary armies, and although the dramatic intrusion of the French state in the form of the army produced much resentment and fear among the French Cerdans, the French side of the Cerdanya was never formally "occupied" between 1793 and 1795, unlike the Spanish Cerdaña.

A military map drawn up for the campaigns of 1794 and 1795 sheds light on the French state's conception of the occupation (illustration 7). The map gives the name of "Cerdagne" only to that portion of the valley on the Spanish side of the boundary; the "French Cerdagne" itself was not named on the map. And indeed, the French revolutionary ar-

34. On the process of the "democratization" in village revolutionary leadership more generally, see L. Hunt, *Politics, Culture, and Class in the French Revolution* (Berkeley, Los Angeles, London, 1984), 166–176.

35. ADPO L 393, no. 28; Brunet, *Le Roussillon,* 297–388; on emigration from the French Cerdagne during these years, see below this chap., sec. 3.

36. AHP Revolució francesa: "Tercios del corregimiento de Puigcerdà, de paysanos auxiliares armados para la defensa de la provincia," May 1795, and below, this chap., sec. 3.

37. Brunet, *Le Roussillon,* 389–407.

38. Fervel, *Histoire des campagnes,* 2: 126–131 and 315–318.

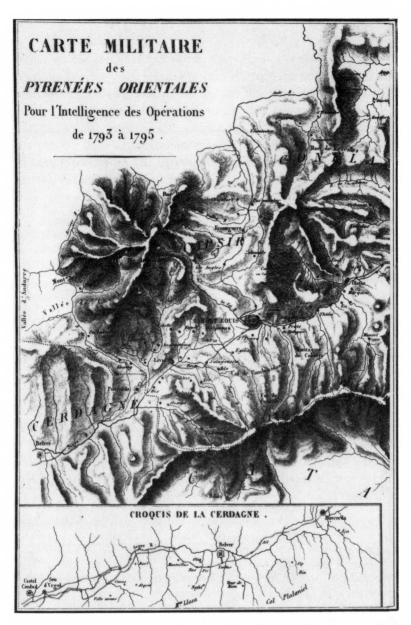

Illustration 7. "Military Campaigns of 1794 and 1795." This French military map of the Pyrénées-Orientales department used in the campaigns of 1794 and 1795 represents the "Cerdagne" as the portion of the valley from Puigcerdà to the Seu d'Urgell, thus identifying the Cerdagne as part of a military frontier extending beyond the political boundary into Spain. During the military occupation of 1793–1795, the French command treated the Spanish side of the Cerdanya as an occupied territory, while the French side remained indistinguishably part of France.

mies only militarily occupied the Spanish Cerdaña, stationing 8,500 troops in villages on the Spanish side.[38] In the later seventeenth century, the Cerdanya valley itself was the military frontier; by the Revolution, the military frontier was reduced to the Spanish Cerdaña, a zone that extended from the political boundary southward into Catalonia.

The young military genius, Carnot, conceived the state's policy toward the Cerdanya; Couthon, Robespierre, and others supported it. Carnot outlined a defensive plan to protect the French frontier against Spain, and the Committee ordered its mission representatives to

> move back the limits to their extremities in the mountains, and thus establish yourselves in the Cerdagne, take the valley of Aran, and in a word, all that which is on this side of the mountains, all that which gives our enemies a foothold on our territory, all that which can assure the inviolability [of French territory].[39]

Yet there was great pressure, coming from the revolutionary generals, mission representatives, and fervent local revolutionaries, to expand into Catalonia. General Dugommier had first proposed the annexation on May 12, 1794. Convinced that the Catalans "with their love of liberty" would join in the cause of the glorious revolution, he believed an uprising would provide France with an opportunity to acquire a "boulevard [into Spain] which forms a better guarantee than the Pyrenees."[40] Couthon and the majority of the Committee in Paris saw it otherwise. Echoing Richelieu, the architect of French absolutism a century and a half earlier, Couthon the revolutionary saw how Catalonia could become "a small independent republic which, under the protection of France, will serve as a barrier where the Pyrenees end."

> It is toward this end, dear colleagues, that we want your political and military occupations directed. . . . Catalonia once French would be as difficult to govern as was the old Roussillon and, on the contrary, if we make it a free country interested in its own defense and in the conservation of its independence, it will fortify itself and form an eternal barrier between Spain and France.[41]

But Catalonia was not ripe for revolution. The thousands of leaflets printed in Catalan which preceded the revolutionary armies in the cam-

---

39. Quoted in E. Vigo, *La política catalana del gran Comité de Salut Publica* (Barcelona, 1956), 37.

40. FMG AAT B4 15: "Mémoire sur la Catalogne," by General Dugommier; ibid. (letter of Delcasso to Committee of Public Safety, 22 floréal an II [May 12, 1794]); and FAN AF (III) 61, no. 3 (Delcasso to the Committee on Public Safety, March 1795).

41. FAN AF (II) 264, no. 2227 (Couthon, for the Committee of Public Safety, to the mission representatives Milhaud and Soubrany, 7 prairial an II [May 26, 1794]).

paign of 1794 found no audience in Catalonia.[42] Far from welcoming the French as they penetrated across the boundary into Catalonia, the Catalans manifested defiantly anti-French sentiments. With poetry, song, weapons, and men, Catalonia mobilized in the *Gran Guerra* against France.

Throughout Catalonia, and in media ranging from the daily conservative *Diari de Barcelona* to popular songsheets sold on the streets of towns and villages, a flood of poetry and popular songs relayed the news of heroic resistance to the "Godless heretics."[43] Some of this literature was clearly propaganda, material produced by the Catalan elites, supported if not written with the help of the central government, and designed to encourage enrollments in the local militias. But much of it was native and autochthonous, expressing the inherited antipathy toward the French in a new idiom. An illiterate poet from Prats in the Spanish Cerdaña, for example, dictated an anti-French diatribe in nineteen verses which encapsulated the history of the Revolution.[44] The poet sang of the Revolution as seen from Paris, but most of the anti-French discourse translated the Revolution into the local experience.

The printed call to arms from the *junta* organized at Berga—to which many Cerdans, French and Spanish, responded—is a good example of this anti-French ideology. It described the horrors of the French occupation—"they burnt religious images to light their chimneys" during the winter in the Cerdanya, and waged "a ferocious and destructive war" that left the Cerdans without homes. The junta evoked the defense of "Religion, Faithfulness to our August Monarch, a love of the Patria, and [that of] ourselves, our wives, children, and goods." A strong localizing tendency of anti-French rhetoric equated, within a religious idiom, the defense of the land (*la terra*) with that of the patria, which was just as likely to be Spain as Catalonia. In this way, the defense of one's family or village was merged with the defense of Spain.[45]

In the Cerdanya, the call to form companies of miquelets produced impressive enrollments in 1795. Several lesser nobles of the Cerdanya—

42. Herr, *Eighteenth-Century Revolution,* 282–293.
43. Ossorio y Gallardo, *Historia del pensamiento político,* passim; M. E. Cahner and E. Duran, "La guerra gran a través de la poesia de l'època (1793–1795)," *Germinabit* 58 (1959): 1–5.
44. BC MS 186, fols. 28r–29r: "Cansó de la nasió françesa," by Josep Bertran, 1792; and more generally, P. Vilar, "Quleques aspects de l'occupation et de la résistance en Espagne en 1794 et au temps de Napoléon," in *Occupants–Occupés, 1792–1815* (Brussels, 1969), 221–225.
45. BC FB 943: *La junta y comisionados para el armamento y tercios de Berga* (Barcelona, 1795).

among them Don Josep de Esteve i Girves, a notary of Puigcerdá, and Don Francisco Salsas of Llívia, an "honored citizen" of Barcelona—came forward as captains. They and others raised at least six companies of miquelets in the Cerdanya; over 570 peasants and townsmen from the Spanish side of the valley signed up to fight against the French—alongside, we have already seen, a large number of volunteers from the villages of the French Cerdagne.[46]

To captain or enlist in a company was hardly an act of pure self-sacrifice to an abstract cause. There were clear material benefits to becoming a miquelet: pay and daily rations of bread, booty from any eventual victory, and, for residents of Spanish Catalonia, exemption from paying the catastro tax for ten years.[47] But there were also concrete reasons to hate the French, beyond the inherited antipathy. The experience of the occupation could only reinforce hostility to the French.

At the request of the mission representatives, the Committee of Public Safety extended the "abolition of feudalism" to the Spanish Cerdaña in August 1794. Yet it hardly mattered, since the purpose and practice of the occupation repeated the age-old policy of treating the valley of Cerdanya as a breadbasket and a garrison. The representative Cassanyes ordered the seizure of all grain, hay, and vegetables of the Spanish villages and confiscated the properties of Spanish lay and clerical landowners.[48] The revolutionary ideology only seemed to increase the brutality of the French armies. So many died in the town of Puigcerdà, according to its municipal record, that a new cemetery was necessary in order to bury them.[49] Throughout the Spanish Cerdaña, but especially along the valley's limits where miquelets and French troops skirmished constantly, the French burned villages, seized cattle, and murdered resisters. Indeed, even if the incidents and damage reported by the Spanish communities to their own government after 1796 are exaggerated, they are nonetheless remarkable. The Spanish Cerdans claimed that the French had destroyed 22 churches, burnt almost 300 houses and devastated another 400, taken almost 10,000 horses

46. AHP Revolució francesa, esp. the list of "Tercios del corregimiento de Puigcerdà."
47. BC FB 943.
48. Vigo, Política catalana, 72–73; FAN AF (II) 134: "Compte-Rendu par Cassanyes," printed report to the Convention, July 15, 1793; see also Cassanyes' reports to the Committee of Public Safety, FAN AF (II) 256 (2169), nos. 60–62; and ADPO L 927, no. 85.
49. AHP MA 1795–1798, fol. 6v (August 8, 1795).

and cows and 25,000 sheep, not to mention grain, hay, and other foodstuffs, totaling more than a million and a half (French) livres in damages.[50]

Many Spanish Cerdans, both nobles and peasants, had withdrawn to the interior of Catalonia at the first signs of trouble—although it is impossible to gauge their numbers. Those who remained during the occupation occasionally demonstrated some signs of resistance. Ramon Carbonell of Llívia wrote to the governor of Puigcerdà, established in the neighboring valley of Ribes during the French occupation, that he and others "had convened up to 600 armed peasants, and that all agreed to take up arms spontaneously against the French," although nothing concrete came of the plan.[51]

Yet some Spanish Cerdans did support the French. Truth be told, only a handful collaborated, and most of them were from Puigcerdà. According to short lists drawn up by the Spanish armies, after they eventually retook the town, a dozen or so inhabitants from Puigcerdà, and others from the French Cerdagne, were incarcerated as having "whispered against the Spaniards" when they returned in April 1795. Nearly all of these men were from artisanal and merchant families, as none of the Puigcerdà nobility was implicated directly.[52] Petitions sent to the Spanish government for reparations after 1795, however, include denunciations of two nobles accused of trying to recover their incomes during the two years in which the French occupied the town. Others complained that "many were those who got rich, especially from Puigcerdà, in the commerce and traffic of goods which they did with the license and under the shelter of the French." These men—no names are mentioned—were clearly outsiders and in the minority; the story that the municipality of Puigcerdà welcomed the French invaders and asked for "the title of French citizens" is probably apocryphal, although it is often repeated in military histories of the war.[53]

Most of the Patriots active in Puigcerdà during the two years of oc-

50. AHP Revolució françesa: "Estado gen. de los daños y perdidas que han sufrido los pueblos y particulares de la Cerdaña española." The petitions were presented to the "Junta de la Frontera," established to make reparations for the damages, which treated the destroyed border villages as "new settlements" and exempted the inhabitants from paying the catastro for ten years: see ACA CA RA, legs. 174–175.

51. AHP Revolució françesa (correspondence of the governor of Puigcerdà to the capitain-general of Catalonia, January 1794).

52. Ibid.: "Lista de los franceses y alguna gente de Puigcerdà y Serdaña que se conocen por sospechosos," July 26, 1795.

53. Fervel, *Histoire des campagnes,* 1: 22; the complaints of enrichment during the occupation may be found in ACA CA RA, leg. 174, no. 1.

cupation were French, some from far beyond the French Cerdagne. A certain Doppet, a doctor from Savoy and an ardent Jacobin, presided over the newly reorganized municipality and founded a "Popular and Mountain Society of Puigcerdà." In May 1794, the society organized a victory celebration of Dugommier's reconquest of Roussillon in the Temple of Reason—the rechristened church of Santa María of Puigcerdà—at which time the society members read the Constitution and Rights of Man in Catalan. According to an official account of the victory feast, the society "had the sweet joy of setting fire to the emblems of the Castilian tyrant's pride and the emblems of fanaticism." Most of the active members of the Puigcerdà municipality were French natives, several from the French Cerdagne—among whom was none other than Joseph Ramonatxo, purchaser of "national goods" in France and soon to become a widely feared "boss" in the borderland.[54]

That French troops never managed to penetrate beyond the towns along the flank of the Pyrenees—Seu d'Urgell and Ripoll—was due in large part to the widespread and popular resistance of the Catalans; it was also due to the central government's neglect of the southern frontier. The major French battlefields in 1794 were along the northern and eastern fronts, sites of the struggle against the principal members of the First Coalition, Prussia and Austria. The Committee of Public Safety largely abandoned the Cerdanya and the Pyrenean frontier, leaving fewer than 5,000 soldiers between Bellver and Puigcerdà. In the borderland, the tide turned. In April 1795 a Spanish army of 18,000 entered the Spanish Cerdaña and, following a bloody battle for the Puigcerdà hospital, took control of the town.[55] The Spanish forces would undoubtedly have penetrated into the French Cerdagne, had the news not soon arrived that a preliminary peace had been signed—four days before their reconquest of Puigcerdà.

THE NATIONALIZATION OF NATURAL FRONTIERS

"The boundaries of France are marked out by Nature," proclaimed Danton to the Convention in February 1793, "We shall reach them at their four points, the Ocean, the Rhine, the Alps, and the Pyre-

54. On the victory feast, Vigo, La política catalana, 67–68; on Ramonatxo, ADPO L 927, no. 55; and on Ramonatxo's activities during the Napoleonic episode, see below, chap. 6 sec. 4.
55. Fervel, Histoire des campaigns, 2: 318–336.

nees."[56] Danton's sweeping assertion culminated months of debates over France's natural boundaries and the limits of its expansion. At stake were the occupied territories of Savoy, Nice, Belgium, and part of the left bank of the Rhine. Still faithful to the "no conquests" formula of 1791, and not entirely confident in the doctrine of popular sovereignty and the freely expressed desire by the occupied territories to be united to France, the Convention needed a justification that would allow France to appear as if it were not embarking on an expansionist policy abroad. Sorel, Mathiez, and others argued that the revolutionary state merely completed the monarchy's goal of establishing France's "natural frontiers," but it was precisely *not* to look like the Old Regime monarchy that the revolutionaries adopted the doctrine. Yet once introduced into revolutionary discourse, the ideas served less to justify existing policies than to establish new claims of the revolutionary state.[57]

The politicization of natural boundaries was possible because the complex of ideas had changed its fundamental character since the seventeenth century. Whereas seventeenth-century statesmen and apologists never abandoned the historical determination of France's boundaries, as exemplified in the Treaty of the Pyrenees, the enlightened conception of Nature stripped the idea of boundaries of their historical determination. With respect to Savoy, Grégoire had announced that "one must consult the archives of nature, to see what rights are permitted, what duties are prescribed." With respect to the Pyrenees, eighteenth-century commentators reinterpreted the Treaty of the Pyrenees in terms of an ahistorical natural law: "that the points of watershed, which is to say the summits of the mountains, would serve as limits." More generally, from Montesquieu, Hume, and Rousseau to royal geographers such as Buache de la Neuville, enlightened thought took it as axiomatic that different nations ought to have as their boundaries natural divisions of the landscape.[58]

Enshrining the principle of natural boundaries, the French–Spanish peace treaty of Basel, signed on July 22, 1795, ordered the definitive demarcation of the Pyrenean boundary. Article 7 dictated that

56. *Le Moniteur Universel,* vol. 15, no. 32 (February 1, 1793).

57. J. Godechot, *La Grande Nation: L'Expansion révolutionnaire de la France dans le monde de 1789 à 1799,* 2 vols. (Paris, 1956), 1: 76–80; Sahlins, "Natural Frontiers Revisited," in press.

58. Grégoire's speech in *Le Monteur Universel,* vol. 14, no. 333 (November 28, 1792); on the Treaty of The Pyrenees, see FMG AAT MR 1084, nos. 41 and 43; and on eighteenth-century thought, see N. Pounds, "France and 'les limites naturelles,'" 51–53; and Sahlins, "Natural Frontiers Revisited," in press.

[b]oth sides . . . name commissioners to work incessantly toward drawing up a delimitation treaty between the two powers. Concerning the lands which were in litigation before the present war, the commissioners should take as the basis of this treaty, as far as possible, the summits of the mountains which form the watershed between France and Spain.[59]

Spain and France failed to delimit the boundary line before 1868. Indeed, the Basel commissioners never even met to resolve the issue, as other needs proved to be more pressing. Yet an outpouring of memoranda by statesmen, military officials and engineers, and political representatives of local society suggest how the idea of natural boundaries was further transformed by the Revolution. Introducing new terms into the definitions of identities and boundaries, the Revolution had nationalized the idea of natural frontiers, with profound implications for the states and the communities of the borderland.

One strain of military thinking in France continued to reject the arguments that mountain crests ought to form the division of Spain and France. Grandvoinet, batallion chief in the Army of the Pyrénées-Orientales, recognized that the Pyrenees were a "limit which Nature has posed" but argued to the central government that such a boundary "should not be idealized. The Pyrenees are not an impenetrable barrier and we do not own them: the summit belongs to the two powers, and in several places, it belongs to Spain." Grandvoinet went on to propose that the Republic profit from its "rights of conquest" and extend its limits, 3,000 *toises* (about five miles) into Spanish territory:

If the border line were placed elsewhere, the Republic would draw no advantage from the Pyrenees; these mountains are a long and very high curtain which would only hide the enemy if it were preparing hostile actions.[60]

In effect, he was proposing to redefine as a territorial boundary line the military frontier in existence since the fifteenth century. In times of war, France had sought to conquer the natural frontier and gain control of the territory on the opposing watershed; now Grandvoinet sought to institutionalize that frontier as a boundary.

But Grandvoinet's opinion was the minority view; more widely, military officials and engineers became the principal spokesmen for the strictest reading of the idea of natural boundaries. "There is nothing more vague than such a phrase, 'the summit of the Pyrenees,'" com-

59. Text in Martens, *Recueil* 6: 542–547.
60. FAMRE Limites vol. 459, no. 113: "Observations du Citoyen Granvoinet," n.d., ca. 1795.

plained Jean Bourgoing, French ambassador to Madrid in 1792, who was given the authority to begin the delimitation talks in 1795. Military men agreed, since they subjected the concept of "watershed" to a new kind of scrutiny. Arguing that watersheds could be misinterpreted through errors "of custom and convention," a civil engineer in the School of Mines urged the determination be made along scientific lines.[61]

Such debates over the location and determination of "true" watersheds were reproduced in the Cerdanya itself. Delcasso wrote to the Committee of Public Safety that Spain was separated by high mountains from the Cerdanya and that the Treaty of the Pyrenees had unfairly given up a portion of the valley, which should have been entirely annexed to France.[62] Military officials in the Cerdanya also urged the French government to retain the Spanish Cerdaña. Although they cited the abundance of grain and hay, as well as its role as a strategic passage, they argued more generally that the Pyrenees Mountains dividing Spain from France were comprised of that range which stretched "from Mont-Louis to Bellver," separating the Cerdanya from the rest of Catalonia.[63]

Yet such an insistence on the "true watershed" implied a boundary that would have forced France to cede the entire valley to Spain. In fact, a few proposals—including one by the town of Puigcerdà—suggested the exchange of the French Cerdagne, on the Spanish watershed, for the Aran Valley, located on the French side. Such an exchange had been debated in the early 1720s; but the revolutionary discourse added a new meaning to the proposal. Whereas earlier it had been a question of jurisdictional compensations to ease problems of administration, after 1795 the issue became one of exceptions to a natural law.[64] Brossier and Chrétien, the cartographic officers of the Caro–Ornano meetings, claimed that "Nature has established a strong equality in the rights of

61. FAN AF (III) 61 254, no. 4: "Quelques reflections sur le traité de Paix avec l'Espagne," Jean Bourgoing, 22 thermidor an III (August 9, 1795); FAMRE Limites, vol. 442, no. 91: "Mémoire sur la fixation des limites entre la France et l'Espagne, d'après les versants," Muthuon, 5e jour complémentaire, an III (September 21, 1795).

62. FAN AF (III) 61 245, no. 3 (March 1795).

63. Ibid. (Adjunt General Delort to Cappin, Representative of the People, 28 ventôse an III [March 18, 1795]).

64. FAN AF (III) 62 (250) 1 (Junker to Committee of Public Safety, 28 thermidor an III [August 15, 1795]); ibid. (Fournier, retired director of fortifications, 4 messidor an III [June 22, 1795]); AHP MA 1795–1798, fol. 48v (January 15, 1796); on the earlier proposals, see FAMRE Limites, vol. 459, no. 1: "Mémoire sur la vallée d'Aran, le val Carol, et la Cerdagne française," n.d., ca. 1720; FMG AAT MR 1221, no. 50 (September 12, 1721).

the two nations, since the exchanges to be performed to conform to its laws give quite equal results."[65] But Adjunct General Junker, head of the Military Archives in 1795, opposed the exchange:

> If we make the slightest exception to the law of watersheds . . . I fear that the demarcation of this boundary will never be achieved, and perpetual quarrels will continue among the frontier inhabitants. As soon as we make an exception to the general rule, the arbitrary exception allows private interests to tread on the general interest, and it is the general interest which must be the basis of the Republican regime.[66]

Such arguments derived from natural law or the public good had less weight than an even more fundamental element of French revolutionary discourse—the ideology of national territory itself. The Directory refused to support the foreign minister's proposal to exchange the Cerdagne for the Aran Valley, arguing:

> The Constitution formally forbids the alienation of any part of French territory, and for a stronger reason still the alienation of our fellow citizens who have exercised the fullness of their rights in the acceptance of this act, and who promise after long sacrifices to live freely and happily under the only empire to which Reason admits, that of Law. . . . The patrie is an indivisible property, and a Frenchman cannot allow it to be taken away.[67]

The revolutionary character of this argument was threefold. In the first place, the Directory assumed that the patrie had a spatial and territorial dimension. It is true that since the later middle ages, the patria had been associated with the French monarchy, as opposed to a mere valley or region, thus implying the appearance of a "limited, territorial" idea of loyalty. But in the seventeenth and eighteenth centuries, under the influence of enlightened thought, the word came to refer to a sense of attachment to a place—an "attitude of mind" or "experience of belonging"—and was tied to the idea of "liberty."[68] The Revolution both

65. FMG AAT MR 1221, no. 11: "Mémoire et projet sur la démarcation des limites," by Brossier and Chrétien, 23 frimaire an IV (December 14, 1795); and FAMRE Limites, vol. 446 (memoir of 1 thermidor an IV [July 19, 1796]).

66. FAMRE Limites, vol. 446 (Adjunct General Junker to the Committee of Public Safety, 26 thermidor an III [August 13, 1795]).

67. FAMRE Limites, vol. 446: "Rapport au Directoire Exécutif sur la démarcation des limites," n.d., ca. 1796.

68. On medieval and Renaissance conceptions of the patrie, see Kantorowicz, *The King's Two Bodies*, 247 et seq.; and G. Dupont-Ferrier, "Le sens de mots 'patria' et 'patrie' en France au moyen age et jusqu'au début du XVIIe siècle," *Revue historique* 188–189 (1940): 88–104. For later usages, see J. Godechot, "Nation, patrie, nationalisme et patriotisme en France au XVIIIe siècle," in *Patriotisme et nationalisme en Europe à l'époque de la Révolution française et de Napoléon*, XIIIe Congrès international des sciences historiques (Paris, 1973), 7–27; and F. Brunot, *Histoire de la langue française*, 7: 132–136.

politicized and territorialized the patrie by firmly identifying it with a national territory.

In the second place, the patrie was made into a "property" of the Republic. Royal apologists for Louis XIV consistently identified the royal domain as a property of the king, and military officials of the later seventeenth century had occasionally identified a territorial conquest as a property right as well. But it was not until the eighteenth century that jurists such as Emmerich de Vattel defined a state's territory as its "property," and only during the Revolution did statesmen and administrators begin to write consistently of territory as a "property" of the Republic. It was a long road from Louis XIV's proprietary claims to his royal dominion, to the Revolutionary notion of territory as property, but the bourgeois state had finally arrived.[69]

Finally, national territory was to be defined by the nationality of its inhabitants—or so it would seem. "The people of the Cerdagne, having accepted the Constitution, are an integral part of the Republic and cannot be alienated," wrote a member of the Committee of Public Safety.[70] In truth, this was a passive acceptance based on the spatial dimensions of national territory: their territory defined them as Frenchmen, while as Frenchmen they defined their territory as French. This self-fulfilling logic was to have far-reaching consequences in local society; for the state, it meant that the Directory could reject the concept of a plebiscite so useful in the annexations of Savoy, Belgium, and Nice—where there had been some body of opinion favorable to France. Thus in 1796, the Directory argued that the nationality of a territory could not be defined by the expressed declarations of the inhabitants, for to seek their consent would be "an offense against the rights" of citizenship, making it a "servitude."[71]

This affirmation of the territorial determinants of nationality created a paradoxical situation in the French Cerdagne: although most of the inhabitants actively resisted the French state during the war years of 1793 to 1795, most were to end up affirming their own French nationality in the years after the 1795 Treaty of Basel. The years after 1795 mark a period of the gradual resettlement of many of those who had emigrated from the French Cerdagne. The experience of the Cerdanya

69. Rowen, The King's State, esp. chap. 4; E. de Vattel, The Law of Nations (1758), bk. 2, chap. 7.

70. FAN AF (III) 61 243, no. 2 (Committee of Public Safety, with a note by Murlia to mission representatives with the army of the Pyrenees, 4 floréal an III [April 23, 1795]).

71. FAMRE Limites, vol 446 (report to Directory, with a marginal note by Reubell, n.d., ca. 1796).

during these years suggests how, despite the failure to physically delimit the territorial border, the Revolution placed new—and more rigid— boundaries on national and territorial identities.

EMIGRATION AND RESETTLEMENT
(1792–1800)

The first to emigrate from the French Cerdagne were the clergy, who found a cool official reception but were generally supported among their Catholic brethren in Catalonia and Spain. As the revolutionary regime was consolidated, more and more lay emigrants fled across the boundary. A register from the municipality of Puigcerdà from December 1792 lists approximately 1200 names of emigrants from France. Most were from the towns and departments of the south, including thirty-seven French Cerdans.[72] Yet it was the war, beginning with the withdrawal of Spanish troops in August 1793, which produced the greatest emigration.

It is difficult to evaluate the total number of emigrants. Between 1792–1793 and 1798–1799, the total population of the French Cerdagne fell from 6584 to 5926 inhabitants—mostly due to emigration. For Roussillon as a whole, it has been estimated that there were 3854 emigrants, mostly humbler families from the frontier districts of Prades and Ceret.[73] In the French Cerdagne, sequestration rolls and the slow sales of emigrant properties suggest that emigrant families dominated numerically. Certainly they did in the village of Err, where twenty-four families had members who were emigrants. For the district of Cerdagne as a whole, it became nearly impossible after 1795 to find families without emigrants from whom municipal officers could be drawn.[74]

But it was more than the predominance of families with émigré relatives which preoccupied military officials and local representatives of the Patriot party in the valley. One official complained in May 1796 that "the inhabitants of the district of Sallagosa entertain daily rela-

72. ADPO L 740. The list dramatically underestimates the number of French Cerdans, since most would not have registered at Puigcerdà. On clerical emigration to Spain more generally, see Herr, *Eighteenth-Century Revolution*, 298–302; and the older but still useful work by J. Contrasty, *Le clergé français exilé en Espagne, 1792–1802* (Toulouse, 1910). Emigration statistics for France have been compiled by D. Greer, *The Incidence of Emigration during the French Revolution* (Cambridge, Mass., 1951).

73. Vidal, *Histoire de la Révolution*, 2: 236 et seq.; Brunet, *Le Roussillon*, 237–253.

74. ADPO Q 122: "Séquestrations des biens, parents des émigrés; ADPO L 388, no. 119: letter of Alart, "Commissaire du Directoire Exécutif, près l'administration municipale du canton de Saillagouse," 27 germinal an V (April 14, 1797).

tions with the émigrés and deported priests, who bustle on the adjoining Spanish territory." Seventeen of 122 households in the town of Llívia harbored French émigrés in 1796. Complaints and protests to the Spanish ambassador in Paris led Charles IV in December 1795 to order French emigrants to be kept at a distance of fifteen leagues from the border.[75] But the edict went unenforced for several years, and Puigcerdà and Llívia became principal emigrant centers, rivaling others in Catalonia.

The weekly and sometimes daily letters to the departmental administration in Perpignan of a loyal Patriot, François Alart of Palau, reveal how, "crackling with an irrepressible audacity," the émigrés remained on the frontier. Supported by their families and friends, they proved more than willing to participate in the traditional festive life of the valley. In August 1796,

> fifteen or sixteen émigrés from the hoard of the commune of Err, all of them armed, came into the commune and held a *veillée de dances* in several houses, and the following morning they went with their families and friends to the Hermitage of Núria, where a procession of both sexes goes every year on August 15.

Soldiers were sent up the mountain after them, but the locals easily eluded them. Turning a revolutionary dance *against* the revolutionary state, the supporters of the émigrés demonstrated their opposition to the repressive forces of order:

> Their families, incensed, started to dance the *carmagnola,* and acted depraved in the streets; they made a great show of their presence in clear contempt of the laws, and nobody stopped them. . . . Few were those in the commune who did not take part in this horseplay.

Nor was it only festive occasions that brought the emigrants back to French territory. During the harvest of 1796, many returned to work their lands, just as they later disappeared in the interior of Catalonia "to earn their keep" during the winter months.[76]

The practice of religion proved the focal point of opposition to the revolutionary regime and of divisions among the Cerdans themselves. The "refractory" priests, those who refused to sign the oath of loyalty, were "like the stone which draws iron toward itself," according to

75. AMLL MA 1789–1823, fols. 78–80; and AHP MA 1795–1798, fol. 54 (decree of the Real Audiència of Barcelona, January 14, 1796).
76. Reports of Alart to the Communal Administration, July–December 1796: ADPO L 388, nos. 26 (16 messidor an IV); 77 (29 thermidor an IV); and 132 (6 nivose an V).

Alart, "forcing a diminution in the number of Patriots and good Republicans."[77] The bishop of Urgell and priests of the Spanish Cerdaña gave orders to the clergy loyal to Spain, while many of the French Cerdans went to hear mass across the boundary or plotted with their neighbors in the Spanish Cerdaña to sneak the "true clergy" into French villages.[78] According to Alart, these religious divisions split the inhabitants in civil war. He wrote that those who bought up their properties were in danger of their lives, as local society suffered an extreme polarization: "Everything is in upheaval, quarrel upon quarrel, insult upon insult, one against another all because of the émigrés."[79]

Alart's view was a minority one. The Revolution did not permanently fracture local society into distinct camps, nor did it break apart the inherited and traditional solidarity of the village communities.[80] It is true that the communities were changed by the Revolution: the status of privileged families was irreparably damaged, and "new men" subsequently entered into political life. But during the 1790s, the emigrants clearly dominated the Cerdagne communities. Indeed, they provided the village communities with a cause and a symbol by which to assert their local autonomy against the French state.

Yet during the later 1790s, this antistate ideal coincided with the return of the emigrants to French territory. Even if they resisted the authority of the central government, their homes lay on French territory, and given the opportunity and/or the necessity, the emigrants sought to return. The necessity came from Spain, as its government in 1797 began to enforce its laws concerning the restriction of the French to fifteen leagues inside Spain. At the same time, the opportunity was provided by the Directory, which relaxed its laws on emigration, encouraging Frenchmen to reclaim their residences. Confiscated estates valued at fewer than 20,000 livres were returned to the families of emigrants. As early as 1795, the Convention, faced with economic woes, passed a law permitting artisanal and laboring emigrants to return to work in France.[81] In subsequent years, the Directory conceded

77. ADPO L 388, no. 23 (letter of Alart, 20 ventôse an IV [March 10, 1796]).

78. Ibid., no. 25 (12 germinal an IV [April 1, 1796]); ADSU Cerdanya, no. 2 (letter of Mauri and Vila, priests of the French Cerdagne, denouncing the bishop of Urgell for his interference); and ADPO L 688 (denunciations of municipal officers of Osseja for plotting to bring the priest of Aja to say mass in the French village, fructidor an VII [August–September 1799]).

79. ADPO L 440, no. 19 (Alart, 11 fructidor an V [August 31, 1797]).

80. Cf. Brunet, Le Roussillon, 261 et seq.

81. Greer, Incidence, 96–105; AHP Revolució françesa (dossier on internment of French emigrants); AMLL MA 1669–1800, fol. 112v (royal order of April 22, 1798, expelling French emigrants within 30 days).

that the names of those who could prove that they left France before 1790 were to be removed from the lists of émigrés. It was clear to the Directory that even those who had taken an oath to the Spanish king—required by Floridablanca in 1791—and who remained in Spain after 1795 were French, even if they fulfilled the requirements of having established residency in Spain.

> The birth of these Frenchmen, and the care with which they demonstrated their desire to return by abstaining from identifying themselves as natives by letters of naturalization . . . assures the conservation of their political rights, and thus the right that they have to be inscribed in the common book of the French.[82]

Many were those in the French Cerdange who took advantage of such openings. Considerable doubt could be raised about their loyalty to France; the dozens of petitions to Paris seeking to be removed from the lists of émigrés are filled with a rhetoric of loyalty and suspicious excuses. Jacques Carcassonne of Err claimed an "involuntary absence" when he was not permitted to return from visiting his sick brother, a priest in Ripoll; Joseph Marranges of Ur claimed to have been living in Puigcerdà since he was thirteen. Clearly not all French Cerdans who had emigrated *wanted* to return to the French Cerdagne: Joseph Gallard, released from prison, settled in Puigcerdà and became a Spanish notary instead of going back to his practice in Sallagosa, while Jean Fabre, a shopkeeper from Angostrina, bought a house in Llívia and established himself there.[83] But most of the French Cerdans who emigrated did attempt to return to their hearths, and most were successful.[84] Even if doubt could be cast on their political loyalty to the revolutionary government, their identity as Frenchmen could not be questioned.

The irony of the Revolution was that despite their political opposition to France, the French Cerdans found their fate tied to France. And if the French Cerdans were Frenchmen, they might as well take advan-

82. FAN AF (III) 63 250: "Rapport au directoire exécutif sur les réclamations des français en Espagne," n.d., ca. 1796.

83. FAN F(7) 5497 (petitions of French Cerdans seeking their names removed from the lists of émigrés); ADPO L 777 (petition of Jean Fabre, 2 ventôse an VII [February 20, 1799]); ADPO L 389, no. 41 (Alart to Communal Administration, 26 pluviôse an VI [February 14, 1798]).

84. AMLL: "Sobre franceses emigrados," (1798), declaration that all Frenchmen, except four widows and "six or seven field workers" left the town of Llívia. By 1806, according to declarations of various municipalities of the Spanish Cerdagne, following an inquiry by the governor of Puigcerdà, responding to a request by the prefect in Perpignan, there were no French emigrants or deserters living on the Spanish side of the boundary: see AHP Internacional (declarations of April 1806 and February 1807).

tage of the fact. In the late 1790s and the first years of the nineteenth century, individuals and communities petitioned the French government for benefits and "privileges" which they believed were due them as Frenchmen. Thus the merchants of La Tor de Carol complained about the burdens of the new customs administration, underscoring that "they are French, that they pay taxes to the Republic," and that they were due the "privileges deriving from that quality." Communal struggles were renewed as well, both among communities of the French Cerdagne and between French and Spanish villages. In the latter case, the mayors and municipal councillors petitioned their departmental and national governments concerning their claims to disputed boundaries, emphasizing their status as national citizens and urging their governments to delimit, once and for all, the territorial boundary of Spain and France. During the renewed communal struggles in 1798 between the French village of Angostrina and the Spanish town of Llívia over their boundaries, Angostrina insisted on a French arbitrator, "who knows the idiom and can appreciate the force of their titles," and demanded that the judgment over a disputed seizure take place in the French courts. "Llívia has always petitioned the courts in France, so the town must recognize the courts as natural judges of this land, and it thus results that the land in question is taken to be situated on French soil."[85]

The uses of French language and nationality were even more evident to the twenty-two household heads from Sallagosa who, in 1796, sent a petition to the district administration requesting the establishment of a grammar school to teach French. The school would give their children "the benefit of learning the Rights of Man and of the Citizen so that they can profit from these rights given to them by Nature." Profit, it would seem, was meant quite literally, since the petition went on to describe the utility of French in commercial affairs as well as in the performance of civic duties.[86]

The inhabitants of the village communities thus took advantage of their status as Frenchmen, using the language and rhetoric of nationality to justify their own local interests. In settling once again in French territory in the late 1790s, they identified their own, local interests as

85. ADPO 33W 6 (petition of year X); ADPO L 393, no. 97 (town council of Angostrina to the Departmental Administration, 28 frimaire an VII [December 18, 1798]); and AMLL VII.5 (copy, in Catalan, of claims of Angostrina); for other examples of renewed communal struggles across the boundary, see Sahlins, "Between France and Spain," 785–790.
86. ADPO L 388, no. 75 (n.d., ca. 1796).

French ones. By nationalizing their local interests and setting themselves off against their Spanish neighbors across the boundary, they asserted their distinctive identity as national citizens. The Revolution thus crystallized national identities, but not by uniting individuals and communities to a common purpose—as most accounts would have it. Indeed, the French state remained notably unable to recruit young men to fight for the patrie. In 1797, Alart decried the fact that "it has been a while since a single [draftee] showed up," and eighteen months later he was still complaining about "the obstinancy of the draftees who should have begun their service: only one presented himself, and none of the others, nor their family members, bothered to show up."[87] Despite their refusal to serve the armies of the Republic, and despite their continued political opposition to the French government, the Cerdans nonetheless affirmed their national identity as Frenchmen. Both opposition and affirmation took place within a new kind of state, one organized around the coherent administration of a national territory. National territory had replaced royal jurisdictions as the idiom of sovereignty in France. And by creating a specifically national territory, the revolutionary state created a specifically territorial nationality. The boundaries of territory and nationality were to be affirmed, both from the state and local perspectives, over the next half-century.

87. Ibid., no. 6 (Alart to Communal Administration, 18 germinal an V [April 7, 1797]); and ADPO L 389, no. 62 (report of Alart, 16 frimaire an VII [December 6, 1798]).

6
Territory and Identity During the Spanish Crises (1808–1840)

The Old Regime in France fell suddenly; in Spain, it fell continuously in a series of political, economic, and cultural crises occasioned by French expansion into the Peninsula under Napoleon in 1808 and, in the 1820s and 1830s, when "liberals" and "royalists" fought a series of civil wars in Spain.[1] From these years of political turmoil was to emerge the legal and administrative framework of a national, centralized state in Spain, and—in France's reaction to Spain's political crises—a national boundary line between France and Spain. The Spanish crises introduced a specifically political content to the national territorial boundary which emerged from the Revolution. In the borderland, the Cerdans were directly engaged by these political crises and developments. The peasants and village communities responded by defining and shaping their own communal and national boundaries and identities, consolidating the boundaries drawn during the French Revolution.

THE NAPOLEONIC OCCUPATION, 1808–1814

In February 1810, the French Marshall Augereau issued a proclamation to the Catalans from his headquarters in the occupied city of Girona:

1. For background on the crisis of the Old Regime in Spain, see J. Fontana, *El crisis del Antiguo régimen, 1808–1833* (Barcelona, 1979); and Carr, *Spain:* 79–209.

198

The French have always embraced you and supported you in your struggles. Charlemagne saved Catalonia from the tyranny of the Moors. The magnificent cathedrals which still exist recall their founder. In 1641, your ancestors asked France to govern you, and you remained several years under its ambit. Your industry, activities, and customs are so similar that some politicians justly call you *the French of Spain.*[2]

Apart from a handful of collaborators, the Catalans would have been stung by the insulting epithet bestowed on them. But propaganda does not have to be successful to be significant: the episodes of French expansion into Catalonia from Charlemagne to Richelieu were indeed precedents to the Napoleonic occupation and eventual annexation of northeastern Catalonia between 1808 and 1814. So too were more recent events unmentioned in Augureau's proclamation, episodes of French expansion across the eastern Pyrenees which would hardly have induced a warm historical memory—the Bourbon alliance and occupation of Catalonia during the War of Succession between 1707 and 1714, and the French Revolution itself. The Napoleonic expansion of France thus fit a pattern; yet by the character of its boundaries, the episode was unique; and so too were the reactions it provoked among the populations of the borderland.

The concept and practice of national territory as it appeared during the events of the French Revolution indelibly marked the boundaries of nationality and territory under the Napoleonic Empire. Such was the case of France's military frontier. Napoleon's expansion across the Pyrenees reproduced a pattern of control and conquest long practiced in the borderland, but one which recognized the existence of the national boundary of France and Spain put into place during the French Revolution. Under Napoleon, the French state simultaneously "conquered" the Pyrenees, extending its military frontier into Catalonia, and attempted to uphold the distinctiveness of its national territory in opposition to Spain's.

The internal contradictions of such a policy became evident in the War Ministry's attempt to establish "a new line of demarcation in the Pyrenees between the French Empire and the Spanish monarchy," proposed in the winter and spring of 1811 and decreed in March and May 1812.[3] Far from an agreement of two sovereign powers, Imperial

2. Quoted in J. Mercader i Riba, *Catalunya i l'Impéri napoléonic* (Barcelona, 1960), 110.

3. FMG AAT MR 1221, no. 23: "Projet d'une nouvelle ligne de démarcation dans les Pyrénées," January 19, 1811; and the "Rapport au comité central des fortification sur la nouvelle délimitation," May 6, 1812.

France imposed on the Spanish government of the Intruder King Joseph Bonaparte the "delimitation" of a specifically military frontier desired by France. But there remained an essential contradiction within French policy. On the one hand, France established an "offensive frontier" that would secure the strongholds and provisioning centers on the backside of the Pyrenees: the age-old "conquest of natural frontiers" and the control of key passes was the basis of this policy. On the other hand, the Central Committee on Fortifications sought to establish the empire's own natural boundaries, to give France "the defensive frontier of the barriers posed by Nature, and consider the districts situated beyond them as battlefields in which it will be sufficient to hold one or two strongholds or depots."[4]

Military considerations were foremost among the concerns of the French government during the Napoleonic episode. Spain's uprising against Napoleon began in the spring of 1808, following his invasion of the kingdom and the replacement of Ferdinand VII with Joseph Bonaparte on the Spanish throne. Throughout the countryside, armies of "insurgents"—irregular companies of Catalan miquelets—waged guerrilla warfare on the French troops. The French established themselves as an occupying power in Barcelona in 1808 and remained there until 1814.[5] But in the borderland, French troops stayed within French territory until September 1810.

The same was generally true of the Catalan insurgents, who tended, in the words of the mayor of Estavar in the French Cerdagne, "to respect French soil in this part of the territory."[6] True, many of the rebels—sometimes barely distinguishable from the bandits and smugglers who infested the eastern Pyrenees during these years—sought refuge in the border villages of the French Cerdagne, from where they occasionally raided villages and redirected convoys of contraband goods toward the Catalan interior.[7] But in the first phase of the "War of Independence," both the French troops and the Catalan counterrevolution-

4. Ibid, no. 27 (registers of the Central Committee of Fortifications, May–June 1812, including the imperial decree of March 1, 1812, which ordered "the incorporation of all territory on the French watershed" and declared the need to secure "all military positions protecting the Empire, even if they are on the Spanish watershed.")

5. General sources include J. R. Aymes, *La guerre d'independence espagnol, 1808–1814* (Paris, 1973); on the "captivity of Barcelona," see P. Conard, *Napoléon et la Catalogne* (Paris, 1910); and more recently, J. Fontana et al., *La invasió napoléonica* (Barcelona, 1981); and the collected articles in *L'Avenç* 113 (1988): 21–47.

6. ADPO Mnc 833 (letter to the Perpignan police commissioner, March 8, 1810).

7. ADPO Mnc 833 (correspondence of mayors of the French Cerdagne with the subprefect of Prades, 1810).

aries took account of the national boundary of France and Spain in the Cerdanya.

The national boundary was defined, first and foremost, as an economic one. By a decree of 1807, Napoleon established the "Continental Blockade" to prevent the entry of British colonial goods into France and its allied states.[8] This economic boundary of empire crossed through the center of the Cerdanya plain, and the neighboring towns and villages in Catalonia quickly came to serve as stockpiling points for what the French authorities suspected were large quantities of British colonial goods.

In September 1810, French troops in the Cerdagne crossed the border, entering Puigcerdà. The suspected stockpiles turned out to consist of 4 sacks of flour and 259 bales of cotton, which were promptly confiscated.[9] But the French penetration into Spain drew a response from the Spanish General Campoverde who, with several companies of Catalan miquelets, drove the French from Puigcerdà on September 30 and continued into the French Cerdagne. Their ostensible aim was to assure the entry of "colonial goods" and to return with cattle and grain with which to supply their armies in Catalonia. These goals were accomplished, but only with a great deal of gratuitous pillage and destruction of the French Cerdagne villages.[10]

After the events of September 1810, the two armies willfully effaced the territorial division. But neither the Catalan insurgents nor the French imperial troops managed to secure more than a fleeting control over the opposing side of the valley. The two armies thus entered into a frenzied exchange of dominion in the Napoleonic military frontier.[11] It was as if the seventeenth-century struggles for military hegemony in the valley were replayed at high speed, and with the same result: France in control of the military frontier of Cerdanya.

In truth, the French state secured its control of the military front only after it had annexed northern Catalonia. Napoleon himself had first proposed the annexation as early as 1804. In February 1810, the French state established a set of military governments and the

8. G. Lefebvre, *Napoleon: From Tilsit to Waterloo, 1807–1815,* trans. J. E. Anderson (New York, 1969), 107–146.

9. ACA Gi caja XXX, no. 2 (confiscations of September 18, 1810).

10. ADPO Mnc 831/1 (commander of National Guard, Girves, to subprefect of Prades, September 26, 1810); see below, sec. 4, for the reactions of the French Cerdans to these invasions.

11. The confusing events of these two years have been reconstructed from the administrative and military correspondence in ADPO Mnc 831 and 833: see Sahlins, "Between France and Spain," 298–300.

framework of a civil and fiscal administration along the southern flank of the Pyrenees from Catalonia to Navarre. Then, by a decree of January 1812—as the French Empire was already falling apart, drained by Napoleon's "Spanish Ulcer" and the Russian campaigns—France formally annexed northeastern Catalonia, creating four new departments. Puigcerdà became the capital of the Segre department, which included its corregimiento minus the Valley of Ribes but including the Valleys of Andorra, along with the districts of Solsona and Talarn.[12]

The French established a civil administration in the newly created departments, abolishing the "feudal system" and introducing French civil law and administrative institutions. Most of the personnel in the new Puigcerdà administration were recruited from France, including many individuals from the French Cerdagne and Roussillon; there were few collaborators from the Spanish Cerdaña.[13] Yet despite the identity of civil administrations and the continuity of personnel in the two Cerdanyas, France maintained the territorial boundary between the annexed department of the Segre and France itself.

It struck the new prefect of the Segre as both odd and unreasonable that the two Cerdanyas should be separated by the political division. In June 1812, he urged the intendant of Catalonia to push for a new delimitation of the frontier, incorporating the Spanish Cerdaña into France.

> It is a delimitation fixed by Nature itself. It is only a question of joining to the Pyrénées-Orientales department several communes; but it is most necessary, and the inhabitants, who share the same language, mode of industry, clothes, and who are linked together by kinship, demand [the union] themselves.

After all, France had incorporated the Aran Valley instead of allowing it to form part of the newly created departments. But the proposal to unite the two Cerdanyas—and others to eliminate the enclave of Llívia

12. Mercader i Riba, Catalunya, 203–325, and the map on 293; J. Nadal i Farreras, "L'organizació administrativa i la vida material en la Catalunya napoleònica," in Fontana et al., La invasió napoleònica, 81–98; on the annexation in the Pyrenean region, see J. Mercader i Riba, "Puigcerdà, capital del departament del Sègre," Pirineos 9–10 (1948): 413–457; L. M. Puig i Oliver, Girona francesa, 1812–1814 (Barcelona, 1976); and B. Faucher, "Une annexion éphémère, ou le val d'Aran de 1812 à 1814," in Actes du 2e congrès international d'études pyrénéennes (Toulouse, 1962), 39–49.
13. On the customs personnel, see ACA GI caja II, no. 10: 127 of the 148 employees came from France, and, of these, 27 were born in the Pyrénées-Orientales, including 11 in the French Cerdagne. For further details on collaborators from the Spanish Cerdaña and personnel from the French Cerdagne, see Sahlins, "Between France and Spain," 659–661; Mercader i Riba, "Puigcerdà," passim; and Mercader i Riba, Catalunya, 281–284.

and to "mark the limits of the two Cerdanyas"—fell on deaf ears.[14] For it involved a delimitation not just of two departments but of two national territories—even if both were part of the Empire.

The persistence of the national territorial division of France and Spain despite French expansion into Catalonia meant that traditional conflicts between the two Cerdanyas continued to plague both administrations. Thus disputes over jurisdiction along the "neutral road" linking Puigcerdà and Llívia, and conflicts over rights of passage in the borderland, threatened to disrupt the harmony of the two allied administrations. The attempt by the Puigcerdà prefecture in August 1812 to tax French proprietors with lands in the Spanish enclave of Llívia also created dissonance both among officials and local residents of the two Cerdanyas.[15] But the greatest source of conflict arose from jurisdictional disputes between the two customs administrations.

The new customs administration of the Spanish Cerdaña was created in February 1812, a month after the annexation. Yet because Catalan counterrevolutionary forces controlled the Cerdaña during the winter and spring of 1812, it was not until April that customs posts were established along the Spanish side of the boundary. In the intervening months, the customs posts of the Segre department were erected in French territory, much to the displeasure of the French customs, military, and civil administrators.[16] Even after 1812, the two customs administrations came into conflict as if they were protecting the territories of separate states. As the director of customs of the Segre department explained in April 1813, referring to his inability to pursue four smugglers into France, "our jurisdiction cannot . . . be exercised in French territory." Disputes over the "neutral road" between Llívia and Puigcerdà, and over rights of passage in the borderland, suggest that the annexed department of the Segre was still Spanish territory.[17] Despite a return to past patterns of French domination in Catalonia, the imperial annexation could not rid itself of the Revolution's contribution of a specifically national territorial boundary.

14. FAN FI (E) 72 (prefect to interior minister, July 18, 1812); other proposals are found in ACA GI caja XLVII, no. 1 (Segre prefect to intendant, June 26, 1812); and ACA GI caja LV, no. 8 (February 10, 1812).

15. ACA GI caja LV, no. 8; and ACA GI caja XXIII, nos. 14–15 (August 1812).

16. ACA GI caja XIV, no. 5: "Rapport . . . sur l'organisation des douanes de la Cerdagne espagnole," February 2, 1812; ACA GI caja XXX, no. 2; and ADPO Mnc 836–837 (customs director of the Spanish Cerdagne to prefect, February 28, 1812).

17. ADPO Mnc 836–837 (police commissioner of Perpignan to prefect February 1812); ACA GI caja VIII, no. 1 (customs director to intendant of Upper Catalonia, April 12, 1813); ACA GI caja LV, no. 8 (February 10, 1812).

Viewed from the state's perspective, the Napoleonic episode thus reveals the enduring impact of the French Revolution. As France expanded its military, economic, and administrative boundaries beyond those posited by the revolutionary treaties, that expansion was forced to recognize the national territory inscribed by the Revolution itself. The tension and contradiction were ultimately ephemeral. In March 1813, Napoleon, defeated on the eastern fronts, ordered the simplification of the civil administration in Catalonia and, in the spring of 1814, Spanish troops entered the valley of Cerdanya.[18] The Spanish invasion of the Cerdanya coincided with the end of the empire: the Treaties of Paris in July 1814 and June 1815, separated by the "100 Days" of Napoleon, restored to France its boundaries of 1792, thus affirming not only France's defeat but also the spatial dimensions of French national territory.

REDEFINING THE FRENCH MILITARY FRONTIER (1820–1840)

In January 1820, an army rebellion in Spain led a colonel and a major to declare for the Cadiz Constitution of 1812, the "liberal codex" that was to provide the legal framework of bourgeois society in nineteenth-century Spain. Backed by a provincial uprising, the two army officers sanctioned their actions by establishing a government that began Spain's first experiment with liberalism, the "Constitutional Triennium."[19] France, under an increasingly beseiged royalist government, looked upon Spain as a serious threat—especially when civil war broke out in Spain in 1821 and fighting in Catalonia came in close proximity to French territory.

"The eyes of France and all of Europe are focused on this frontier," wrote the French journalist and future minister Adolph Thiers, as he described the Cerdanya borderland in the winter months of 1822.[20] The establishment of the Royalist Regency in the nearby town of Urgell during August 1822 drew the attention of the Spanish constitutional forces

---

18. On the simplification of the administration, ACA GI caja XXI, no. 13 (copies of ordinances); and Mercader i Riba, "Puigcerdà," 449–451; on the Spanish invasion of the Cerdanya, ADPO 3M1 20B, no. 2 (mayor of Carol to sub-prefect, March 21, 1814).

19. R. Carr, *Spain, 1808–1975*, 129–146; and J. Comellas Garcia-Llera, *El trienio constitucional* (Madrid, 1963); for an overview of the events in Catalonia, see J. Carrera i Pujal, *Historia política de Cataluña en el siglo XIX*, 3 vols. (Barcelona, 1957), 2: 115–246.

20. A. Thiers, *Les Pyrénées et le Midi de la France pendant les mois de novembre et décembre 1822* (Paris, 1823).

toward the Cerdanya. Throughout the year, Puigcerdà and the Spanish Cerdaña had been the battleground between constitutional troops and the royalist armies of miquelets. The struggles of the "two Spains" were fought out in close proximity to French territory.[21] Indeed, the proximity was too great for the French royalist government, which attempted to prevent the civil wars in Spain from spilling onto French territory.

As France affirmed its goal of territorial integrity in opposition to Spain, the distinction of national territories became politicized in both France and Spain. The event that crystallized this process—almost exactly a century after the Spanish sanitary cordon of 1721 against the Marseilles plague—was France's own sanitary cordon, established in September 1821. The cordon was a "line" of barracks within a wider "zone" watched over by the military administration, assisted by the brigades of customs guards; it was established with the ostensible reason of preventing the spread of cholera, already present in Barcelona during the summer of 1821. Its purely sanitary functions were, as Thiers himself saw, a vain attempt to seal off an otherwise permeable boundary. "One hundred thousand men," he wrote, anticipating the same number of Frenchmen who were to invade Spain in the spring of 1823, "would not succeed in closing down the boundary, because they would have to guard the smallest rock crevices, and what is more difficult, to know them all." More important, as Thiers among others was quick to point out, the quarantine functions of the cordon quickly yielded to more political ones.[22]

Alongside its efforts to prevent the spread of cholera to France, the government in Paris tried to prevent the plague of liberal politics from contaminating French territory, defining and defending its national territory in opposition to Spain's. This was a project informed, partly, by narrow political motives, since the French government protected its territory but allowed Spanish royalists to take refuge on French soil and sometimes supplied the Spanish "Army of the Faith" with arms and munitions.[23] But the political defense of national territory transcended

21. On the counterrevolutionary uprisings in Catalonia, see J. Torras, *Liberalismo y rebeldía campesina, 1820–1823* (Barcelona, 1976), the second chapter of which has been translated as "Peasant Counterrevolution," *Journal of Peasant Studies* 5 (1977): 66–88. A succinct chronology of the events in the borderland is sketched out by the Constitutional General Espoz y Mina, commander of the Army of Catalonia during these years, in his *Memorias,* in Biblioteca de autores españoles, 146: 332–414; and 147: 5–105.

22. Thiers, *Les Pyrénées,* 182; J. F. Hoffman, *Le peste à Barcelone* (Paris, 1964), 27–49; and ADPO Mnc 2100/1–3 (ordinances and registers of the "sanitary intendancy" established at Perpignan).

23. Thiers, *Les Pyrénées,* 156–157; Mina, *Memorias,* 377.

the political groupings of liberal and royalist. When the French government disbanded the cordon in September 1822, French troops remained in place as an "observation army"—over 12,000 strong in the Pyrénées-Orientales in 1823. According to the royal ordinance, the point was "to prevent the political dissensions of our neighbors from disturbing the profound peace we are enjoying." Customs guards and constabularies joined these troops in protecting national territory by stopping the actual fighting between civil factions from spreading to France. Thiers described their difficult task in his account of battles between the Constitutional General Mina against the royalists near Llívia:

> The horsemen were not always sure of being on Spanish territory. The French sentinels would alert them, and make them return to within their territorial boundaries. One of them, caught up in the chase, didn't hear the warning cry, and continued his gallop. The sentinel shot and missed. The horseman reeled around, and came after him; but seeing that he was French, he stopped and, without firing back, returned to Spanish territory.[24]

In the early spring of 1823, the "army of observation" became an army of occupation. The French ministry declared the invasion a "political necessity" dictated by "the blood ties of the monarchy" and the threat of invasion of French territory by armed bands. While "100,000 Sons of Saint Louis" accompanied the Duc d'Angoulême into Spain along the western Pyrenees, one French batallion chased out the Spanish constitutional troops from Puigcerdà. French troops remained in Barcelona and as near the border as the Seu d'Urgell until 1828. But in the borderland itself, the French withdrew ten days after their "invasion." They remained stationed in the French Cerdagne, however; alongside 900 National Guardsmen, the French army attempted to prevent the renewed struggles between liberals and royalists from spreading to France.[25]

24. Thiers, Les Pyrénées, 148; SAHN Estado, legajo 6228: (consular reports from Perpignan on troop strength, 1822–1823). The royal ordinance is quoted in J. Sacquer, "La frontière et la contreband avec le Catalogne dans l'histoire et l'économie des Pyrénées-Orientales de 1814 à 1850," (Diplôme d'études supérieures, Université de Toulouse, 1967), 46; on the use of customs guards and police to prevent territorial violations, see ADPO Mnc 1961; and AN F(7) 11984 (reports of the gendarmérie, 1822–1823).

25. On the French invasion of 1823, see Geoffrey de Grandemaison, L'expédition française d'Espagne en 1823 (Paris, 1928); and R. Sanchez Mantero, Los cien mil hijos de San Luis (Seville, 1981), esp. 57–82; and FAMRE MD Espagne, vol. 147, fols. 328 et seq., including the Occupation Treaty that left 24,000 troops in Spain. On the Cerdanya, see the dossier in ADPO 4R 34; and M. Martzluff, "Aspects socio-politiques du Roussillon dans le contexte de l'expédition d'Espagne en 1823" (Mémoire de maîtrise, Université de Montpellier, 1977), esp. 63 et seq.

This short occupation of the Spanish Cerdaña differed radically from those earlier occupations of the late seventeenth century, the Revolution, or the Empire. The valley served none of its traditional roles as garrison, breadbasket, or mountain pass toward Barcelona. Instead, the Spanish side of the valley figured in the 1823 "invasion" as just one more site of a political spectacle, a symbolic enactment of French domination in Spain.

The sanitary cordon of 1821–1822 politicized the national territorial opposition of France and Spain; yet the boundary line defined by the cordon was ultimately ephemeral. Not only was the boundary again transgressed with the French invasion of Spain in the spring of 1823, but there was no permanence given to the politicized national territory. True, the soldiers placed in the French Cerdagne in 1819 remained stationed there throughout most of the nineteenth century. When for the first time they were removed in 1851, the local mayors clamored for their return, for they had come to depend upon the soldiers for protection against the "vandalism" and "thefts" of the "disorderly and undisciplined" Spanish troops garrisoned across the boundary.[26] But the permanent politicization of the national boundary occurred only within the French reaction to the events of the First Carlist War (1833–1840), as it was fought out in both the eastern and western borderland, and especially in the Cerdanya.

The Carlist revolt was at once a political and popular reaction to the liberal centralizing state in Spain. With its first stronghold in Andalusia, the rebellion spread north to the Basque districts and to Catalonia, where in some cases it drew its strength from the defense of regional privileges and liberties.[27] But while most of the Catalan towns in the Pyrenees supported the Carlists, Puigcerdà and the Spanish Cerdaña affirmed their liberal loyalties. During ten separate Carlist "invasions" of the Spanish Cerdaña between March 1836 and December 1837, there is little evidence of sympathy or collaboration among the local inhabitants. Indeed, Puigcerdà's impressive self-defense against the Carlists in 1837 won the town the title of "Heroic and Invincible."[28]

26. ADPO Mnc 1883/1 (mayor of Bourg-Madame to prefect, June 7, 1851).
27. On the Carlist uprising in relation to the liberal reform movement of the 1830s, which restructured municipal and provincial administrations, dividing Spain into "provinces" in 1833 (and dividing the Spanish Cerdanya between Lérida and Girona), see Carr, *Spain,* 184–195; and C. Maréchal, *Spain, 1834–1844* (London, 1977). On the Carlist struggles in the Basque countries, see J. Coverdale, *The Basque Phase of Spain's First Carlist War* (Princeton, 1984).
28. On the Carlists in Catalonia, see R. Grabolosa, *Carlistes i liberals: Historia d'unes querres* (Barcelona, 1974); on the invasions of 1836 and 1837, see the dossier in ADPO Mnc 1918; and AHP MS, "Relación del sitio que los carlistas pusieron a Puigcerdà

Strictly a Spanish affair, the Carlist revolt was restricted in space to the Spanish side of the boundary. The Carlists saw in the Cerdaña a granary, a garrison (usually the village of Alp and those close to the passes toward Berga), and a strategic point near the French border. Levying taxes, confiscating cattle, and taking prisoners for ransom, the Carlist soldiers were periodically chased out by the arrival of Spanish constitutional troops, only to return to the Cerdaña when the government forces left.[29]

In 1836, Adolph Thiers held the portfolio of foreign minister in the newly formed French cabinet. And once again, the attention of France was directed toward the Pyrenean borderland. Thiers' own experience in the Cerdanya fifteen years earlier, combined with the failure of an Austrian alliance and the growing civil war in Spain, made the borderland a site for the forging of national policy. Thiers himself came to favor French intervention in support of the Bourbon Infanta, María Cristina; he may have envisioned an army of "100,000 Liberals of France" which would replay the scenario of 1823, with the political signs of liberal and royalist reversed. But lacking the support of the Chambers and king, Thiers was unable to convince even his fellow ministers, and France sent no troops beyond some hastily assembled regiments of the Foreign Legion.[30]

The council of ministers debated possible intervention, but they did not question the need to defend national territory. French policy in sealing its territory evolved less from ministerial directives than in response to the demands of the Cerdagne municipalities, and in reaction to the events in the Cerdanya itself. During the night of February 6, 1837, a band of some 200 Carlist guerrillas descended from Ribes to Llívia through the Err valley and the French Cerdagne; after looting the Spanish enclave, the column attempted to pass along the "neutral road" back to Puigcerdà but was stopped by French troops, who arrested 176 Carlists and confiscated the pillaged goods.[31]

en noviembre 1837" by Don Juan Capdevila y Oliva. For additional material on Puigcerdà's growing liberal identity in the nineteenth century, see Sahlins, "Between France and Spain," 675–681.

29. ADPO Mnc 1918/1: "Rapports concernant les invasions carlistes en Cerdagne espagnole"; and ADPO Mnc 1921 (correspondence of sub-prefect of Prades, 1835–1838).

30. FAMRE PA Desage, vol. 29, fols. 29–94 (correspondence between the minister of foreign affairs and the French ambassador in Madrid, esp. letters of July 22 and November 18, 1835); and J. Allison, *Thiers and the French Monarchy* (New York, 1968 [1926]), 232–239.

31. ADPO Mnc 1981/1; FMG AAT E(4) 43: "Rapport sur le passage des troupes carlistes pendant la nuit du 5–6 fevrier 1837." Only one of the 176 was from the Cerdanya: F. Gallard of Puigcerdà.

The double violation of French territory and the "neutrality" of the international passage produced heady debates at the French ministerial level. These debates drew attention to the special quality of the Spanish town completely enclosed by French territory; and from the debates actually emerged a general French policy toward Spain. Even as he favored intervention, Thiers knew that it was important to keep the Carlists from penetrating into France, "to push them back beyond the boundary." The war minister complied, placing nearly three-quarters of the French troops in the Cerdagne along the neutral road, at once to prevent such territorial violations and to guard the town of Llívia—for safeguarding Llívia would be defending French territory. The consensus was to react swiftly to any territorial violation, which included movements of Carlist forces along the so-called neutral road.[32]

The configuration of troops deployed around the enclave of Llívia, the international road, and an imaginary boundary line separating two territories in 1837 formed part of a fundamental shift in the conception and practice of France's military frontier. For one, France instituted its defensive policy at a time when the two states were not at war, nor even threatening a confrontation. The difference between the new military frontier and the traditional one was apparent to a military engineer writing—with some disgust—in 1836. He described the enclave of Llívia.

> I know that the declaration of war [between France and Spain] would immediately rectify this monstrous result of politics; that once the treaties are torn up by the sword, our army will take up the military position which it needs, and enough will have been said both about the enclave of Llívia and its annoyances. But in times of peace, and given our actual position vis-a-vis Spain, these annoyances are particularly troublesome. We have formed an observation cordon the length of the border, charged with watching the movement of bands which infest the neighboring country, and with forbidding them entry into our own territory. The duties which we have to accomplish would be clear and simple faced with a boundary line neatly traced, but on disputed lands, these duties are no more well defined than the line itself.[33]

Yet France and Spain were not to "tear up their treaties," and the French military frontier became redefined as the strategic defense of the national boundary line.

32. FMG AAT E(4) 43 (correspondence of February 1837, quoting letters of war minister and foreign minister, April 9 and 16, 1836). Maps of the borderland drawn up by military engineers during these years suggest that France simply took over the jurisdiction of the neutral road: see, for example the map in FMG AAT E(4) 43 (1837).

33. FMG AAT MR 1349: "Mémoire sur l'enclave de Llívia," by Dumontet, May 7, 1836.

The shift in French military strategy was hardly a sudden one: since the later seventeenth century, the military frontier had increasingly recognized a territorial division of the Cerdanya. The "traditional" military frontier had taken no account of the division, as the army in control occupied the entire valley, but by the later seventeenth century, the military administrations came to recognize the existence of two Cerdanyas. During the Revolution and especially in the Napoleonic Empire, the military frontier protected French territory by extension: it became a zone beyond France into Catalonia. But after the 1823 "invasion" of the Spanish Cerdaña, the military frontier took full account of the political difference of national territory; indeed, it became organized around the defense of national territory.

This shift in French military strategy was shaped by military engineers themselves, whose memoirs and maps of the period both reflect and inform the military consolidation of the borderland. Previously, military engineers and commanding generals made a distinction between the military frontier of the Pyrenees and that of the political division of territories, "the limits of France and Spain." The latter were considered irrelevant in any strategic sense, since French armies sought to secure the towns of Bellver and Puigcerdà and to control the major passes from the four sides of the valley. Thus the attention paid to roads and transports, and especially to the topography of access to the valley on all sides.

The reports and maps of the 1830s and 1840s, by contrast, broke down the topography in a manner unthinkable even a half-century before. Detailed maps of the "neutral road" and of the small depressions and hills immediately coterminous with the political boundary appeared. The military frontier was redefined to approximate the national boundary of Spanish and French territories; it took that political division as its point of departure (illustration 8).[34]

Yet, less a clear shift from an old to a new military frontier, the memoirs of the 1830s and 1840s reveal a certain tension between them. On the one hand, there were attempts to meet the goals posed by the definition of the new boundary. The result was an international "buffer zone," a neutral area on Spanish territory designed to protect France from straying bullets. In 1837, the representatives of the French government received assurances from both Carlists and Liberals to respect

---

34. FMG AAT MR 1223, nos. 48–49: "Reconnaissance militaire de la Cerdagne française," by Labadie and Demesnay, 1840.

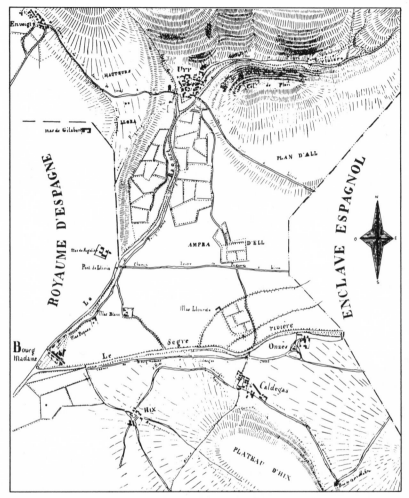

Illustration 8. "Military Reconnaissance of a Portion of the Spanish Frontier" (1840). This map and accompanying memoir includes a detailed depiction of the terrain in France located between the Spanish enclave of Llívia and Spain itself. During the 1830s and 1840s, the military frontier became equated with the defense of national territory, designed to prevent the spread of Spanish political struggles into France. Representing Spanish territory as blank, this map illustrates the completion of this evolution.

French territory, and in 1874, during the last Carlist War, they reached a formal accord creating a "neutral zone" on Spanish territory so that stray bullets would not cross the streets of Bourg-Madame.[35] Thus the military frontier continued to be represented as a zone, not a line. Since the Napoleonic occupation the zone had shrunk substantially, but it still remained distinct from the national boundary line. On the other hand, military engineers remained faithful to an older notion of the frontier, and they often stressed the need to secure Puigcerdà as a strategic point in guarding the French border—even though the town was fully recognized as Spanish. Other memoirs, while including detailed plans to defend the French Cerdagne, still entertained fantasies about taking control of the entire valley.[36]

Beyond this nostalgia for the old military frontier, the transformations were apparent. Despite the oscillation of royalist and liberal governments in France and Spain, French policy in the borderland of Catalonia displayed a profound continuity. Transcending its ideological position toward Spain, the French state sought to confine Spanish civil struggles to Spanish territory. Toward that end, the state deployed a new military frontier that explicitly defined as it protected the boundaries of French national territory.

CROSSING THE BOUNDARY: SPANIARDS IN FRANCE

The disengagement of French and Spanish national territories during the Spanish political crises of the early nineteenth century was the context for a paradoxical transformation within local society in the Cerdanya: as the boundary was affirmed politically and nationally, and a new military frontier designed to defend the territorial distinction was put into place, the boundary line became more permeable socially and locally with the increasing imbrication of the people of the two Cerdanyas. This linkage was above all demographic, and its origins were political. In the same way that many French Cerdans became emigrants during the French Revolution, seeking refuge in Spain, so too did Spanish Cerdans—and Catalans more generally—cross the border into France during the political crises beginning under Napoleon.[37]

35. ACA Diversos Comandante Ingenieros no. 2413 (royal order of June 4, 1875).

36. For example, FMG AAT MR 1222, nos. 112–113: "Mémoire militaire sur la Cerdagne française" March 1836; and ibid., 1223, nos. 50–51: "Mémoire sur la reconnaissance militaire de la frontière d'Espagne," October 5, 1740.

37. Studies of Catalan and Spanish immigration in nineteenth-century France include M. Isern, "L'évolution de l'immigration catalane en Roussillon à partir du XIXe siècle"

By 1810, French civilian and military authorities in the Pyrénées-Orientales had become aware of the wave of emigrants from the Spanish Cerdaña and Catalonia. There were 200 Spaniards—or off-spring of Spaniards—in the French Cerdagne in May 1810, a proportion as high as 10 percent in the village of Ur. Of those living in the villages of Sallagosa and Osseja, 14 had come from the Spanish Cerdaña and 23 from other parts of Catalonia, and almost half had arrived since 1806.[38] Some had come to France to escape debt or military service. Others were more clearly political refugees and collaborators loyal to the French-supported government—*Josefins* or *Afrancesats*. But most were Catalans simply fleeing from the material dislocation and civil discord of the Old Regime crisis in nineteenth-century Catalonia.

The French government only slowly developed a consistent policy for the treatment of these refugees. In 1816, the police commissioner of Perpignan wrote to the prefect proposing the creation of a special bureau to monitor these "foreigners" who, he believed,

> took part, more or less, in the revolutionary events of their country. Pushed by their need, they run after fortunes, or run from the punishment of their crimes. Traces of their immorality are everywhere visible. What is more, the Spanish Revolution has thrown into this country a mass of families and individuals: a few of which are waiting for the remission of their so-called political failings in order to return to their patrie; others, without means of subsistence, only live here from brigandage, theft, or begging. The latter could become a sinister instrument under a malevolent influence.[39]

Each successive political crisis in Spain produced its contingent of immigrants. In November 1822, at the height of the civil war during the Constitutional Triennium, the Roussillon and French Cerdagne served as a royalist refugee camp. An estimated 4000 men, women, and children, "all without means of subsistence, and who receive help in bread and money," were living in the department; two months later their number had doubled.[40] The French government gave them some finan-

---

(Mémoire de maîtrise en espagnol, Université Paul Valéry, Montpellier, 1972); and more generally, R. Sanchez Mantero, *Liberales en exilio: La emigración política en Francia en la crisis del Antiguo régimen* (Madrid, 1975); and J. Rubio, *La emigración española a Francia* (Barcelona, 1984), 64–114.

38. ADPO Mnc 831/2 (tabulation of resident Spaniards by the sub-prefect, March 23, 1810); ADPO L 777 (Spaniards in the commune of Ur).

39. Quoted in Sacquer, "La frontière et la contrabande," 42.

40. FAN F(7) 11984 (prefect's report, May 31, 1822, reporting 580 Spanish refugees in the Cerdagne); in February 1823, the Prefecture reported over 8,000 refugees in the department; see also SAHN Estado, legajo 6228 (consular reports from Perpignan, November 1822–February 1823).

cial assistance; but it was only during the 1830s, which witnessed an important increase in the number and proportion of immigrants, that the French government developed a consistent policy. The idea was to provide "humanitarian" aid while at the same time instituting a system of surveillance and internment of political refugees, keeping them isolated in the French heartland at equal distance from the frontier and the political center.[41]

Such a policy was not, however, applied in the peculiar circumstances of the Cerdanya. During each appearance of the Carlists in the Spanish Cerdaña, peasants and townspeople from the district fled "with their most precious effects, and even grain," into French territory. During the first invasion of March 1836, the local administrations welcomed them and arranged for their food and lodging; yet by June 1836 the prefect of the Pyrénées-Orientales ordered the customs guards to accept only women and children, and to redirect the men to Llívia. Undertaken in cooperation with the Puigcerdà governor, who sought to stop the emigration of able-bodied men from the Spanish side, this policy attempted to limit the number of political refugees seeking French assistance. But the Spanish Cerdans were not sent into the refugee camps in the interior, for, as the prefect explained in 1835:

> There is an established tradition and explicit instructions which should not be forgotten. The inhabitants [of Puigcerdá and the Spanish Cerdaña] who withdraw into France in such circumstances are not at all political refugees: they come temporarily and for their security. We must welcome and protect them, without requiring their internment.[42]

But many of Spanish Cerdans and Catalans who took refuge during the 1830s in the French Cerdagne did not return across the boundary. The Carlist struggles were frequently cited by the immigrants as the foremost reason for leaving Catalonia. According to a police report of 1840, two-thirds of the sixty-four Spaniards living in La Tor de Carol—almost half with their families—had only arrived since 1833, the first year of the Carlist revolt. Politics may have pushed them into France, but economic factors help to account for why many immigrants chose to stay.

41. Sanchez Mantero, *Liberales en exilio,* esp. 37–63 and 120–141; M. Larrieu, "Les réfugiés espagnoles à Toulouse lors des guerres carlistes" (Diplôme d'études supérieures, Université de Toulouse-le-Mirail, 1971); and Sahlins, "Between France and Spain," 323–332.

42. ADPO Mnc 1921 (letter to sub-prefect, May 4, 1835); ADPO 1918/1, including reports of prefect to interior minister during "invasions" of March 1836); FMG AAT E(4) 43 (division commander to war minister, May 20, 1837).

Most of the Spaniards in the French Cerdagne were agricultural laborers or artisans; the lists consistently report their occupations as "field workers" (more than half in 1810), "laborers," and "domestic servants," with most of the others living by their trades—a tailor, two bakers, three shoemakers, a potter, a blacksmith, a few masons, and an innkeeper in 1840.[43] In the nineteenth-century French Cerdagne, as a century earlier, semiskilled artisans were few and Spanish immigrants found a ready market for their trades. In addition, the French Cerdagne depended heavily on this "foreign" labor force in the exploitation of their agricultural estates; as during the eighteenth century, the larger farms were labor intensive, requiring a year-round labor force as well as seasonal harvesters. And increasing numbers of lands tended to be rented out to "farmers," many of whom came from Catalonia. "Their sons born in the French Cerdagne," wrote the prefect in 1866, "generally avoid military service in Spain by staying at home." He insisted that to arrest these Spaniards as deserters "would deprive agriculture . . . of numerous workers, and would be highly detrimental to the general interest."[44]

Catalan immigrants in the French Cerdagne, moreover, often married native French women, as soundings in parish registers for several villages of the French Cerdagne reveal (see app. B and table B.4). One-fifth of the spouses who married in the village of Odello in the 1850s came from the Spanish Cerdaña, and that percentage was twice as great in the village of La Tor de Carol during the same period. In Càldegues, 44 percent of the marriages between 1853 and 1871 involved Spanish Cerdans, up from 32 percent in the late eighteenth century. In Santa Llocaya between 1853 and 1867, nearly a third of all spouses came from villages in the Spanish Cerdaña.[45]

By the 1860s, the French Cerdagne villages included a high percentage of Spanish inhabitants, both men and women (see tables 3 and 4, and map 4). Over 16 percent of the population in the French Cerdagne was born in Spain or of Spanish parents. In the border villages, the percentages reached 34 percent (Palau) and 36 percent (Hix, Santa Llocaya). In Palau and Err, two-thirds of the immigrants were from vil-

43. ADPO Mnc 1840; tabulations of the gendarmérie brigades of Osseja and La Tor de Carol, May 1840; and ADPO Mnc 831/2.

44. Quoted in Isern, "L'évolution de l'immigration," 84. As early as 1804, the prefect wrote that "many foreigners, especially Spaniards, establish themselves in this department without formal declarations; the mayors keep quiet since most are concerned with agriculture, which lacks workers": quoted in Brunet, *Le Roussillon*, 238.

45. See below, app. B.

TABLE THREE   FRENCH RESIDENTS IN THE SPANISH CERDAÑA, 1857

	Residents	French (% of Total)	Percentage of French Women	Percentage of French Men
Alp	648	12 (1.9)	1.7	1.6
Baronia de Bellver	2,012	7 (.3)	.2	.5
Bolvir	385	9 (2.3)	.5	4.5
Das and Tartera	348	26 (7.5)	4.6	10.3
Ellar	254	—	—	—
Ger	631	4 (.6)	—	1.3
Grus	239	—	—	—
Guils	394	37 (9.4)	7.6	11.1
Isobol	417	18 (4.3)	2.5	6.0
Llívia	968	37 (3.8)	2.1	5.6
Maranges	391	—	—	—
Montella	1,212	1 (.1)	—	.2
Aransa	520	—	—	—
Prats and Sampsor	268	—	—	—
Prullans	654	—	—	—
Puigcerdà	2,342	136 (5.8)	5.1	6.5
Queixans	274	—	—	—
Talltendre	390	1 (.2)	.5	—
Urtx and Vilar	544	—	—	—
Villallobent	271	6 (2.2)	1.5	2.9
Villec and Estana	406	—	—	—
TOTAL	13,568	294 (2.2)	1.6	3.0

lages in the Spanish Cerdaña, and the remainder from Catalonia. Two-thirds of the heads of household established in the two French villages had married French women from the commune or from other neighboring French communes.[46]

The contrast with the Spanish Cerdaña is striking. In 1857, only 2.2 percent of the Spanish Cerdaña inhabitants were either French-born or had French parents, and the border settlements of the district—especially the towns of Puigcerdà and Llívia—accounted for nearly three-quarters of the French inhabitants in the Spanish Cerdaña. By the

46. ADPO Mnc 2620 (1866 Census); ADPO Mnc 1925/1 (lists of resident Spaniards in Palau and Err, May 1866).

Territory and Identity during the Spanish Crises 217

TABLE FOUR   SPANISH RESIDENTS IN THE FRENCH CERDAGNE, 1866

	Residents	Spaniards (% of Total)	Percentage of Spanish Women	Percentage of Spanish Men
Angostrina	448	33 (7.4)	5.4	20
Càldegues	187	65 (34.8)	31.9	37.6
Dorres	300	32 (10.7)	12.3	8.9
Enveig	424	118 (27.8)	26.7	28.8
Err	738	99 (13.5)	12.8	14
Estavar	309	32 (10.4)	8.8	12
Hix/B-Mme	289	106 (36.7)	38.3	34.5
Nahuja	172	19 (11.0)	8.0	14.3
Odello	461	9 (2.0)	3.9	1.7
Osseja	1,053	232 (22.7)	18.4	26.3
Palau	310	105 (33.9)	36.3	32.9
Sta. Llocaya	128	46 (36.0)	34.7	36.7
Targasona	198	4 (2.0)	2.8	3.9
Ur and Flori	311	53 (17.0)	13.4	20.4
Vilanova	157	10 (6.4)	5.9	6.8
Carol Valley	1,375	182 (13.2)	9.3	11.5
TOTAL	7,725	1,267 (16.4)	14.8	17.3

midnineteenth century, then, the movement of the population across the border in the Cerdanya had definitively shifted direction along its north–south axis, even if out-migration from the French Cerdagne and Roussillon continued to be directed toward the principality.[47]

DRAWING THE BOUNDARIES OF NATIONAL IDENTITY

The settlement of so many Spaniards in the French village communities in the course of the nineteenth century is but one indication

47. *Censo de la población de España, 1857* (Madrid, 1858). Although Puigcerdà had a population of 136 French-born inhabitants (5.8 percent of its population in 1857), a disproportionate percentage had established themselves as permanent residents (*vecinos*)—some 49 of 505 household heads in 1866, or almost 10 percent: see AHP Populació; and AHP MA 1855–1860, fol. 55v (November 20, 1856, including complaints against French residents in Puigcerdà). On the continued patterns of out-migration toward Barcelona into the twentieth century, see Vila, *La Cerdanya*, 200–202.

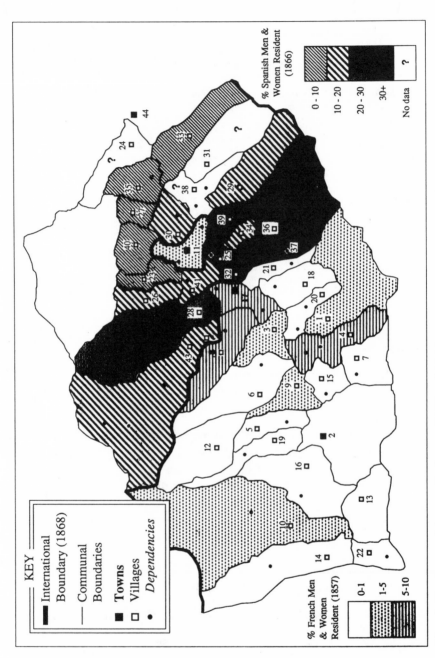

Map 4. Foreigners of the two Cerdanyas (1857 and 1866).

SPANISH CERDAÑA

1 Alp
2 **Bellver**
3 Bolvir
4 Das
   Mossol
   Sanabastre
5 Ellar
   Olopte
6 Ger
   Saga
7 Grus
   Riu
8 Guils
   Saneja
9 Isobol
10 Lles
   Coborriu
11 **Llívia**
   Sareja
   Gorguja
12 Meranges

13 Montella
14 Musser
   Aranser
15 Prats
   Sampsor
16 Prullans
   Ardovol
17 **Puigcerdà**
   Ventajola
   Rigolisa
18 Queixans
   Les Pereres
19 Talltendre
   Orden
20 Urtx
   Vilar
   Estoll
21 Villalobent
   Aja
22 Villec
   Estana

FRENCH CERDAGNE

23 Angostrina
24 Bolquera
25 Càldegues
   Onzès
26 Dorres
27 Egat
28 Enveig
29 Err
30 Estavar
   Bajanda
31 Eyna
32 Hix/Bourg-Madame
33 Llo
   Rohet
34 Nahuja
35 Odello
   Via
36 Osseja
   Valsebollera
37 Palau

38 Sallagosa
   Ro
   Vedrinyans
39 Santa Llocaya
40 Targasona
41 Ur
   Flori
42 Vilanova
43 Carol Valley
   Carol
   La Tor de Carol
   Porta
   Porté
   Ques
44 Mont-Louis

that the communities were changing demographically. In comparison to two generations earlier, they were becoming more and more "open" to all outsiders, not just Spaniards. Local endogamy rates fell dramatically: in Odello, only a third of the marriages between 1853 and 1862 involved both spouses born in the village, half the percentage of eighty years earlier. In Càldegues, endogamous marriage fell from 50 percent in 1770–1785 to 34 percent in 1853–1871. The increased percentage of foreign-born spouses in the French Cerdagne villages of the mid-nineteenth century was part of a more general opening up of the village communities to "outsiders."[48]

Yet the village communities tended with greater insistence to affirm their local boundaries and identities, despite this flow of personnel across social and territorial boundaries. There may even have been an inverse correlation between the opening up of the communities and the defense of local boundaries against encroachments by "outsiders." This seemingly paradoxical process has been described by the Norwegian anthropologist Fredrik Barth in his study of ethnic groups and boundaries:

> Boundaries persist despite a flow of personnel across them. In other words, categorical ethnic distinctions do not depend on an absence of mobility, contact, and information, but do entail social processes of exclusion and incorporation whereby discrete categories are maintained despite changing participation and membership. . . .[49]

A similar assessment could be made not only of village identities but of national identities more generally. Like Barth's definition of ethnicity, the expression of identities in the Cerdanya depended on the maintenance of boundaries of exclusion and inclusion. After the Revolution, the French communities tended to harden these boundaries. And by identifying a set of outsiders, they defined for themselves new identities as national citizens.

The definition and exclusion of outsiders could take place within a political context, as occurred during the Napoleonic episode in the Cerdanya. In September 1810, after French troops had occupied Puigcerdà in search of contraband goods, a ragged army of Catalan counter-revolutionaries chased the French out and continued into the French

48. Below, app. B; on a similar if slightly earlier "opening" of village communities in the Aran Valley during the nineteenth century, see M. Angels Sanllehy i Sabi, "L'afllament a les comunitats araneses," 35.

49. F. Barth, Introduction to Barth, ed., *Ethnic Groups and Boundaries: The Social Organization of Cultural Difference* (Boston, 1967), 10.

Cerdagne. "From the ramparts we sadly watch," wrote a dozen mayors and large landowners who had taken refuge in Mont-Louis, "the cloud of smoke produced by the general incineration of our houses." Over the course of two days, the miquelets destroyed some two million francs worth of property, a sum—with accompanying atrocities toward the local population—similar to that claimed by the Spanish victims of the French revolutionary occupation fifteen years before. The Catalan insurgents had passed with such ease into the French Cerdagne that the interior minister suspected the worst. The failure to organize the National Guard, to arm the population, and to defend French territory led the Paris government to inquire as to whether the French Cerdans were not actually siding with the Spanish counterrevolutionaries. In late November, a commission of inquiry dispelled such rumors:

> These brave folk, whose reputations have been blackened in the eyes of the government, have nothing but French sentiments. It was these sentiments which they dared to show among the Spaniards and which drew their hatred and madness: why else would all their villages have been burned when nothing was even taken away were it not to quell the rage and resentment of the invaders against the poor folk of the French Cerdagne?[50]

The "French sentiments" so graciously attributed to the French Cerdans, however, remained ambivalent. After all, if "the first virtue of a good Frenchmen," according to the interior minister, was "to participate in saving his country from foreign domination," the Cerdans showed little passion to save the Empire.[51] Far from offering a strong contingent of enlistments for the armies to defend the Empire, the district of Sallagosa produced a greater percentage of deserters than nearly any other district in the department—perhaps as large a proportion as anywhere in France. In February 1813, the prefect complained that the Cerdans were following "their ancient custom" and not presenting themselves for the draft. In May of that year, the police commissioner described the "opposition and spirit of disobedience which shows itself greater each day among the conscripts of the frontier communes." The recruiting captain Lemaire wrote to his colleague about his latest tour in the French Cerdagne that he "would rather have gone to hell."[52]

50. ADPO Mnc 831/1, including the "Rapport de M. Essanyé au Prefet . . . sur les pertes de l'invasion du 29 septembre 1810."

51. ADPO 3M(1) A/C 2, no. 88 (letter to prefect, December 28, 1813).

52. Quoted in Brunet, Le Roussillon, 367; on the failures of conscription in the Pyrénées-Orientales and especially the frontier districts, see Brunet, 297–388; and FAN F(7) 3608 (prefect's reports of 1811–1813).

Yet even though the French Cerdans did not participate in distant or proximate wars under Napoleon, their identity as Frenchmen did find expression in a more local context. Unwilling to enter the life of the nation, they brought the nation into their lives: their "Frenchness" appeared as a set of statements about the boundaries of their communities. For the French village communities, being French meant condemning the others—those few individuals who actually sided with the Spanish rebels, or those who actively manipulated their national identities in the service of their private interests.

The best example of the former is Joseph Ramonatxo, one of the few purchasers of national lands and a municipal councillor in Puigcerdà under the French during the Revolution. Destined for an ecclesiastical career by his clerical godfather, Ramonatxo abandoned the cloth and took up an account book. He owned a modest estate at Hix, but his great fortune was made in contraband trade. Despite his losses during the war and a bankruptcy of some importance, he nonetheless controlled—with two associates, Malibran and Autel—nearly all the contraband trade with Catalonia which passed through the Cerdanya.[53] But Ramonatxo was less a local hero than a tyrant who terrorized the local population. The French General Garreau called him the "protector of the frontier" and claimed that no one would speak against him.

But when the French commission met in November 1810 to inquire as to whether the French Cerdans had actually sided with the Spaniards during the invasion of September, many voices were raised against Ramonatxo. In fact, the French Cerdans universally denounced Ramonatxo for his behavior during the September invasion. His house had been the only one untouched, guarded and protected by the "insurgents." Jean Cot, one of his neighbors, saw the rebel leaders eating and drinking with Ramonatxo as the French Cerdagne burned. Ramonatxo proclaimed his innocence, stating he was "a good and loyal Frenchman, with a total devotion to my patrie, and completely obedient to the laws of my government." And although his government never convicted him, public opinion in the Cerdanya had universally condemned him in 1812. The condemnation was confirmed six years

53. On Ramonatxo's earlier activities, see above, chap. 5 sec. 1. Police files of Ramonatxo's background and activities, from which this information is drawn, can be found in ADPO Mnc 822. Both Malibran and Vedrinyans were from local families but, like Ramonatxo, had traveled widely in their search for profit all over Europe, as far away as Moscow. Both maintained houses in Puigcerdà, and Ramonatxo's daughter married Vedrinyans' son.

later, when the municipal council and the principal inhabitants of Bourg-Madame denounced Ramonatxo for usurping communal lands —of using the public good for his private interest.[54]

The Cerdans thus affirmed their French identities by ostracizing those, like Ramonatxo, who had supported the Spanish revolutionaries. In doing so, they gave expression to a sense of the (French) public good as distinct from the individual private interests of individuals. The French Revolution hardened the boundaries of communities, making it more difficult for individuals to choose their identities on the basis of their private interests. Indeed, the French Cerdans ended up excluding from their communities those who blatantly manipulated their national identities, nobles such as Don Vicens de Travy and Don Bernandino de Mir.

Both men held estates in the French Cerdagne, but their eighteenth-century forebears had taken differing stances on the national question. The Travy family of Sallagosa had a long history of collaborating with the French crown, while the Mir family, whose ancestral lands were in the French village of Hix, nonetheless identified themselves as Spaniards—and gained exemption from the capitation tax.[55] With the outbreak of the Revolution, both Bernandino de Mir and Vicens de Travy were drawn toward Spain. Travy, who had pursued his legal education in Catalonia and was to later join the Real Audiència in Barcelona, became a naturalized Spanish citizen in 1795; Mir, a youngest son who had not inherited property in the French Cerdagne, had moved to the Seu d'Urgell, but he moved back to the family estates in Hix and Bajanda when his brother died without heirs.[56]

Though the two nobles acted differently during the Napoleonic occupation, both tried to claim a French nationality. They did so in part to regain possession of their estates in the French Cerdagne sequestered in 1808 with the outbreak of Spanish–French hostilities. Until 1811, Travy held an official position in the revolutionary junta in Puigcerdà and acted as public attorney for many villages in the Spanish Cerdaña. When the French annexed the Spanish Cerdaña, the minister of police ordered Travy's arrest, causing him to flee to the interior of Catalonia.

54. ADPO Mnc 831/1 (letter to prefect, October 26, 1810); ADPO O Bourg-Madame, no. 8 (sub-prefect to prefect, September 6, 1819).

55. For background on these families, see Lazerme, *Noblesa catalana;* on Mir's exemption, see the dossier in ADPO 864, esp. the letter of Don Antoine de Mir to intendant, May 1743.

56. The Real Audiència recognized all the sons of Don Antonio Travy as Spaniards, "by direct male line, despite having been born circumstantially in the kingdom of France" (FAN F[7] 5497). On Mir, see his petition in ADPO O La Tor de Carol, no. 8D.

Travy explained that, notwithstanding his having gained employ in the new intendancy of Upper Catalonia, he was "drawn from [his] apathy to take a position" and, cutting off all relations with family and friends, requested French citizenship.

The inhabitants of the French Cerdagne raised an outcry. "The public rumor is that Travy only wants to be French so that his goods confiscated in France will be released," according to the mayor of Osseja. Another mayor described Travy as

> an egoist: his party spirit is neither for the French nor for the Spanish, but directs him according to the profit he can extract from the events. . . . He is faithful only to his egoism.

In June 1812, the inhabitants of the Osseja Valley accused Travy of sending around a false petition demanding the division of the valley into two communes—a perfect metaphor for local unity against an outsider's intrigues.[57]

Travy had tried to play the game of selective advantage, manipulating his nationality according to his interests. Mir, too, sought to affirm his French nationality, arguing the principle of *jus soli*, that he was born on French territory. In 1810, however, a number of unnamed individuals denounced Mir for serving in official capacities at Puigcerdà and for being "a revolutionary entirely devoted to the Spanish government against France." Mir admitted to holding municipal office before the war broke out in 1808 but claimed that he became a private individual afterward and was even excluded "as a nativeborn Frenchman" from an assembly in Puigcerdà which met to send deputies to the League of Ferdinand VII. To the prefect, he wrote,

> It would be most unfortunate to be recognized neither in France nor in Spain, and since I am not recognized in Spain, at least I should be in France, where both I and my father were born, and I should be allowed to enjoy the rights inherent in my status.

Mir elaborated upon his loyalty and devotion to the French government, but it was his birth certificate that sealed the case, as his lands were returned and he was recognized as French.[58]

If the case histories of Ramonatxo, Travy, and Mir suggest the possibilities of defining nationality in the borderland under Napoleon, they also reveal the boundaries within which manipulations of nationality could take place. Each man took a different path to the status of

57. ADPO 3M(1) 20/B 2 (dossier on Travy, esp. nos. 29, 104, 111, and 237).
58. ADPO 0 La Tor de Carol, no. 8D.

outsider. But in each case, the other members of the Cerdagne communities affirmed their identities in opposition. Public opinion denounced Ramonatxo, a powerful supporter of the Spanish cause. The village communities excluded Travy and Mir—both from previously privileged families—who had sought to claim unjustly their status as Frenchmen. Just as the Revolution had marked the limits for the state's construction of national boundaries under the Empire, so too had the experience of the Revolution hardened the boundaries of identity and loyalty in the borderland.

Truth be told, the French state of the early nineteenth century was less concerned about the manipulation of identity than were the local communities. But as the system developed of policing, assisting, and incarcerating political immigrants in the borderland, both states tended to increase their surveillance of the uses and abuses of citizenship. According to the civil governor of Barcelona in 1835,

> The right of nations has never allowed a foreigner to enjoy promiscuously the quality of being a citizen of his own and of another nation where he is living temporarily. Neither should the public order permit a stranger to change his nationality when his interest so dictates, according to the principle of utility. . . . This abuse is prejudicial to both nations; it interrupts the free flow of business and distracts the authorities of both countries from other, more pressing and arduous matters.[59]

The states became especially concerned with the growing number of abuses stemming from their laws of universal conscription. The entire district of the French Cerdagne, according to the prefect in 1813, "wants to follow attentively French or Spanish laws, according to which ones will exempt them from personal obligations."[60] The claim of "foreign" ancestry was a common one among those who managed to escape their military service, especially in France. The Cerdans also sought to affirm one citizenship or another in order to escape from national taxes, especially those levied for "extraordinary purposes." For example, two French industrialists established in Puigcerdà in 1835 claimed French citizenship while refusing to pay a tax for urban militias: they were imprisoned then forced to leave, while the prefect raged that it was "against the rights of nations to subject Frenchmen to pay a purely national tax."[61]

59. ADPO Mnc 1883/2 (letter to French consul in Barcelona, April 23, 1835).
60. Quoted in Isern, "L'évolution de l'immigration catalane," 48.
61. AHP Correspondencia (letter to Puigcerdà governor, February 11, 1835); see also the examples of "mistreatment" of French citizens listed in ADPO Mnc 3818/1 (subprefect to prefect, March 16, 1831).

The case of Marc François Garreta (or Marco Francisco Garreta, or Marc Francesc Garreta, depending upon whom he petitioned) illustrates the permeability of the boundaries of nationality. Born in the French village of La Tor de Carol in 1771, Garreta moved to Llívia after the revolutionary war, where he married a local woman, acquired property, and came to hold political office in the late 1820s and early 1830s. Then came rumblings of civil war in Spain, and Garreta decided to move back to France. He may have been pushed by the increased tax levies or by fears about his physical safety as a prominent citizen. In November 1834 he requested confirmation of his French nationality:

> I remained so long in Llívia because of business affairs and for no other reason; as a result, the Spanish authorities think that I am a Spaniard. I was elected mayor, and I accepted because I was ignorant that it was against French law to hold office in Spain.

In June 1835, the French minister of justice granted Garreta his petition, returning to him the "quality and rights of a Frenchman," which he had lost by holding public office in Spain.[62]

But Garreta continued to live in both France and Spain. He was elected to the municipal council of Santa Llocaya in 1837, but less than a year later the town of Llívia confiscated some of his goods and fined him, claiming

> that he had not returned to France to live as a Frenchman, but continued to live in this town behaving the same way and exercising the same functions as always, and he only left Llívia when the rebels invaded the Cerdaña, in order to flee from the hardship and taxes which good Spaniards had to suffer.[63]

Such manipulations proved disturbing to the Spanish authorities, who thought Garreta was typical of the Cerdans. "The Cerdaña abounds," wrote the civil governor of Barcelona, "in men who could be called amphibious in political society, because they will just as quickly wish to be considered French as Spanish"—complaints echoed by the French authorities as well. But the local response to Garreta's behavior suggests otherwise. In point of fact, the designation of "amphibious"— neither Spanish nor French—came from the municipal councillors of Llívia, who thereby denounced Garreta's "refined egoism."[64]

62. ADPO Mnc 1883/2 (dossier on Garreta, esp. the decree of June 17, 1835).
63. AMLL (municipality of Llívia to Puigcerdà governor, July 16, 1838).
64. ADPO Mnc 1883/2 (civil governor to French consul, April 23, 1835); AHP Correspondencia (municipality of Llívia to civil governor, March 18, 1835).

Already in 1820, Garreta had succeeded in alienating the town council by diverting water from the town's canal. In March 1835, as Garreta's desire to regain French citizenship became known, the town council deliberately imposed upon him the official responsibility of levying the *quinta* or military draft for that year. According to the minutes of the town council meeting of March 17, Garreta started to scream and shout, calling the councillors "shits" and "rogues," insulting them individually and collectively. The next day, the councilmen attempted to arrest him, but he shut himself in his house and later fled out the back window. Nor did the antagonism between Garreta and the community of Llívia end with that incident. Three years later, Garreta was again living in Llívia and entered into further altercations over his administration of the annuities of the Marqués de Vilana, and over his refusal to pay extraordinary local taxes raised for the town's defense against the Carlists. And in 1842, his son Paul took up the role of outsider, refusing to pay a recruitment tax while claiming his French nationality.[65]

The case of Garreta suggests that it was the local communities, much more than the national states, which decried those "political amphibians" who manipulated their nationalities in the service of their interests. The French communities tended to develop a sense of nationality by drawing local boundaries of inclusion, excluding those they saw as "selfish" and "egoistical." By the 1830s, when Spain instituted its national military service, the Spanish communities began to do the same.[66] In both countries, the burden of supplying another village son to replace the one who was drafted and escaped fell back onto the village communities. They saw their "public interest" threatened by "those who do not fulfill their municipal obligations, but nonetheless want to enjoy all the advantages."[67] This attitude of a local mayor had many echoes in the mid-nineteenth-century communities. The national governments tended to support those who petitioned for citizenship, taking the rhetorical claims of citizenship as solidly grounded. But the communities thought otherwise and, by identifying these outsiders, the communities affirmed their own national identities.

65. AMLL MA 1789–1823, fol. 373 (June 28, 1820); AMLL MA 1832–1843, fols. 119–122 (March 17–18, 1835); AMLL I.3.3.1 (October 5, 1838); and ADPO Mnc 1883/2 (French consul to prefect, September 27, 1842).

66. On military service in Spain beginning with the 1837 law, see N. Sales i Bohigas, "Servicio militar y sociedad en la España del siglo XIX," in *Sobre esclavos, reclutas, y mercaderes de quintos* (Barcelona, 1974), 207–277.

67. AMLL VII.I (municipality of Llívia to civil governor, September 23, 1842).

It is important to underline that such local affirmations of national identity took shape according to local terms and frequently contradicted the definitions of national sentiment advanced by political and military officials. When, for example, the Spanish Constitutional General Espoy y Mina and his troops crossed into French territory in June 1823, many French Cerdans fêted the Spanish troops, leading the sub-prefect of Prades to denounce "the sentiment of patrie which no longer exists among the inhabitants of La Tor de Carol devoted to the Spanish revolutionary cause." And during the 1830s, the French police and military reports consistently accuse the French Cerdans as "taken to serving the Spanish Carlists." Police reports from both episodes suggest some of the motives of the French Cerdans, most notably the profit that they drew from supplying Spanish causes with contraband goods and weapons.[68] But the profit motive alone cannot account for the reception and appropriation of Spanish political causes in the French Cerdagne. Rather, the permeability of the boundary and the proximity of Spanish territory were such that the French Cerdans became more engaged in the national political events of Spain than in those of France.

Thus one can search in vain for indices of political "assimilation" and "integration" in the Cerdans' reactions to the political revolutions of 1830 and 1848 in France. Following Maurice Agulhon's seminal study of the Var department, social historians have been concerned with the "descent" of politics and its reception in peasant society during the Second Republic, stressing either the ("traditional") local or the ("modern") national content of political participation.[69] By these standards, the Cerdans were barely even primitive rebels. The French Cerdans largely ignored the electoral politics of the Second Republic; instead, taking advantage of the breakdown of political authority, the peasant communities chased out the customs officers, as they had during the Revolution, but also, led by an ex-forest guard, took over the state forests. Indeed, the municipal council of Osseja went so far as to

68. ADPO Mnc 1832/3 (sub-prefect to district attorney, July 12, 1823); ADPO Mnc 3818/5 (sub-prefect to prefect, November 6, 1831); and ADPO Mnc 1921 (prefect to sub-prefect, January 22, 1835).

69. M. Agulhon, *The Republic in the Village: The People of the Var from the French Revolution to the Second Republic*, trans. J. Lloyd (Cambridge, 1982 [1970]); E. Berenson, *Populist Religion and Left-Wing Politics in France, 1830–1852* (Princeton, 1985); and, for the argument that national political parties were simply masks of local concerns, see the two articles by E. Weber, "The Second Republic, Politics, and the Peasant," *French Historical Studies* 11 (1980): 521–550; and "Comment la Politique Vint aux Paysans: A Second Look at Peasant Politicization," *American Historical Review* 87 (1982): 357–389.

claim that the February Revolution "meant that communes could now take control of their woods."[70]

Yet the refusal of the Cerdans to partake in the formation of a national, political movement in 1848 has less to do with traditionalism and backwardness than with difference: the Cerdans invented a set of national identities, but they did so in the context of Spanish political life, and from their local experiences and sense of place. They affirmed their respective nationalities by defining their communal identities: by drawing boundaries of inclusion and exclusion, the village communities of the French Cerdagne isolated and ostracized individuals who crossed too many boundaries. But in addition, and more dramatically, disputes over territorial boundaries between French and Spanish village communities provided the mechanism that affirmed the local basis of national identity.

COMMUNAL STRUGGLES AND NATIONAL IDENTITIES

The self-definition of national identity in the Cerdanya took shape by the exclusion of outsiders from the village communities—people who indiscriminately manipulated identities in the service of their private interests. But national identity also found expression as collective statements of opposition between communities divided by the national boundary. In the eighteenth century, French village communities had identified their counterparts as "Spaniards" while not yet describing their own nationalities. After the French Revolution, communities in both France and Spain underscored their own identities as French and Spanish.

Intercommunal struggles reemerged in the years after Napoleon, just as the early eighteenth century had witnessed a resurgence of local

---

70. ADPO U 3071 (judicial dossier on 1848 in the French Cerdagne); and ADPO 0 La Tor de Carol, 1848–1850, including a denunciation of the recently dismissed forest guard, Pierre Olive, by the municipal council of La Tor de Carol, March 10, 1848. Olive seems to have been supported by the inhabitants of the nearby village of Carol. P. McPhee, "The Seed-Time of the Republic: Society and Politics in the Pyrénées-Orientales," (Ph.D. dissertation, University of Melbourne, 1977), argues on the basis of such incidents that the Second Republic brought only "family factionalism" in the Cerdagne, due to the "isolation of small and often dispersed communities, poor communications, widespread if insufficient land ownership, stagnant polycultural economy, absence of 'urban' ownership . . . " (532 et seq.). On the questions of rights to firewood and pasture in the midnineteenth century which surfaced in 1848, see the report by the Departmental Council in ADPO 1 N 20 (procès-verbal of the séance of December 2, 1848); and Assier, *Le peuple et la loi,* passim.

conflicts following the dismantling of the military frontier. Between the 1820s and the 1860s, communal struggles tended to increase in duration and intensity regardless of the boundary. There were sound ecological reasons why such conflicts tended to increase toward the middle decades of the nineteenth century. The growing population of the valley, as it neared its maximum density before out-migration, put added strain on essentially limited resources—pastures, waters, and collective grazing rights to meadows and fields. Throughout the Cerdanya, village communities sought to restrict the presence of "foreign" cattle on their territories, excluding the herds of nonresident landowners. They further decried the "usurpations" of communal lands by other communities and the diversions of waters from their territories. Beginning in the 1820s, communities entered into disputes over the construction and control of irrigation canals, which transformed arid into arable land. Although there is no statistical evidence of increased agricultural yields in the first decades of the nineteenth century, there are signs of an intensified interest in water rights, suggesting a sign of increased agricultural production on both sides of the valley and a movement—more visible in the French than the Spanish Cerdanya—away from stockraising.[71]

Throughout the Cerdanya, irrespective of the political division, conflicts among communities tended to increase in the course of the nineteenth century. Yet those disputes between communities divided by the political boundary came to differ significantly from quarrels among communities in each of the Cerdanyas. For one, disputes between French and Spanish communities almost inevitably evoked the presence of one or both states. Part of the conflict may even have been with the state, as was the case of Llívia's claims to harvest wood in a forest called the "Bac" which lay within the territory of Bolquera, in the French Cerdagne. The Old Regime government had long prohibited the export of wood, yet Llívia managed to have its claim validated by the Sovereign Council of Roussillon in 1723, and again by First Minister Necker himself in the fall of 1788. After the Revolution, Llívia

71. Sahlins, "Between France and Spain," 805–808, and app. 5; and Assier-Andrieu, *Le peuple et la loi*, 158–168. Increases in disputes and seizures reflected in part the growth in number of indigent peasants; as Pau Cot Verdaguer explained in his "Notes folklorqiues pirinenques" (1928): "There are always men who are ready to take advantage of some mistake, to seize heads of cattle, and to receive greater or lesser quantities of money; there are many who live from these seizures . . . so that friendship and harmonious peace are not always to be found between the closest neighboring communities," in BC MS 1535.

found its usufruct "right" classified as a "privilege." The town argued instead that the right "has become one of its properties," and the claim was upheld by the central government despite the opposition of the customs director at Ax.[72]

Llívia's struggle with the French state was but part of its problems, since it also had to contend with the village of Bolquera where the forest was found. In the fall of 1816, peasants of Bolquera seized several mule-loads of wood gathered by inhabitants of Llívia. In response the mayor of the Spanish town appealed to the French departmental administration in Perpignan:

> The Treaty of the Pyrenees divided the Cerdanya into two Cerdanyas, French and Spanish, and isolated the town [of Llívia] by surrounding it on all sides with French territory; it also gave to France the commune of Bolquera and its territory. Before the division, the two peoples were united [and] shared the same sentiments. After the division, they looked upon each other as enemies, and Bolquera tried to take away the previously recognized rights of Llívia. Frequent disputes followed, and Bolquera was strengthened by its emerging domination of Llívia, with which it used to be closely united. Bolquera then used its ties to the forest and customs administrations, and succeeded in taking away from Llívia the enjoyment of its most legitimate claims.[73]

The Treaty of the Pyrenees did not itself cause the struggles, as Llívia and Bolquera's rights were already at issue in 1574, and probably much earlier. But it is revealing that the inhabitants of Llívia said that it did. By attributing its enmity with Bolquera to the Treaty of 1660, Llívia offered a view of the past colored by a sense of the national difference. This sentiment of difference continued to inform both communities in their struggles over the Bac de Bolquera, a struggle not resolved until 1832, when a French appellate court awarded to Llívia one-third of the surface of the Bac as its full and unqualified property.[74]

As the mayor of Llívia pointed out, the village communities tended to "use the ties" with their national administrations in their struggles against each other. What distinguished communal conflicts across the boundary from ordinary village disputes was the existence of two sepa-

72. ACA RA, vol. 137, fols. 41–42 (letter of November 24, 1724); SAHN Estado, libro 6673d, fols. 160–185 (correspondence and memoirs of 1788); ADPO Mnc 3818/3 (dossier, including letter of inhabitants of Llívia to French interior minister, May 1800; and attorney of Llívia to prefect, July 2, 1800).

73. ADPO Mnc 3818/3 (petition of Llívia, November 30, 1816).

74. On the earlier disputes, see ADPO C 2052 (dossier of renewed struggles in 1761–1762; for the nineteenth-century struggles and their resolution, see the dossier in SAMRE TN 221–222 (copies of court decisions concerning the Bac de Bolquera, 1817–1911).

rate authorities to which appeal might be made. Representatives of the village communities empowered themselves by locating themselves in their respective nations—literally, by seeking out their state's authority; and rhetorically, by identifying themselves as members of a national community, as Frenchmen or Spaniards.

But such appeals to the power and authority of the nation were rarely successful. As the French customs director wrote to his finance minister in 1827,

> Contiguous communities often have interests to fight out. When they are not under the same superior authority, their rivalries cannot be contained, their territorial encroachments cannot be stopped, and their usurpations of pasturing rights cannot be impeded. When local demands for protection remain without effect, the communities resort to armed struggle.[75]

At the local level, political authorities often attempted to quell these local "armed struggles." For example, renewed struggles over the Raour River led in June 1817 to a joint French–Spanish commission of inquiry headed by the mayor of Puigcerdà and an army captain at Mont-Louis. With a copy of the 1750 accord in hand, they proceeded to the site of the disputes, where they found it

> impossible to finish verifying the boundaries, because of the obstinancy of the Dominican monks of Puigcerdà, the principle riverfront proprietors, who are demanding that the French proprietors push their walls back. . . . Since 1750, the river has shifted course a great deal, and the localities have changed completely; there is a great and urgent need to draw up a new convention taking into account the actual situation.

But the central government in Paris was unwilling to take up the issue, preferring to wait for the "general demarcation" of the frontier.[76] Nor was that immediately likely, since the relations between the two governments in the early 1820s decidedly worsened.

What further distinguished village struggles across the boundary from communal disputes within each Cerdanya was the intersection of state and local territories. In the early 1820s, the French state's definition and defense of national territory during the Constitutional Triennium in Catalonia were matched by the local communities' po-

---

75. FAMRE Limites, vol. 461, no. 31 (January 27, 1827).

76. ADPO Mnc 3817: "Procès-verbal de la vérification des limites," June 1, 1817; FAN F (2)I 447 (prefect to interior minister, June 19, 1817); FAMRE Limites, vol. 459, no. 119 (response of interior minister, July 10, 1817); and SAMRE TN 221–222: "leg. B: Raour, 1820–6," esp. the "Memoria sobre el establecimiento del Reur," May 13, 1820.

liticization and defense of claims to their local territories. In 1821, the sanitary cordon had politicized the opposition of national territories for the two states; it simultaneously politicized an ongoing set of local struggles between French and Spanish communities in the borderland.

Village communities in Spain and France adopted and developed a rhetoric of nationalism, identifying themselves with their respective nations against both soldiers and peasants of the other. The sanitary cordon reproduced the uncertainties of communal limits: it thereby overlaid and transformed an ongoing process of local conflicts in the borderland. The placement of the quarantine houses on the boundary sparked the extant disputes between French and Spanish communities over their boundaries and rights.[77] Llívia protested that the French communities "took advantage of their soldiers to usurp lands in contention." In a letter of February 1822, the town council of Llívia sent a petition to the Puigcerdà governor:

> The French are beginning to place border markers on the cordon and dividing line between this commune and the ten French communes around it. . . . They are planting these markers at their discretion, without asking this town to participate, and what is worse, violating in the process the dividing line of the two kingdoms. . . . This town has been long enough under French claims. Now we can be promised the support of our own armies, now we are in a happy period that our constantly buried rights are reborn. . . . This town has always sought to protect the property rights of its fellow citizens and to uphold the integrity of Spanish territory.[78]

Disputes over boundaries helped forge the ideology of national difference at the political centers of power. Already in January 1820, Colonel Diaz had addressed the new Cortes in Madrid, urging Spain to defend "the Spanish citizens who tremble under the tyranny and despotism" of the French in the Cerdanya. Yet the most elaborate expressions of nationalism occurred at the extreme periphery, where the opposition of France and Spain was most salient and a vituperative militarism dominated local discourse. As the municipality of Llívia wrote in May 1822,

> The Spanish government should announce that its Nation will not let itself be subjugated by any other one, and that it seriously guards its rights. Does not the French Nation recall that the Spaniards are valiant warriors, that

77. SAHN Estado, libro 672d, fol. 303 (French prefect's report of November 16, 1821); AMLL VII.5 (Angostrina and Llívia, November 1821; Llívia and Càldegues, May 1822); ADPO Mnc 1722/3 (Guils and Carol, 1821–1822).
78. SAHN Estado, libro 672, fol. 331 (February 8, 1822).

the Spaniards beat back the wings of the eagle [i.e., Napoleon]? . . . Have the French forgotten that the Spaniards created the most powerful army of warriors, that our Nation is now resplendent, and that it shall always know triumph despite treachery and intrigue?

The Spanish nation did rise up—at least that part of it in the commune of Guils, where the mayor and Spanish troops, supported by the peasants, took the municipal council of La Tor de Carol prisoner for a day.[79] But the Spanish nation—the political nation—did not respond. The logic of events in Paris and distant European capitals led to the royalist Congress of Verona in 1822 and to French intervention in Spain.

The experience of the 1823 invasion through the Cerdanya sharpened rivalries whose origins lay in the ongoing struggles over limited ecological resources. Puigcerdà, in particular, had renewed its struggles with the French communities of La Tor de Carol and Enveig, which the Spanish town claimed were cutting off its water supply provided by a canal that crossed French territory. The struggle had been ongoing since the Revolution; under the Empire, officials and engineers of the neighboring departments had tried to draw up binding regulations.[80] But nothing had been resolved, and in 1825, local animosity between Puigcerdà and La Tor de Carol found expression in ritual confrontations of local youth groups. On May 22, during the local feast of La Tor de Carol, the youths of the French village had exacted an unusually high price from their Puigcerdà counterparts for the latter's right to dance in the village square.[81] On July 3, during the Feast of the Rosary, the major festival of the Spanish town—and of the valley as a whole—a similar encounter between groups of young male musicians had greater repercussions.

In the narrow streets of Puigcerdà, an angry crowd of several hun-

79. SAMRE TN 222–223 (memoir sent to the Cortes, August 11, 1820); see also the "Memoria sobre las limites de la Cerdaña española," by J. B. de Ponsich, sent by the captain-general of Catalonia to the first secretary of state, November 12, 1820; AMLL VII.5 (letter of May 26, 1822); see also the municipality's letter to the military commander at Puigcerdà (October 3, 1821); and on the events of March 1822, see ADPO Mnc 1722/3 (sub-prefect to prefect, March 25, 1822); SAHN Estado, libro 673d, fols. 240–249 and 250–256, esp. the account by the municipality of Llívia, March 26, 1822; and ibid., libro 674d, fols. 65–67 (French Foreign Minister Montmorency to Spanish ambassador, April 13, 1822).

80. ADPO S XIV 112 (dossier on disputes over the canal, 1806–1817, esp. the petition of the municipality of Puigcerdà, July 17, 1817); ACA GI XLV, no. 1 (Segre prefect to intendant of Upper Catalonia, March 13, 1813); and ACA GI XXIII, no. 5 (dossier on canal repares, January 1814); and AHP Asequia.

81. AHP MA 1825, fols. 38 and 41 (June 1825).

dred men and women began to stone the musicians, then turned on the French villagers gathered for the feast. The town's gates were closed and the crowd, armed with clubs, screamed—in Catalan—"Kill the gavatxos, they have ruled Spain for too long."[82] Identifying the inhabitants of the French Cerdagne as gavatxos, the derogatory term for a Frenchman, meant pushing back the ethnic boundary of French and Catalan in order to match the political division of the valley.

Was the revolt merely another instance of communal rivalry, as the intendant of Catalonia concluded, or was there more to it than that, as a French commission of inquiry sought to prove? Certainly local communities had a propensity to fight during village festivals, and police reports frequently commented on the "noisy affairs" that often ended in blows. The principal protagonists, young men, represented an age group that customarily defended and defined ritually the distinctive identity of the village community.[83] At the same time, the political context of an ecological struggle gave to the local conflict a national significance: not only did officials read the event as partaking of politics and nationalism, but so too did the participants themselves, claiming that "national hatred ruled this ugly scene." Those who thought otherwise, added a seasoned brigadier long stationed at Bourg-Madame, were recent arrivals, "people who do not understand the district."[84]

Though officials and administrators soon forgot about the events of 1825, the Cerdans themselves did not; the Feast of the Rosary acquired a certain mythology and was used later on to elaborate upon national differences. In 1866, the principal proprietors of La Tor de Carol and Enveig protested to the Bayonne commission that had awarded the town of Puigcerdà full property ownership of the canal. The French landowners claimed a property right, relegating Puigcerdà's own claim to the status of usufruct right. The reason for their enmity, the Frenchmen argued, lay in the Treaty of the Pyrenees:

After the Treaty of 1660 had united the communes of La Tor de Carol and Enveig to France, national rivalries gave birth to continuous struggles be-

82. ADPO Mnc 1943: "Rapport sur le rixe du 3 juillet 1825," anon., n.d., ca. August 1825; letter of mayor of Carol to sub-prefect, July 4, 1825).
83. FAN F(7) 9496 (prefect's report of March 24, 1819); ibid. 8480 (annual report of 1802); and P. Sahlins, "The 'War of the Demoiselles in Ariège, France, 1829–1867" (Honors thesis, Harvard College, 1980), chaps. 3 and 6.
84. ADPO Mnc 1943 (Gelaro, replacing the sub-prefect, to prefect, July 14, 1825); the commissioner in charge of the administrative inquiry, the mayor of Mont-Louis, was of the opinion that "political principles" and "sentiments of national hatred" were at issue, but the principal causes lay in "imprudence," "exasperation," and the "effervescence of passions."

tween those [communities] which enjoyed use-rights between the divided countries. These struggles were resolved by specific conventions of which tradition alone is the witness. The accords permitted Puigcerdà the use of the canal, while imposing on the town two obligations.

In exchange for its use of the water, Puigcerdà was obliged to donate wood each year to the Mont-Louis hospital and to recognize the prerogatives of the young men of La Tor de Carol to use freely the town square of Puigcerdà during the Spanish town's Rosary festival. Linking their festive rite with their juridical rights, the French proprietors argued that the former was "drowned one day in the blood of the young men of La Tor de Carol."[85]

The revolt of the two Cerdanyas in 1825 and its reinterpretation a half-century later speak to an intersection between local and state conceptions of territory and identity. In the early 1820s, the local adoption of national identities coincided with the territorial definition of national sovereignty: national territory was politicized at once from above and from below. Viewed from the center, France's territory emerged from the opposition to Spain and the attempt to limit the spread of Spain's domestic strife into France. Viewed from the periphery, French territory emerged as it reproduced and exacerbated village conflicts over boundaries and rights in the borderland. But such a historical coincidence was rare; indeed, not until the delimitation conferences of the 1860s did both state and local society intersect in their claims to define a national boundary line. And then it was too late, since local claims were more entrenched and local representatives less willing to give up territory than were either of the two states.

Instead of repressing the disorder occasioned by local disputes in the nineteenth century, French and Spanish authorities tended to encourage it both indirectly and directly. Indirectly, their own contestations over disputed contraband seizures in the borderland were grafted on to local disputes over boundaries and rights. To take but a single example, a seizure by French customs guards in 1845 set of another round in the increasingly bitter conflict between Angostrina and Llívia.[86] Directly, customs officers, the police force, and even the soldiers stationed in the border villages joined together with local communities to defend their

85. FAMRE CDP 11 (petition of December 31, 1867). The petition originally read "Carnival of Puigcerdà," which was crossed out and annotated "Fête majeur."

86. ADPO Mnc 1951/3 (police commissioner at Bourg-Madame to subprefect, March 18, 1845); and ADPO Mnc 1924/1 (mayor of Angostrina to sub-prefect, December 6, 1845).

territories. The customary practices of *penyores*, the seizures of "foreign" cattle in defense of local territory and usufruct rights, became forms of rivalry engaged in by official representatives of state political authority. For example, the border war that erupted between Guils and La Tor de Carol in August of 1848 involved seizures and counterseizures of hundreds of head of sheep, with the inhabitants of each village backed by their own customs guards and constabularies.[87] By collapsing the distinction of local and national territory, the village communities also collapsed the distinction between themselves and their states: identifying themselves as national citizens, they defined themselves in opposition to each other. It only remained for them to define the territorial boundaries of their national identities.

87. SAMRE TN 221–222 (correspondence of 1848–1849).

7
The Treaties of Bayonne and the Delimitation of the Boundary

The treaties of Paris in 1814 and 1815 restored France to its boundaries of 1792 and mandated that France and its neighbors formally "settle their limits"—delimit and demarcate their respective territories—thus completing the work begun by the late-eighteenth-century "treaties of limits." The Restoration government in France quickly formed two commissions in 1816 to delimit the northern frontier (from Dunkerque to the Rhine) and the eastern frontier (from the Rhine to the Mediterranean.)[1] The Northern Commission completed its work in 1820, when France and the Low Countries signed a delimitation treaty for their boundary, demarcated between 1821 and 1823. The Eastern Commission took longer but, by 1831, France had agreed on the demarcation of its boundary line with Prussia, Bavaria, and Baden. After France's annexation of Savoy, the French–Italian boundary was delimited in 1860–1861. Except for the loss and return of Alsace and Lorraine, the delimitation and demarcation of France's northern and eastern boundaries have remained, despite minor modifications, unchanged until today.[2]

Yet the French–Spanish boundary was all but ignored. The treaties of May 30 and July 20, 1814, ordered its delimitation, and the French

1. FAMRE Limites, vol. 7: "Note pour son excellence sur la démarcation des frontières," n.d., ca. 1826; and ibid., vol. 446: "Notice sur l'exécution . . . du traité de Paix," August 1, 1814.
2. C. Rousseau, *Les frontières de la France* (Paris, 1954).

foreign minister actually named a commission of engineers and military experts to "fix the demarcation of the limits" of Spain and France.[3] But the minister soon adopted the view that the French–Spanish boundary was of minor importance. It was not simply because France's interests lay to the north and east, where territories were to be ceded and exchanged, but also because local disputes along the Spanish frontier were of little cause for concern.[4] Such was *not* the opinion of his successors, or their contemporaries, or the local populations. In the course of the next decades, both French and Spanish officials recognized the need to "define the true line of division of the two kingdoms," as the director of the Royal Corps of Engineers in Spain wrote to the secretary of war in 1830. The local populations as well complained consistently to their respective governments about the need to "define the limits of the frontier," as the community of Angostrina wrote to the prefect. But when the Spanish government was ready to "talk limits," France was not; and when France came around and was prepared to begin negotiations, Spain was unable to treat what appeared then to be an issue of minor importance.[5]

Not until 1851 did the Spanish and French states coincide in the need to delimit the French–Spanish boundary, and negotiations for a "treaty of limits" began in earnest in 1853. Starting with the boundary in the western section of the Pyrenees, the commissioners took fifteen years to complete their task. On May 26, 1866, they signed the third Treaty of Bayonne, which covered the eastern Pyrenees from Andorra to the Mediterranean, "fixing definitively the common frontier of Spain and France." By the further treaty of July 11, 1868, the commissioners modified certain articles of the 1866 accord and laid the official boundary stones in the Cerdanya.[6]

3. FAMRE Limites, vol. 463: "Commission de démarcation des frontières," n.d., ca. August 1814.

4. FAMRE Limites, vol. 35 (*avis* sent to foreign minister, July 14 1814); FAMRE Limites, vol. 7: "Mémoire du Général Maureillan," Northern Boundary Commissioner, September 16, 1814; and "Note sur les traités de 1814," October 23, 1838.

5. SAHN Estado, libro 673d, fols. 328–330 (letter to secretary of war, November 2, 1830); FAMRE Limites, vol. 459, no. 119 (memoir of July 10, 1817); FAMRE Limites, vol. 446: "*Pro-memoria* pour M. le Baron de Damas," minister of foreign affairs, September 1825.

6. Sermet, *La frontière hispano–française,* passim; and C. Fernandez de Casadevanti Romani, *La frontera hispano–francesa y las relaciones de vecindad* (*Especial referencia al sector fronterizo del pais vasco*) (San Sebastian, 1986). The Bayonne treaties involve three principle accords: December 2, 1856, delimiting the western Pyrenees corresponding to the Spanish provinces of Guipuzcoa and Navarre; April 14, 1862, from Navarre to Andorra; and May 26, 1866. These conventions, along with their annexes and additional acts, have been reprinted as *Acuerdos fronterizos con Francia y Portugal* (Madrid, 1969); see below, app. A.

The French geographer Roger Dion remarked that the image of the Pyrenees was "the first boundary which the French national consciousness, at its awakening, clearly perceived."[7] Yet the Pyrenean frontier was the last of France's boundaries to be delimited and the one which presented the most difficulties. The delay in delimiting the boundary during the nineteenth century resulted in the entrenchment of both local and national competition in the French–Spanish borderland. By the time the commission reached the Cerdanya in the 1860s, the border was at once more inscribed and more contested than either state wanted. The boundary through the Cerdanya was the most disputed site along the whole Pyrenean frontier, and perhaps of any section of France's frontiers in the nineteenth century.[8] It was not the lack of a natural frontier in the Cerdanya but the absence of a historical one which posed such great difficulties for the commissioners, and for the village communities themselves.

POLICING THE BOUNDARY: SMUGGLING AND COMMUNAL STRUGGLES

If the Pyrenean boundary became relatively "fossilized" or "cold" in the course of the early nineteenth century as compared to France's northern and eastern frontiers, it nonetheless became the object of political debate and contention in much the same way. The politicization of national boundaries and national territory was the final stage in a development that led from the emergence of territorial sovereignty in the late eighteenth century through its nationalization during the French Revolution. In the nineteenth century, the repeated claims of "rape," "pillage," and "usurpation" of national territory against the "rights of nations" fill hundreds of dossiers in the archives of Paris and Madrid and of the Cerdanya itself. The concern to prevent "violations of territory," so evident in France's border policy during the Spanish civil wars of the 1820s and 1830s, became an obsessive preoccupation of French and Spanish local and national officials throughout the nineteenth century.

In addition to the threatened spread into France of Spanish civil struggles, two principle sources of territorial violations stood out: contraband trade and local struggles. The repression of contraband trade

7. Dion, *Les frontières de la France,* 56.
8. FAMRE CDP 10, fol. 386–415 (report of the French Commissioner General Callier to foreign minister, November 7, 1867).

was part of a larger attempt on the part of both states to affirm and consolidate national territorial sovereignty. In part, officials were reacting to changes in the volume and character of trade in contraband goods through the Cerdanya during the first half of the nineteenth century, changes largely in response to economic and political transformations in Catalonia. The industrial renaissance in northern Catalonia, for example, provided the opportunity to "export" new goods, including men. In 1818, the customs director reported to the prefect that industrial looms and even skilled workers were being smuggled into Catalonia through the Cerdagne. Several smugglers were named, including an important merchant established at Mont-Louis.[9] The Carlist Wars in Spain further provided the opportunity for many in the French Cerdagne to profit from the sale of grains, weapons, or uniforms to the rebels. François Durand, a merchant at La Tor de Carol, received an order for six hundred Carlist uniforms, a third of which were confiscated in his warehouse by the police commissioner of Bourg-Madame.[10]

Changes in the content of trade were matched by changes in the attitudes of military and civilian officials toward that trade. In the late eighteenth and early nineteenth centuries, despite concerns for the public order and exceptions to national law, officials could still see some good resulting from smuggling operations in the Cerdanya, especially as concerned the balance of trade and bullion in France's favor.[11] By the end of the 1830s, however, most officials denounced smugglers as enemies of the social and moral order. Thus a military official wrote that

> contraband trade has organized itself, it has penetrated in the mores of the people and corrupted them, it has destroyed honor and created fraud, and caused the abandonment of work. There is not a single inhabitant of these frontier districts who would accept for honest and easy work the salary that he so hastily accepts for a contraband enterprise. From the one, he sees only drudgery; from the other, merit and pleasure.

9. ADPO Mnc 1978/5 (customs director to prefect, November 6, 1818).
10. ADPO Mnc 1921 (police commissioner at Bourg-Madame to subprefect, March 3, 1847).
11. FMG AAT MR 1223, no. 99 (anonymous memoir of 1806); ADPO 2J 81/1–3: "Statistique," fol. 43; ADPO 1J 160: "Statistique de la Cerdagne," n.d., ca. 1800; ADPO 3M1 21–22 (report on the state of commercial activities in the départment by Jaubert de Passa, sub-prefect of Perpignan, 1814: "Contraband trade . . . takes from us an excess of products and gives us an enormous mass of currency").

As part of their attack on contraband trade in the 1840s, both the Spanish and French governments grew increasingly intolerant of any local exchanges across the boundary and attempted to police the most insignificant aspects of economic life in the borderland. The French customs director in Perpignan proposed a weekly fair day in Bourg-Madame in 1846 in order to encourage French Cerdans not to cross the boundary; and in these same years the Spanish state spent a great deal of energy attempting to stop the enclave of Llívia from provisioning itself with meat from the French Cerdagne.[12]

But the point of delimiting the boundary line was less to repress contraband trade itself than to avoid the international conflicts that stemmed from its repression. For even if the two governments cooperated occasionally after 1815 in the mutual repression of contraband activities, as they had before the Revolution in accordance with the formal convention of 1786, they generally competed against each other in attempting to prevent "territorial violations" of opposing soldiers and customs guards.[13] Archives of both states suggest that the number of violations grew after 1830, as did official evocations of the need to delimit the national boundary. By the late 1840s, the French state was so committed to defending its national territory that it protected its smugglers, claiming "territorial violations" when "French" merchandise was seized by Spaniards on lands believed to form part of French territory.[14]

The contested zone in which disputed seizures were made could be as wide as several kilometers, as was the case of a seizure in 1845 on

12. FMG AAT MR 1221, nos. 63–64: "Mémoire sur la délimitation de la frontière et sur la contrabande," April 10, 1835; ibid. 1222, nos. 112–113: "Mémoire militaire sur la Cerdagne française," March 1836; and AMLL VII.5 (correspondence concerning the Llívia's meat provisions, 1845–1849). One example of the French state's growing intolerance of exchanges across the boundary was the ruling in 1840 forbidding proprietors of the French Cerdagne to frequent the blacksmiths in Puigcerdà for repairs of their agricultural tools. The case was complicated because the Puigcerdà blacksmith was French, and used French iron, whereas the ones established in the French Cerdagne, who supported the prefect's prohibition, were Spaniards, or sons of Spaniards: see the dossier ADPO Mnc 1921.

13. ADPO Mnc 1944 (district attorney to prefect, May 9, 1820, reporting that between 1815 and 1830, approximately eighty Spaniards were extradited in fulfillment of the December 24, 1786, accord).

14. Cases include ADPO Mnc 3818 (mayor of Osseja to prefect, August 27, 1846); AHP Correspondencia (mayor of Osseja to Puigcerdà governor, February 18, 1836); and FAN F(7) 9692 (prefect to interior minister, September 1, 1830, citing the fifty infantry troops he had sent to the frontier in order to "shelter from destruction" the "entrepôts of merchandise at Bourg-Madame, Palau, and Osseja.")

the mountains above Osseja;[15] or, it could be restricted to a couple of meters of land, as was often the case on the floor of the valley, and especially along the disputed "neutral" paths between Puigcerdà and Llívia. Thus an incident ensued in February 1835, within the confines of Càldegues and Llívia, in which the mayor of the French village supported the French customs guard's seizure of two mules near the "neutral road." On the site of the seizure, in the midst of discussion between the guards and smugglers, a crowd of peasants from Llívia arrived, disarmed the guards, and surrounded the French mayor. In his report, he wrote:

> I was forced to shout: "respect authority," and my voice was heard. "We shall respect authority, but since we are in Spain, it will be Spanish," they replied. "If you are in Spain," shouted Isidro Arro [a peasant of Onzes], "we are in France." Again I shouted, "Respect authority, for if you don't respect it today, you will have to respect it tomorrow, since we know who you are."[16]

To both governments, such scenes were of more than local interest: in the midnineteenth century, territorial violations were matters of high diplomacy. A disputed seizure of contraband goods set off dozens of letters to all levels of administration, mobilizing officials from the administrative districts to the foreign ministry. A badly defined boundary not only encouraged smuggling, it encouraged the violation of territorial sovereignty in the course of its repression. No single section of the French–Spanish frontier produced as much contention as the Spanish enclave of Llívia. Thus to Dumontet, a French military engineer stationed in the Cerdagne in 1836, Llívia was a

> bizarre enclave which an ill-willed act seems to have thrown into the heart of our territory. An obstacle to our commerce and to our agriculture, it protects smuggling enterprises against the watchfulness of our customs guards, destroys the regularity of this part of the frontier, and complicates the problem of defense. This fragment of foreign land lost on our soil is the enclave of Llívia, which remained Spanish in the midst of a district which has become French.[17]

Such opinions were seconded by the demands of the General Council of the Department of the Pyrénées-Orientales, made almost annually

15. ADPO Mnc 1921 (report of police commissioner at Sallagosa to sub-prefect, October 30, 1845).
16. ADPO Mnc 1828/1 (procès-verbal of mayor of Càldegues, February 27, 1835).
17. FMG AAT MR 1349: "Mémoire sur l'enclave de Llívia," by Dumontet, May 7, 1836).

from 1825 until years after the delimitation, to incorporate the Spanish enclave into France.[18]

The second source of territorial violations came from the competition and struggles among local communities themselves. Both Spanish and French officials insisted on seeing the boundary as a disputed and badly demarcated zone: "No inhabitant can flatter himself in saying he knows the boundary well," wrote the customs director at Perpignan in 1825. Yet there is much evidence to suggest that the village communities knew the boundary all too well. In court testimony concerning disputed seizures, peasants described the borderline in detail.[19] More striking still were the uses made of the boundary by individuals and communes alike.

The boundary, as it took shape during the decades before the delimitation conferences, wandered across the floor of the valley. This was true not just of the Raour River, which periodically left its established bed and overflowed onto the French bank, leaving new islands and streams with which to be reckoned. It was true of the border stones between neighboring communities as well. On March 29, 1826, the mayor of Angostrina went with the local customs official and various notables of the commune to a site called "dels Cormers" on the path between Angostrina and Sereja (Llívia). According to his official report, they found that

> one of the stones serving as the boundary of the territories of the two kingdoms of France and Spain had changed places; that the said stone had been pierced in a way to prove that attempts had been made to destroy it, which evidently proves the bad faith of our Spanish neighbors, with whom we have always tried to live in the best of relations. We went about 500 meters into French territory, and found a large stone with a newly engraved cross on it, the customary way in which limits are marked . . . this new boundary takes away a considerable part of French territory . . . it is my duty not only to defend [the rights of] the proprietors of the commune of Angostrina but also the rights of the kingdom and the crown.

But was it in fact a boundary stone separating the two kingdoms? The mayor of Llívia quite adamantly denied it, in a letter to the governor of Puigcerdà, asserting that the true demarcation of Spain and France

18. FAMRE Limites, vol. 461, nos. 39 (1826), 47–48 (1828), 70–72 (1829), 91 (1831), 103–104 (1836), 106 (1839).

19. FAMRE Limites, vol. 461, no. 41 (letter to prefect, March 19, 1825); for examples of testimony, see SAHN Estado, libro 673d, fols. 343–348 (testimony following a seizure on the boundary of Guils and La Tor de Carol, June 1830); and SAMRE TN 221–222 (inquest of February 3, 1850, concerning the location of a border stone between the same two communities).

lay at a distance of "one hundred paces, more or less, toward the territory of Angostrina," and claimed that the inhabitants of the French commune had "violated Spanish territory" in their attempts to verify the boundary.[20]

There were other incidents of wandering boundary stones in the nineteenth century, as there were indications of other kinds of manipulations of the boundary. Writing in 1830, the proprietors of Sant Pere de Cedret, a hamlet forming part of the commune of La Tor de Carol, claimed that Laurent Vigo, the mayor of the French village of Carol in 1807, had testified that certain properties in the borderland were part of Spain. Vigo's motive, the petitioners claimed, had been to favor his brother-in-law from Puigcerdà, who was stopped by Spanish customs guards with two oxen while he was coming back from France. The seizure was nullified, and the inhabitants of the village of Guils in Spain had moved the boundary stones to assure the extension of their territory—or so the landowners of Sant Pere argued in August 1830. They contended further that Vigo's son, Bonaventure, had conspired with the "Spaniards" to maintain the new boundary—although Bonaventure Vigo himself was on record in June 1830 denouncing "a violation of French territory which provoked the inhabitants of this commune and the customs guards of Carol to carry out just reprisals and seize two Spanish herds pasturing on French territory, or at least on disputed land."[21] It is impossible to say who was telling the truth: the important point is that the lands of neighboring communities were openly contested, and identified in national terms, just as the exact location of the boundary was openly disputed and subject to constant manipulation throughout the nineteenth century.

In the late 1850s in the Cerdanya, as throughout the Pyrenees, struggles among communities divided by the political boundary actually intensified under the impact of the delimitation proceedings. As news of the delimitation reached the valley, local communities insistently

20. FAN F(1) I 447 (procès-verbal of mayor of Angostrina, March 29, 1826; and French Foreign Minister Damas to captain-general of Catalonia, August 25, 1826); ADPO Mnc 3818/1 (municipality of Llívia to Puigcerdà governor, May 30, 1826); SAHN Estado, libro 673d, fols. 303–310 (correspondence among Spanish officials, March–July 1826). For some reflections on wandering boundary stones as symbols of conflict, see J. Fernandez, "Enclosures: Boundary Maintenance and Its Representations in Asturian Mountain Villages [Spain]," in E. Ohnuki Tierney, ed., Symbols in Time (Stanford, in press).
21. ADPO Q supp. 18 (petition of landowners of Sant Pere de Cedret to prefect, August 1830; mayor of La Tor de Carol to prefect, August 18, 1830); the more complete Spanish dossier is in SAHN Estado, libro 673d, fols. 327–398.

evoked the powers of the state to draw official attention to their local interests, in the process nationalizing those interests and interesting the nation in their struggles. "The inhabitants of La Tor de Carol cannot take their herds across the lands usurped by the Spaniards, who threaten to seize their livestock," wrote the mayor of the French commune in 1858.

> The position of Frenchmen along the frontier is always worse than that of Spaniards; the latter are encouraged by their governors to undertake more or less arbitrary actions, which our governors forbid us. We do not wish to be abandoned to the mercy of the Spaniards.

In fact, the mayor failed to mention that three days earlier about 150 inhabitants of all the French villages in the neighboring district, joined by the police commissioner of Bourg-Madame, had penetrated into Spain—or at least into disputed lands—and destroyed a canal built by the villagers of Guils which they claimed amounted to "a permanent violation of French territory."[22]

The desire to prevent such incidents lay at the heart of the delimitation proceedings, since local struggles over waters, pastures, and communal boundaries were the source of the politicization of national territory. Unless such struggles could be eliminated, the integrity of national territory would be constantly threatened. In his final report to the foreign minister in 1868, the French commissioner of the Bayonne negotiations, General Callier, explained that the purpose of delimiting the boundary was "to resolve definitively the secular litigation of the frontier communities and to abolish all future motives of collision in the borderland." The Spanish first minister, writing to the French ambassador, agreed that the essential motive for the delimitation of the boundary was "to abolish all the germs of discord and defiance generated by the long-extant controversies" over local boundaries.[23]

THE END OF NATURAL FRONTIERS AND THE DEMILITARIZATION OF THE BOUNDARY

The French–Spanish commission's focus on local struggles during the delimitation conferences meant taking into account the historical

22. ADPO Mnc 1924/1 (mayor of La Tor de Carol to sub-prefect, April 23, 1859); FAMRE CDP 5, fol. 371 (Spanish first secretary of state to French ambassador in Madrid, June 24, 1850, and the additional correspondence in fols. 351, 354, and 385–393, as well as the dossier in SAMRE TN 221–222. I count nearly forty documents generated by the incident).

23. FAMRE CDP 10, fol. 208 (report of January 19, 1867); ibid. CDP 1, fol. 27 (letter to French ambassador, September 11, 1853).

claims and contemporary opinions of local society. It was no longer possible to evoke such abstract principles as "natural frontiers." After the French War Ministry had persuaded its government to adopt the idea of dividing mountain crests as the basis for the delimitation in 1851, the Spanish government withdrew from the talks—using as its excuse the renewed struggles in the western Pyrenees. The Spanish commissioners argued instead that the people of the borderland were "far too intertwined, through their interests and by their histories, in their respective nationalities" to separate them by "natural frontiers."

> In vain can we say today to men: this or that is your nationality, this is your fatherland, that is your history, only because here or there stands a mountain which defines a watershed; in vain can we tell them, "we will change your name, because there is a barrier which divides you," . . . these people will not believe us.[24]

The French Foreign Minister Boislecomte eventually consented. Though conceding that the "dividing crests indicated by Nature herself" would serve strategic and economic interests, he nonetheless stated that "considerations of a higher order prohibit the complete and absolute application of this principle." Exceptions were thus to be made for certain districts in the western Pyrenees and for the Cerdagne and Carol valleys, "which long-standing habits unite to France and which we could never consent to see pass under foreign domination."[25]

During the actual demarcation of the boundary, the commissioners sought to use topographical features—streams and rivers—whenever possible. But the possibilities were extremely limited, since both parties had agreed to structure their discussions around other concerns. As the Spanish secretary of state wrote to his commissioner, Monteverde, the commission "should prefer a demarcation by natural limits," but only in the interest of "reconciling mutual interests and not causing any further dissension." Callier explained that "we always tried to find natural boundaries, but without losing sight of the principle goal which was to satisfy with equanimity the recognized needs and rights."[26] Natural frontiers had, in short, been entirely subordinated to the principles of eradicating local struggles.

The French military establishment had been the principal spokesman for the doctrine of natural frontiers. "Natural frontiers are the

---

24. FAMRE Callier 4 (propositions of the Spanish commission, June 28, 1852).
25. FAMRE CDP 1 (instructions from Foreign Minister Boislecomte to Bayonne commissioners, May 1, 1852).
26. FAMRE CDP 1, fol. 91 (Monteverde to Spanish commission, April 21, 1852); FAMRE CDP 11, fols. 189–209 (Callier report of August 5, 1868).

true military frontiers," wrote a general about the Alps in 1819, and the same was often said about the Pyrenees. The idea of the Pyrenees as a natural frontier had become a platitude among nineteenth-century military officials and engineers, who underscored in their memoirs the need to "mark the limits of the frontier" along the dividing line of crests.[27] Abandoning the principle of natural frontiers thus entailed the decline of the military establishment itself. Indeed, as early as 1851 the French war minister protested to the minister of foreign affairs, echoing complaints of his predecessor in 1772, about his lack of influence on the recently formed commission. In all other cases of delimitation, he argued, the military had been represented in the negotiations. The point was not simply about jurisdictional competency, it was about the nature of the border itself. "It seems to me beyond the point of discussion," he wrote,

> that political and military considerations, and I might say exclusively military considerations, dominate all questions concerning state boundaries. Before anything else—before the customs barrier is erected, before commercial relations are regulated, before the questions of taxes can be debated—the line of military defense must be established . . . to protect French interests against perfidious and duplicitous neighbors. The military position of the boundary is necessarily the key point, because all other interests are grouped and arranged under the aegis of military power.[28]

Military interests, in short, should dictate the shape of the boundary. But military concerns had been displaced during the decades before the delimitation talks even began.

Since the Spanish Carlist struggles of the 1830s, the French military frontier had shrunk to the dimensions of national territory: no longer a zonal extension beyond the political boundary into Spain, the military frontier became subordinated to the defense of the political division. More generally, the decades after the Carlist War in Spain

---

27. FMG AAT MR 1207, no. 1: "Rapport sur la frontière d'Italie ou des Alpes," 1819; FAMRE Callier 4: "Mémoire sur la nouvelle démarcation de la frontières des Pyrénées, par M. Blevec, Colonel du Génie à Bayonne," October 25, 1851; and copies of memoirs of the Fortifications Committee of the Foreign Ministry, 1839. Generally speaking, the idea of natural frontiers was so thoroughly entrenched in nineteenth-century military discourse that it had lost its referent. For example, instead of signifying a boundary marked by a natural topographical feature, a border that was "natural" was one that was well established. Thus the Northern Boundary Commissioner Maureillan wrote that the boundary between the Atlantic and the Mosel River "has only become natural by the creation of fortified sites"; see FAMRE Limites, vol. 7: "Mémoire du Général de Maureillan," September 16, 1814.

28. FAMRE Callier 4 (war minister to foreign minister, July 24, 1851).

witnessed the demilitarization of the national boundary, especially on the French side. Such a demilitarization may be charted in the declining influence of the War Ministry's opposition to road construction in the borderland. Traditionally, during the late eighteenth and early nineteenth century, Spanish and French military officials consistently opposed projects to link the Cerdanya by a "free passage to the interior." Both the French and Spanish Cerdans had since the later eighteenth century requested that paved roads—"on which vehicles could pass"—link them to their respective interiors. In 1791, for example, the mayors of the French Cerdagne argued in a petition that "France has nothing to fear from the Spanish" and that military interests would not be damaged by opening a road from Ax-les-Thermes to the valley of Carol over the Puigmorens Pass. The events of 1793 were to prove them wrong, and at their own expense; but the opposition of the military to that road and others linking the Cerdagne to the interior continued until the late 1830s.[29]

Ironically, during its defense of the boundary in the 1830s, the military establishment itself assisted in the creation of road networks linking the Cerdagne to the Roussillon plain. In part they were spurred on by demands of the French Cerdans, who were concerned that the Spaniards would soon complete the road from Vic and Ripoll to Puigcerdà, over the Tossa Pass.[30] But more generally, road construction emerged from the strategic concern of defending French territory. Previous armies had of course built roads suitable for artillery, and military reminiscences never fail to mention the spectacular road-cutting feats of the Duc de Noailles in the 1690s or the Maréchal de Berwick during the War of the Pyrenees. But their roads, constructed for campaigns, virtually disappeared with the armies themselves. During the 1830s, however, the French army itself assisted in the construction of paved roads into and within the Cerdagne. In its permanent de-

29. On opposition of the military establishment in the late eighteenth century, see Gavignaud, "La frontière pyrénéenne depuis 1659," 61; and FMG AAT A(1) 3719 (petition of French Cerdagne mayors, n.d., ca. 1791); on the continued opposition of the military establishment during the early nineteenth century, see E. Fornier, "Les relations frontaliers en Cerdagne et Haut-Ariège," 2 vols. (Diplôme d'études supérieures, Section géographie, Université de Toulouse, 1966), 1: 157–158; M. Chevalier, *La vie humaine dans les Pyrénées ariegeoises* (Paris, 1956), 963. Demands of Puigcerdà to complete the road include BRP MS 2346, fol. 88 (memoirs of the Economic Society of Puigcerdà, 1784); and AHP MS 1820–1835, fol. 74 (letter to intendant of Catalonia, September 24, 1824).

30. ADPO MS 7469: printed "Mémoire à Messieurs les Ministres, Députés, et Pairs de France, par le Sr. Jh Sans Cadet, Négotiant à Palau," 1832; and FAIG 4.1.6.3: "Notes sur les routes exécutés par les Espagnols," February 5, 1833.

fense of French territory, the French military command constructed a road network of equal permanence. By the late 1840s, military objections to road construction had largely dissipated, although dissenting voices were still to be heard. The paved road from Perpignan to Bourg-Madame reached Mont-Louis and crossed the Perxa Pass in 1854, and National Road 20, ordered by decree of Napoleon in 1808, was completed in 1866.[31]

The demilitarization of the national boundary formed the backdrop of the military establishment's declining influence in the delimitation conferences during the 1850s and 1860s. Upon seeing the delimitation treaty of 1866 between Spain and France, the French war minister protested angrily to his foreign minister:

> I am astonished and worried to see the definition of the frontier of France and Spain submit itself to the habits, customs, and claims of small Pyrenean communes; the dividing line twists and turns carefully so as not to bother these communities, instead of following a general rule dictated by great interests far superior to local quarrels.

Yet the French commissioners of the Bayonne treaties did have such "great interests" as their concerns; it was simply that military goals, and the corresponding doctrine of natural frontiers, had been definitively displaced. If the commissions paid so much attention, as General Callier responded, to the "habits, customs, and claims" of local society, it was only in the interests of preventing "territorial violations." The boundary was the point at which national territorial sovereignty found its expression, and unless local quarrels could be settled, the political and diplomatic relations of France and Spain would be constantly threatened with disruption.[32]

THE DELIMITATION AND DEMARCATION OF THE BOUNDARY

In sorting through the myriad local claims of village communities in the borderland, the Bayonne commissioners played the contradictory roles of both experts and judges: they represented the national interests of their respective governments, but they also evaluated and judged the

31. FMG AAT MR 1222, nos. 112–113: "Mémoire sur la Cerdagne française, par le Capitaine Th. Puvis," March 1836, fol. 22 ff.; J. Sermet, "Les communications pyrénéennes et transpyrénéennes," in *Actes du 2e congrès international d'études pyrénéennes,* 59–193.

32. FAMRE CDP 9, fol. 283 (war minister to foreign minister, March 14, 1866); and fols. 297–300 (response of Callier to foreign minister, April 7, 1866).

claims of local communities on both sides of the boundary. According to General Callier, "The role of judge took precedence over the role of expert," since the two commissioners were to make their decisions

> based on law, equity, and the necessities of order. In other words, we had to pronounce judgments between the parties, as would a mixed court with the approbation of our governments. In these judgments without appeal, we deduced the line dividing the two states.

All previous commissions inevitably failed, Callier continued, in reversing the primacy of the two roles of expert and judge.[33] But like their seventeenth-century predecessors, the present commissioners had still to prove themselves experts—competent historians of local society.

The history preoccupying the nineteenth-century commissioners was a very different one from that which had been the object of expert researches in the seventeenth century. It was no longer the ancient history that described the separation of "the Gauls" and "the Spains," nor the history of early administrative divisions of the Roman Empire or the Visigothic counties. The nineteenth-century commissioners were uninterested in the 1850 memoir sent by the archivist of the Pyrénées-Orientales detailing the prehistory of the Pyrenean boundary. They were unmoved by the need to uncover the gigantic iron rings described in the memoir, those "placed along the dividing line of the Gauls and Iberia" that, "like ancient stone markers [menhirs], remain eternal sentinels which would tell each of the two peoples: here is where your nationhood ends."[34] The Bayonne commissioners sought out instead the history of the local communal boundaries that formed the basis of the national boundary of France and Spain.

By 1853, the Spanish and French commissioners had adopted the historical criteria by which they would judge local quarrels. The failure of the 1659 treaty and the 1660 accord to specify the territorial confines of local municipalities led the commissioners to rely upon the documents closest to 1660 which described a community's boundaries.[35] As in 1659, both French and Spanish commissioners prepared their relevant documentation; but in the nineteenth century, their ar-

33. FAMRE CDP 10, fol. 208 (Callier to foreign minister, January 19, 1867); see Callier's remarks on the same subject in FAMRE Callier 12: "Deuxième mémoire français," August 19, 1865.

34. ADPO 1 N 195 (n.d., ca. 1850); see also ADPO Mnc 3817 (archivist Morer to prefect, October 18, 1860).

35. SAMRE Tratados 222–223: "Frontera de Cerdaña," including instructions, memoirs, and lists of documents used by the Spanish commissioners, n.d.; FAMRE CDP 1, fol. 187 (Spanish ambassador to French foreign minister, March 6, 1853).

chives spoke much more to local concerns: a tenth-century ecclesiastical donation of a parish, an administrative ruling from the sixteenth cen ury, a "private" accord signed between neighboring communities of the eighteenth century. As in 1659, the best material lay in Spanish archives; but in the 1860s, the local archivists shared "a common passion for historical research" which ultimately conflicted with their support of distinct national claims.[36] And as in 1660, both French and Spanish commissioners sometimes subverted these criteria in advancing their own claims, although compared to Pierre de Marca, the French commissioner of 1660, General Callier was a model of fairness and conciliation.

The problem of determining the communal boundaries was a two-stage process: first, the commissioners sought out documents describing communal limits of earlier centuries; then they tried to determine where those boundaries could be marked on the ground. The real issue, as General Callier noted, was the "signification" of the descriptions—the relation between the text and the terrain. For example, Llívia produced a 1540 administrative decision giving the town its claim to the disputed lands of "El Tudor" bordering Angostrina. The letters defined a "straight line" between two named boundary stones that had since disappeared. Callier recognized that "all the terrain of the Serra d'Angostrina" might be correctly signified by the description.[37] The problem was especially complicated because many of the toponyms used in the medieval documents had themselves changed: thus the "Serra de la Tor," mentioned in 906 as the dividing topographic feature of Guils and La Tor de Carol was a different one than that noted in 1210—and in the nineteenth century, the two communities claimed both ridges by that name (see illustration 9).[38]

Both the local communities and the commissioners of the two states insisted that "straight lines" be traced between agreed-upon border markers. By adopting the principal of "straight lines," however, the French commission and the French communities were at a distinct advantage over their Spanish counterparts. As became apparent in the early-going, the boundary line established by the French state for the purpose of defining its fiscal limits in the late 1820s and early

36. "Correspondence inédite de l'archiviste Alart et du Général Callier au sujet de la délimitation de la frontière des Pyrénées-Orientales," *Ruscino* 4 (1914): 6–16 (Alart to Callier, May 13, 1864); and ibid. 6 (1915): 142–144 (Callier to Alart, March 31, 1865).
37. FAMRE Callier 16 (letter of March 14, 1865).
38. "Correspondance inédite," *Ruscino* 6 (1915): 129–139 (March 27, 1865).

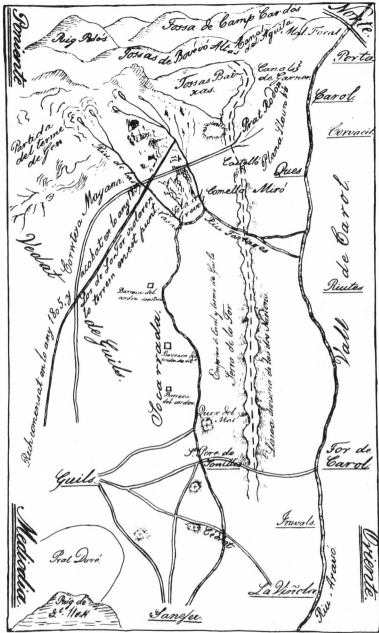

Illustration 9. "Exact Reduced Copy of the Drawing Submitted by the Village of Guils," n.d. ca. 1860. Drawn up for the Spanish commissioners to use during the Bayonne negotiations, this map represents Guils' claim to a boundary with the French village of La Tor de Carol, drawn along the Serra del Tor. The French and Spanish communities disagreed with each other over the location of the ridge by that name, and with the French and Spanish commissioners, who sought to leave the terrain as a joint holding.

1830s became the basis of the international boundary. Except for the half-dozen sharply contested sites between Spanish and French communities, the French cadastral survey formed the foundation of the delimitation and demarcation accords of 1866 and 1868.[39]

Napoleon had ordered a land survey of France in 1804 for the purposes of imposing a uniform land tax in the Empire; interrupted during years of political turmoil and upheaval, the plan was reactivated in the late 1820s.[40] Beginning in 1827, French civil engineers drew up land matrices, rolls, and maps of the Cerdagne communities. They did not, however, request participation by Spanish authorities, either local or national. Local officials of the Spanish Cerdaña protested to the French government, but Spanish Cerdans sometimes went a step further, vandalizing the work of measurement and mapping. The French communities disputed among themselves the boundaries established by the engineers; but disputes between the Spanish enclave of Llívia and the surrounding French communes were even more problematic, lasting almost ten years. The departmental administration in Perpignan waffled on the question of the nationality of property and lands in the contested zones, finally according the cadastral survey the status of a "provisional arrangement" pending the formal delimitation of national territories.[41]

According to General Callier's reports, the problem with the cadastral line was that it "zig-zagged" since it reproduced the actual limits of private possessions owned by Frenchmen and Spaniards. The international line of 1868 thus "straightened out the limits of the cadastre," and the two nations "exchanged equal surfaces on the different sides of the boundary." Yet rarely did the demarcation of the boundary actually follow the limits of private property. In drawing "straight lines" as opposed to the "zig-zags" of the cadastre, the commissioners cut fields and meadows in half—often naming the owner of the property in the formal act of demarcation (see text in app. A). The case of Llívia, as usual, presented special difficulties. The commissioners tried to draw a line that would incorporate within each commune the lands of proprietors of different nationalities; yet the final solution

39. FAMRE CDP 10, fols. 386–415 (Callier to foreign minister, November 7, 1867).
40. R. Herbin and A. Pebereau, Le cadastre français (Paris, 1953); and Konvitz, Cartography in France, 53–62.
41. Sahlins, "Between France and Spain," 353–360; FAN F(2) I 447 (prefect to interior minister, October 3, 1828); ADPO Mnc 3818/1 (prefectoral decrees of 1832 and 1835); see also the correspondence and extracts from municipal council registers in ADPO Q supp. 17.

gave to Spain a certain number of Spanish and French fields so intertwined that after several useless attempts, we were forced to renounce the principle of tracing a line which would have assured each landowner the lands corresponding to his nationality.[42]

The precedence of territorial location over the national identity of the property—itself a function of the proprietor's nationality—reversed the hierarchy that resulted from the eighteenth-century debates over the fiscal limits of sovereignty in the Cerdanya. In the eighteenth century, nationality determined territory. In 1868, the reverse became true. The formal territorialization of sovereignty was constituted in the delimitation and demarcation of the boundary.

As part of the goal of eradicating the seeds of local discord, the task of the Bayonne commission went beyond the demarcation of a linear boundary; the commission also built upon the provisions of the Llívia accord of 1660 while sanctioning social and economic continuity across the boundary. Article 7 of the Additional Act of 1866 not only allowed proprietors "complete freedom" in the agricultural exploitation of lands divided between the two nations, but also guaranteed them the entire "fruits of their lands." Articles 21, 25, and 28 further reaffirmed the "free passage" of French nationals across Llívia, and of Spaniards between Llívia and Spanish territory, explicitly locating each of six "free paths" in common use and further guaranteeing the pilgrimage route across French territory to the chapel of Nostra Senyora de Núria.

Historians have argued that treaties such as those of Bayonne represented a kind of juridical federalism, whereby the autonomous legal order of local communities was incorporated into modern international law.[43] Yet more than simply affirming local practices, the Bayonne treaties included provisions that established the rules and institutions governing communal usufruct and property rights to resources in the borderland. Of great importance were the conflicts over waters that flowed from France to Spain, and especially the canals of Puigcerdà and of Llívia. The rights and access to the water of these two canals were settled in 1868 by an International Commission of Engineers that completed detailed measurements of water availability and the needs of the

42. FAMRE CDP 10, fols. 386–415 (Callier to foreign minister, November 7, 1865).
43. Descheemaeker, "La frontière pyrénéenne," passim; and Fernandez de Casadevanti Romani, La frontera hispano–francesa, passim. For a more general discussion of the nature of a "frontier regime," see Lapradelle, La frontière, 236–264.

communities. The engineers, two French and two Spanish, also completed a new demarcation of the Raour riverbed, which had shifted course since 1820 and was again the object of local disputes. Other waterways included the irrigation canals derived from the Vanera, Err, and Aravó or Carol Rivers and two streams that flowed from Guils in Spain to Sant Pere and La Tor de Carol in France. In the 1660 accord and subsequent Figueres conferences, all these waters had gone unmentioned; in the early 1860s, Callier and Monteverde began to prepare "a general project for the use of waters which flow from one country to another." The two governments established international administrative commissions for each of the canals, designed to enforce the execution of the international regulation governing the amount of water each was assigned. Such an international agreement was the first of its kind in a modern delimitation treaty.[44]

In 1660, the Llívia accord ceded thirty-three villages "and their dependencies" to France, but the modern political boundary dividing two territorial states only appeared more than two hundred years later, in the Bayonne treaties of 1866 and 1868. These treaties between France and Spain delimited and demarcated distinct national territories, while they sanctioned local relations of continuity in the borderland. The movement toward a linear, territorial boundary had been a slow and discontinuous process involving both states and local society. The village communities of the Cerdanya had been demanding the delimitation and demarcation more persistently than either France or Spain; yet when the two states finally agreed to establish their national boundaries, the communities were less willing to settle their differences than either of the two states.

CULTURAL IDENTITY AND THE
EXPERIENCE OF DIFFERENCE

In August 1868, General Callier wrote to the foreign minister expressing his hope that the establishment of the international boundary and the institutional regulation of resources would usher in a new and less contentious era in the history of the borderland:

> The neighboring communities must break entirely with the old spirit of animosity . . . for there are neither motives nor pretexts to dispute. Their

---

44. SAMRE Tratados 222–223 (final report, including maps, of the International Commission of Engineers, September 10, 1866); and FAMRE CDP 10, fols. 96–97 (Callier to foreign minister, September 14, 1866).

common interest is to live as good neighbors and to enter the path of peace and concord which has been opened by the international regulation of the frontier.[45]

In the long run, his hopes were justified: although occasional disputes surfaced during the late nineteenth and twentieth centuries, the "spirit of animosity" that the Cerdanya communities manifested during the delimitation negotiations generally dissipated. But in the short run he was wrong, since the village communities rejected the settlement reached by the Spanish and French commissioners.

It was not just a question of chasing away government officials sent into the Cerdanya to survey the demarcation of the boundary line, as occurred in the spring of 1866.[46] More significantly, the communities refused to countenance the compromises that the Spanish and French commissioners had agreed upon. When in 1866, for example, the commission had awarded the full property of the Puigcerdà canal to the Spanish town, leaving usufruct rights to the French villages of La Tor de Carol and Enveig, the latter had protested so vociferously that the commissioners were forced to redefine the accord two years later. Likewise, when the commissioners were unable to reach a compromise over the limits of La Tor de Carol and Guils and agreed to institute a zone of shared pasturage or "compascuity," the commune of La Tor de Carol protested loudly and the commission was forced to concede the drawing of a dividing line. A similar concession resulted from the protests of the communities of Llívia and Angostrina, which refused to recognize a common pasture as their boundary and insisted on a straight line.[47]

It was not, moreover, always the same issues or the same communities which opposed the settlements offered by the international commission. The construction of canals by Guils and La Tor de Carol, for example, diverting water from the Carol River set those communities against each other, when they had previously only fought over their pasture lands and territorial boundaries. Aja in Spain and Palau in France, neighboring villages, had some history of conflict prior to

45. FAMRE CDP 11, fols. 189–208 (July 3, 1868).

46. FAMRE CDP 10, fol. 87 (Callier to foreign minister, September 4, 1866); and ibid., fols. 166–199v (Spanish ambassador to Callier, November 30, 1866, concerning additional incidents on November 19, 1866).

47. FAMRE CDP 10, fol. 180 (Callier to foreign minister, December 7, 1866), and fol. 298 (June 7, 1867, on Guils and La Tor de Carol); FAMRE CDP 11, fols. 13–16 (Callier to foreign minister, January 23, 1868); and FAMRE Callier 12 (Callier to foreign minister, March 19, 1868, on Angostrina and Llívia).

1868, but they had been united most recently in a common struggle during the 1840s to prevent the French village of Osseja from extending its irrigation canal from the Vanera River.[48] Yet their mutual animosity during the delimitation talks provoked Señor Fivaller, a Spanish deputy to the Cortes who had lived in Palau, to remark:

> I come from the district. I know the feelings of the inhabitants of Palau. If we [Spaniards] had the misfortune of incorporating one inhabitant or one house [into Spain], the other inhabitants of Palau would burn his house down so that he would not have to become Spanish.[49]

Despite its exaggeration, Fivaller's assessment evokes the depth of local opposition that found a national expression. That opposition was derived from the communities' local experience but extended far beyond it: national identity was forged ultimately on the defense not of village but of national territory.

National territory, in turn, defined nationality: the local opposition was framed as an expression of national differences, or at least the perception of cultural and national differences between Frenchmen and Spaniards in the borderland. Decrying the 1866 Treaty, the mayor of La Tor de Carol had written:

> The inhabitants of La Tor would like to see all the rights of compascuity abolished, so as not to have any interaction with the inhabitants of Guils. We want a dividing line to separate from now to eternity *two villages of foreign nations and of different mores.*[50]

That the villagers claimed to belong to different "nations" with different "mores" is somewhat astonishing. After all, it was less than two generations after the town council of Puigcerdà had asserted quite the contrary:

> The peasants follow the same methods in cultivating their lands; the families of both sides are intertwined; landowners and proprietors from one Cerdanya own estates and properties in the other, and vice versa; the people speak a same language, and they dress alike, so that in seeing two indi-

48. ADPO XIV S 108 (councillors of Aja to prefect in Perpignan, August 29, 1847); FAMRE Callier 13 (correspondence and memoirs concerning the canal, 1846–1847); FAMRE CDP 6 (order of French Ministry of Public Works to foreign minister, June 22, 1849); and, on the 1868 conflicts, FAMRE CDP 11, fols. 114–119 (minister of agriculture to foreign minister, May 13, 1868).

49. Quoted in E. Brousse, *La Cerdagne française* (Perpignan, 1926), 68.

50. FAMRE CDP 10, fol. 298 (letter to Callier, June 20, 1867, emphasis added); similar remarks by Angostrina in FAMRE Callier 12 (extracts of municipal record, April 13, 1868).

viduals from the Cerdanya, one Spanish and the other French, it would be impossible to distinguish which is which without knowing them already.

The town council had, of course, a distinct agenda in advancing such a claim. News of the Paris treaty of June 1814 had just reached the Cerdanya, and the town's response was to urge the Spanish government to annex the "upper" or French Cerdagne. Arguing that the Cerdanya had formed a unity "since at least Roman times," and that the county had always been attached to the Principality of Catalonia, the municipal councillors of Puigcerdà had stressed the unity of the valley in support of Spanish claims.[51]

In the absence of detailed descriptions of language, customs, and costumes, it is impossible to verify fully the claims of cultural identity and national differentiation in the borderland during the first half of the nineteenth century. Yet limited evidence suggests that Puigcerdà's claim of 1814 could still have been made in the 1860s—that at the moment of the delimitation of the boundary, the perception of national difference was more important than positive indices of national differentiation.

Had agricultural methods remained the same? It is clear that larger landowners in the Cerdanya had been experimenting for decades with ways of improving agricultural yields. Enclosures had been essayed since (at least) the early eighteenth century; and in the early nineteenth century, some proprietors had made improvements in irrigation and artificial meadows, adding new crops, such as potatoes.[52] But since such innovations took place on both sides of the boundary, they were more a function of class than of nation. Most peasant proprietors still followed the same agricultural methods in the 1860s as they had a century earlier. Firm statistics are lacking, but it does not appear that the local economy of the valley experienced any major structural differentiation before this period. It is true that the French Cerdagne seems to have enjoyed a higher ratio of meadows to arable land in the 1860s than did the Spanish side, suggesting a greater emphasis on larger livestock maintained during the winter months on the fodder from those meadows. But the quantities and ratios of livestock remained more or less steady in the French Cerdagne from the 1730s to the 1860s, suggesting

51. SAHN Estado, libro 674d, fols. 57–64 (petition of municipality of Puigcerdà, July 14, 1814).
52. ADPO 7J 105: "Mémoire à l'occasion de fixer les limites," by Pera, 1803; ADPO 2J 81/1–3: "Essai sur la statistique du département," 1802, fol. 184.

that the French Cerdans had not moved away significantly from pastoral concerns, nor had their Spanish counterparts.[53]

It has already been shown how much more "intertwined" were families from both sides of the valley in the midnineteenth century than they were a half-century earlier; and landownership across the boundary continued to follow its eighteenth-century patterns. The larger estates owned by Spaniards in the French Cerdagne remained intact across the revolutionary decade, throughout the nineteenth century, and even until today. Indeed, the contrast between Spanish and French properties was accentuated. Spanish properties in France were not subject to the equal inheritance provisions of the Civil Code and were thus passed on intact according to Catalan customary law, which guaranteed the prerogatives of primogeniture. Although French proprietors in the Cerdagne undoubtedly followed a range of notarial strategies—from fictitious sales to free gifts—to avoid compliance with the Civil Code, the larger French-owned estates tended to break up beginning after the middle of the nineteenth century.[54]

The continuity of property ownership across the boundary was matched by the continuity of religious practice. The religious geography of pilgrimage sites, for example, continued to ignore the political boundary. A nineteenth-century military engineer—one of the better ethnographers of rural life in the Cerdanya—described how women from both Cerdanyas frequented Nostra Senyora de Núria in Spain, going on pilgrimage "during the fine season to stick their heads in the miraculous pot" in the interests of conceiving, although it is true that the role of Nostra Senyora de Belloch as "patron and protectress of both Cerdanyas" was much weaker in the nineteenth century than a century earlier.[55]

53. Sahlins, "Between France and Spain," app. 5.

54. The effects of the Civil Code on property ownership in the Cerdagne have not been adequately studied. For the neighboring Capcir, however, whose patterns the Cerdagne seems to have followed, see L. Assier-Andrieu, *Coutumes et rapports sociaux: Etude anthropologique des communautés paysannes du Capcir* (Paris, 1981); and for the central Pyrenees, R. Bonnain, "Droit écrit, coutume pyrénéenne, et pratiques successorales dans les Baronnies de 1769 à 1836," in *Los Pirineos: Estudios de antropologia social e historia*, Coloquio Hispano–Francés, Casa de Velazques (Madrid, 1986), 232–240.

55. FMG AAT MR 1223, no. 66: "Reconnaissance de la vallée d'Err, par M. Bragouse de Saint Sauveur, Capitaine d'Etat Majeur," September 29, 1841. In part, Belloch had been displaced by more localized and more generalized sites in the French Cerdagne. More locally, the patron saint of Err in the French Cerdagne underwent a revival in the 1840s. In 1847, a severe drought brought villagers of the Spanish and French Cerdanyas to pray at the site, and the processions and invocations on that saint's day (September 24) continued to be popular. More generally, the chapel of Nostra Senyora de Font-Romeu also revived in the early nineteenth century, due in large part to the efforts of

Nor did religious practice in the two Cerdanyas appear to have suffered much differentiation. The Catalan language remained the idiom of local and popular religion. Although liturgical and devotional books in Catalan appeared less frequently than in the eighteenth century, popular religious drama and *goigs* or hymns suggest a similar religious sensibility. The evidence is impressionistic: but religious practices such as popular drama appeared to an English traveler to the Cerdanya in the 1850s not to have been affected by the boundary.

> Both the French Cerdagne and the Roussillon are as essentially opposed, in the manners of their population, to anything that can be termed French character, as the people of the Spanish Cerdaña, or of Catalonia itself. It is scarcely to be supposed that a population which retains the intellectual traditions of the Middle Ages in reference to [popular morality plays] should become materially affected, in point of social character, by the stipulations of a treaty concluded within the last two centuries.[56]

"The people speak a same language, and they dress alike," wrote the municipality of Puigcerdà in 1814. Given the absence of detailed descriptions of costume in the nineteenth century, the latter claim remains unverifiable. The question of language, however, can be subjected to a more detailed scrutiny.

The Catalan language remained throughout the nineteenth century the language of rural life in the borderland. It does not appear that the boundary had any significant effect on the phonetic pronunciation of Catalan in the two Cerdanyas. Commentary from the late eighteenth and early nineteenth centuries—including the linguistic inquiries of the "Friends of the Constitution" in Perpignan during the Revolution and the musings of military engineers in the French Cerdagne—does not yield much insight about the borderland. The Prefect Martin in 1806 did suggest that the Catalan spoken "in Roussillon" had grown much closer to the "patois" of Languedoc than to the Catalan proper spoken across the boundary.[57] But the only reliable evidence

Msgr. de Laporte, Bishop of Perpignan. By the 1850s, Font-Romeu had become a site of widely publicized miracles, its invocation (September 8) drawing the participation of the Perpignan elite as well as the villagers and townspeople of both Cerdanyas: see B. Cotxet, *Noticia històrica de la imatge de Nostra Senyora d'Err . . . y una curta relació de la seguedat de 1847* (Perpignan, 1855); Tolra de Bordas, *Notices sur Notre Dame de Font-Romeu* (Perpignan, 1865); and Abbé Emilie Rouse, *Histoire de Notre Dame de Font-Romeu* (Lille, 1890).

56. F. H. Deverell, *Border Lands of Spain and France: With an Account of a Visit to the Republic of Andorra* (London, 1856), 185.

57. FAN F(20) 243 (prefect to interior minister, November 3, 1806); see also M. Bouille, "La décadence de la langue catalane en Roussillon au XVIIIe siècle," *Tramontana*, 434–435 (1960): 89–95.

about linguistic frontiers comes from systematic sampling of the midtwentieth century. Henri Guiter's detailed atlas of the boundaries of spoken Catalan, reconstructing phonetic differences within the Cerdanya, suggests a minimal boundary effect. The difference introduced by the political boundary is infinitely smaller than the inherited linguistic boundary of Occitan and Catalan, north of the Cerdanya, or even the linguistic frontier—incorporating the entire Cerdanya valley—which distinguishes the Roussillon dialect from the Catalan spoken in the principality.[58]

The Catalan spoken in the Cerdanya during the late nineteenth and early twentieth centuries was certainly affected by the influence of both Castilian and French. In 1917, Xandri noted how "schools, the press, military service, and relations with the state" were the cause of lexical shifts on both sides of the boundary. "The dialect of the French Cerdagne has suffered notably the influence of the French and Provençal language. On a lesser scale, the Spanish Cerdaña has also experienced the influence of these two neighboring tongues." Around the same time, Jaume Bragulat Sirvent recalled the "corruption" of spoken Catalan by French:

> One has to remember that the Cerdanya was occupied by the French more than 200 years ago, and had since been inundated by gendarmes, customs guards, and other state officials, with their families. Naturally, nearly all of them spoke only French, which rubbed off on the locals, and changed their dialect and pronunciation. These corruptions spread slowly, more or less attenuated, to the Spanish side.

Especially in Puigcerdà, where there were "constant interchanges and daily relations of the people from both sides," French "modalities" could be heard. Yet such differences in spoken Catalan remained lexical and secondary, not structural.[59]

The persistence of spoken and written Catalan in the nineteenth-century French Cerdagne did not preclude the differential adoption of French, as suggested in part by the extension and quality of primary schools. It is true that many sons of wealthier families may have con-

---

58. H. Guiter, *Atlas linguistique des Pyrénées-Orientales* (Paris, 1966); see the maps and discussion of J. Costa i Costa, "Aproximació linguistïca al català de la Cerdanya," in *Primer congrés internacional*, 207–213.

59. J. Xandri, *La Cerdaña* (Madrid, 1917), 206–207; Brugulat, *Vint-i-cinc anys de vida Puigcerdanesa, 1901–1925* (Barcelona, 1969), 26–27. For examples of "gallicisms" and "occitanisms" in the Catalan spoken today in Roussillon, see P. Verdaguer, *El Català al Rosselló: Gallicismes, Occitanismes, Rossellonismes* (Barcelona, 1974).

tinued to frequent schools across the boundary, where—for example, at the Escolas Pias in Puigcerdà—the curriculum was largely in Castilian.[60] But the growth of schools in the French Cerdagne during that time is impressive, especially since a succession of French governments in the nineteenth century failed to build upon the Revolution's experiment in universal cumpulsory primary education.

At its height under the Committee of Public Safety, the French revolutionary state never managed to enforce its policies of "linguistic terrorism" on the periphery. Under the Directory, four schools of the French Cerdagne had all taught in the local idiom, some having "translated the Rights and Duties of Man into Catalan."[61] Throughout the first part of the nineteenth century—until the passage of the Guizot law of 1833 requiring communities to maintain a school—the French state rarely interfered with the educational practices in the French Cerdagne. Yet schools had grown in number and had shifted into French during the first three decades of the nineteenth century: by 1833, there were nine schools with a total of 283 winter students and 160 summer ones, with students averaging five years of instruction in "religion, catechism, reading, writing, and French grammar." By 1863, there were twenty schools, including four "mixed" ones for boys and girls. The number of students had risen to 722, including 73 girls.[62]

The quality of the schools, although difficult to measure, was also striking: in 1833, the canton of Sallagosa had the "best-maintained" schools in the district. According to an inspector, this was due in part to the "large number of families involved in commerce" who had asked for schools and supported them. Others commented on the winter's severity in the region and the lack of farm work during the winter months, thus freeing children to attend school.[63] The result was the extremely high literacy rates by the 1860s—80 to 85 percent of military conscripts from the Cerdagne could sign their names in 1863, the highest percentage in the whole department.[64]

60. Blanchon, En Cerdagne, 19.
61. ADPO L 1115: "Canton de Mont-Louis: Ecoles primaires établies conformément à la loi du 3 brumaire an IV"; see also ADPO L 389, nos. 41 and 127; and FAN F(17) 10459: "Tableau des instituteurs primaires, 4 thermidor an II." More generally, see P. Higonnet, "The Politics of Linguistic Terrorism and Grammatical Hegemony during the French Revolution," Social History 5 (1980): 41–69; and M. de Certeau, D. Julia, and J. Revel, Une politique de la langue: La Révolution française et les patois—L'enquête de Gregoire (Paris, 1975).
62. FAN F(17) 139: "Inspection des écoles primaires," October 1833; FAN F(17) 10459 (inspection report of 1863).
63. FAN F(17) 139; Brousse, La Cerdagne française, 60.
64. L'Indépendent (Perpignan), January 16, 1864.

But what did it mean to be able to speak, read, and write in French? For one, it did not mean abandoning the usage of Catalan, which remained the language of family and community, of culture and religious life. It is generally believed that the use of French remained restricted to the public and official realm—its value lay in facilitating participation in commercial relations with the developing markets and industries north of the Corbières, or of giving a younger generation career opportunities other than those offered in the Cerdanya itself. "A peasant who learns to read and write quits agriculture," wrote the intendant of Provence in 1782, but literacy was as much an effect as a cause of peasants becoming peddlers, merchants, and businessmen.[65] Commerce was an important means of participating in the life of the nation, and the language of the nation was necessary to participate in commerce. It is thus no accident that high literacy rates correspond with a precocious out-migration from the valley, in comparison with neighboring districts.

But French was not just the language of commercial advancement, it was also the language of power, especially of the provincial ruling elites. By the midnineteenth century, the provincial elites—landowners and manufacturers as well as professionals and civil servants—had been won over to France. Among these groups, the Second French Republic was a critical period in forging national identities, as it was throughout France.[66] But the rural population was more belatedly "assimilated" into the nation. When landowners spoke French, peasants resisted in Catalan. As Henri Baudrillart observed in the 1890s,

> The Catalan language has remained the ordinary language of the Roussillon countryside . . . landowners are forced to use the Catalan language: the peasants, even if they understand French, impose it upon them. By speaking in Catalan, they try to keep their distance, and they feel humiliated in hearing orders given in an idiom which they do not see as their own. A certain moral isolation results from this difference of idioms.[67]

In the same way, as the French language was linked to the world of authority and the power of the central state, the use of Catalan could serve members of local society as a means to state their opposition to the state—whether French or Spanish, or both. In 1810, Jean Garreta from Err ("stocking merchant during the winter, carpenter in the summer") was arrested in Puigcerdà, accused of having assisted four Span-

65. Quoted in Febvre, "Langue et nationalité en France," 24.
66. McPhee, "The Seed-Time of the Republic," passim; and the sources cited above, chap. 6, n. 68.
67. H. Baudrillart, *Populations agricoles de la France* (Paris, 1893), 3: 333.

ish prisoners of war to escape back to Spain. Asked if the soldiers were French or Spanish, he replied: "I didn't ask them, they spoke Catalan." Written use of the Catalan language could also express solidarities across the boundary. "Qual trist temps es aquest" [What sad times are these]," lamented in 1822 a cousin of Batholome Carbonell of Gorguja (Llívia), who himself lived in Santa Llocaya in France. Unable to cultivate his lands in Spain because of the sanitary cordon, he asked if a neighbor of Gorguja would be able to lease them out for the year.[68]

The Catalan language, then, could and did serve local society as a language of resistance to the state. But contrary to what the nineteenth century would lead us to believe, there was no *necessary* relationship between language and identity—between the use of French and the abandonment of local identity or between the persistence of Catalan and the lack of a French or Spanish identity. Even the French language could be used in the written correspondence among close family members in the French Cerdagne during the nineteenth century. The creation and expression of national identities depended less on language than on the affirmation of local and national boundaries.

When in 1866, the mayor of La Tor de Carol called the inhabitants of the neighboring village of Guils citizens of "a foreign nation, [with] different mores," he was strengthening his village's claim to a boundary line by evoking a national and cultural distinction. He did not, of course, elaborate on the content of that distinction—if indeed there was one, given the apparent similarity of the Catalan language and customs on both sides of the boundary. What counted was the affirmation of distinctive nationalities as a perceived difference.

In the 1840s, the intrepid Englishman Richard Ford, wandering between France and Spain in the central Pyrenees, was struck by the animosity Spanish peasants expressed toward their French neighbors:

> The [Spaniard's] hatred of the Frenchman . . . seems to increase in intensity in proportion to vicinity, for as they touch, so they fret and rub each other: here is the antipathy of the antithesis; the incompatibility of the saturnine and slow, with the mercurial and rapid; of the proud, enduring, and ascetic, against the vain, the fickle, and the sensual; of the enemy of innovation and change, and the lover of variety and novelty; and however tyrants and tricksters may assert in the gilded galleries of Versailles that *Il n'y a plus de Pyrénées*, this party–wall of Alps, this barrier of snow and hurricane, does and will exist forever.[69]

68. ADPO Mnc 831/2 (police interrogation, September 11, 1810); Carbonell family papers.
69. Ford, *Gatherings from Spain,* 29–30.

Revealing more about the enduring stereotypes of national character than about the people of the borderland themselves, Ford nonetheless noted a significant feature of life on the frontier—the importance of differences *perceived* among villagers in proximity to each other. Such perceived differences were all the more striking in the Cerdanya as compared to elsewhere since in the valley there were no Pyrenees to contend with—the boundary had been drawn through the center of the plain. The inherited antipathy of Frenchmen to Spaniards was matched, moreover, by an equally strong animosity toward Spaniards by Frenchmen, one which had formed during the centuries after the Treaty of the Pyrenees. Finally, the sense of difference was all the more astonishing where the inhabitants shared a common ethnicity distinct from their nationalities as Frenchmen or Spaniards—where, alongside their claims of separate nationalities, they remained Cerdans and Catalans.

8
Conclusion: Identity and Counter-Identity

In 1870, France went to war with Prussia. It was the first national conflict since the Napoleonic Wars to involve the French Cerdans in an immediate and direct way. Throughout the nineteenth century, the French Cerdans had been more engaged by the civil wars in Spain than by political events in Paris: the Constitutional Triennium and First Carlist War in Spain had mattered more than the July Revolution or 1848 in France. Before 1870, the French Cerdans had defined their national identities less by participating in the life of the nation than by using the state for their own ends. In the process, they drew their local boundaries of territory and identity in opposition to Spaniards on the other side of the political boundary, despite sharing with them a common Catalan ethnicity. But the focus of the French Cerdans shifted dramatically in 1870 when they were called upon to serve their own nation at a time of crisis. The Franco–Prussian War forced the French Cerdans to make choices about their attachment to France: 1870, like 1914 and 1940, was a moment that shaped and fixed French identities and loyalties.

The fate of three sons from the village of Palau in the French Cerdagne, called to serve France in 1870, reveals much about the linguistic abilities, political choices, and national identities of the French Cerdans. The eldest son, Raphael, was drafted, joined his regiment, and was soon on his way to the Rhine front. He happened to be in Paris "at the moment the Republic was proclaimed, and I can tell you that

267

everything happened in an orderly way, and not a drop of blood was shed," he wrote to his parents, in the good French of a *bon bourgeois*. To such values he gave his life as, tragically, he died two months later in the Battle of Noyelles. The youngest son, Antoni, chose to flee to Catalonia, and from there he eventually made his way to Brussels and Liège. From Barcelona he had written to his father in a garbled, barely recognizable French, and over the next few years he wrote his mother in much the same style. Yet to his middle brother, Isidor, he wrote in a Castilian decidedly "corrupted" by Catalan syntax and vocabulary. Isidor himself had been living in Barcelona when he was called to serve the French nation, and he too fulfilled his national obligation. Taken prisoner of war in 1871, he wrote to his father in French (not unlike Raphael's), but once released and living in Barcelona, he wrote in Castilian to his family. In 1874, after not having news of his parents for some time—due, he was to learn, to the Carlist uprising—he anxiously wrote to his parents in Catalan.

The Delcor sons came from a relatively prosperous family whose eighteenth-century forebears had already traveled extensively in southern France as "stocking merchants."[1] Yet they were not unlike many Cerdans in their abilities to speak and write in several languages. Their example suggests the need to modify simple assertions about the restriction of Catalan to the private realm of the family, and of French to the language of public affairs.[2] More generally, the Cerdans' chosen languages of expression, and their various degrees of literacy, stood in no necessary relation to their possible identities and chosen loyalties. Throughout Europe in the nineteenth century, that "golden age of vernacularizing lexicographers, grammarians, philologists, and litterateurs," language became the essential element in the definition of national identity, while the recovery of "submerged" languages became the claim of nationalist parties.[3] The example of the Cerdanya suggests how much such claims were political constructions, bearing little relation to the experiences of peasants and others who established and maintained their local and national identities.

One of the three sons chose to escape the draft by fleeing across the boundary and settling in Catalonia, just as many Cerdans before him

1. Delcor family papers.
2. For some interesting reflections on "intrafamilial" boundaries of language use, see Ll. Planes, *El petit llibre de Catalunya Nord,* 121–146.
3. Anderson, *Imagined Communities,* 69; see also H. Seton-Watson, *Nations and States: An Inquiry into the Origins of Nations and the Politics of Nationalism* (Boulder, Colo., 1977).

had done, and as many others were to do later when faced with conscription demands of the First World War. That the other two chose to serve against the Prussians, however, was an index that national sentiment had become entrenched, at least to a much greater extent than suggested in the Cerdans' record of military service under the First Empire. The classical ideal of national identity has exalted the willingness to sacrifice individual or local interests in the service of one's country, the willingness to fight and to die for the fatherland. In its different historical expressions, the ideal of dying for one's country—pro patria mori—can be traced from Virgil and Horace to Machiavelli (via the Christian paradigm of martyr), Rousseau, the French Revolution, and modern nationalism.[4] It is inscribed today in nearly every town, village, and hamlet in France, on war memorials to sons who died for the fatherland, morts pour la patrie, during the First World War.

But the experience of the Cerdanya has suggested that this ideal is only one possible way of affirming national identity: in the first two centuries after the division of the valley, the Cerdans created their own national identities in other ways. One was instrumental, through the use (and abuse) of the nation, whether France or Spain. The Cerdans developed a rhetoric of national identity that masked their own interests and appealed to the ideals of government officials. Yet over the course of two centuries, the Cerdans ended up convincing themselves of their affiliation to France or to Spain; their "national disguises" wound up "sticking to their skin," to borrow a phrase of Michel Brunet.[5] The Cerdans, moreover, affirmed their nationality without abandoning a local sense of place; indeed, their national identities were grounded in the affirmation and defense of social and territorial boundaries against outsiders. Outsiders were either those "political amphibians" who abused the possibilities of choosing different identities in the borderland, or other communities, across the bound-

4. The Roman ideal of pro patria mori had died out by the second millenium, only to be reborn within Christian theology as the idea of the martyr. The "resecularization" of that concept, and the appearance of "sentiments of semireligious devotion" directed toward the regnum, has been beautifully elucidated by Ernest Kantorowicz, in his King's Two Bodies, 232–272; for further documentation of the history of the idea in France, see P. Contamine, "Mourir pour la Patrie, X–XXe siècle," Nora, ed., Les lieux de mémoire, 2 (pt. 3): 11–43; and C. Beaune, Naissance de la nation France (Paris, 1986), esp. 324–335.

5. Brunet, Le Roussillon, 58. Brunet writes that "[by] a long and invisible process, the fantastical and pejorative name-calling would one fine day be claimed with pride by those one sought to humiliate," although he dismisses the idea that such a movement had already taken place in the early nineteenth century.

ary, with whom villagers competed for limited ecological resources. In both cases, national identity involved less the sacrifice of personal or local interests than an affirmation of the local interests of community—often against the interests of the state itself.

More generally, national identity—much like ethnic or class identity—involved the supposition of a boundary between "us" and "them," "friends" and "foes." This arbitrary but very real quality of boundaries was evoked in a *Pensée* by Blaise Pascal (1623–1662):

> Why do you kill me? What! Do you not live on the other side of the water? If you lived on this side, my friend, I would be an assassin, and it would be unjust to slay you in this manner. But since you live on the other side, I am a hero, and it is just. . . . A strange justice that is bounded by a river! Truth on this side of the Pyrenees, error on the other.[6]

Pascal was writing in a period during which rivers and other natural frontiers between states were becoming increasingly visible, highlighted on maps and emphasized in geographical discourse. In the Cerdanya, however, except for the Raour River separating Hix (later Bourg-Madame) and Puigcerdà, there were no natural boundaries to divide the people of the valley. The French and Spanish states had drawn an arbitrary division through the center of the plain. The political boundary did not transform the essential cultural identity of the Cerdans as Catalans between 1659 and 1868, yet around it the Cerdans shaped new national identities as Frenchmen and Spaniards: political constraints became meaningful boundaries of territorial and social identities.[7]

The definition of national identity does not depend on natural boundaries, nor is it defined by a nuclear component of social or cultural characteristics—an essential, primordial quality of "Frenchness" or "Spanishness."[8] National identity is a socially constructed and continuous process of defining "friend" and "enemy," a logical extension

6. Pascal, *Pensées*, ed. L. Brunschvig (Paris, 1937), nos. 293–294.

7. Anderson, *Imagined Communities*, 55, on the "ways in which administrative organizations create meaning"; for similar conclusions about the role of administrative boundaries in rural society, see A. Zink, "Pays et paysans gascons sous l'Ancien Régime," 9 vols. (Thèse de doctorat, Université de Paris, 1985), 2209–2210; and P. H. Stahl, "Frontières politiques et civilizations paysannes traditionelles," in *Confini e regioni*, 459–465.

8. Compare the continuing attempts by historians, sociologists, and psychologists to define "national character" as social-psychological types, including R. Darnton, "Peasants Tell Tales," in *The Great Cat Massacre and Other Episodes in French Cultural History* (New York, 1985), 9–72; D. Bell, "National Character Revisited," in E. Norbeck, ed., *The Study of Personality* (New York, 1968); and D. Peabody, *National Characteristics* (Cambridge, Eng., 1985).

of the process of maintaining boundaries between "us" and "them" within more local communities. National identities constructed on the basis of such an oppositional structure do not depend on the existence of any objective linguistic or cultural differentiation but on the subjective experience of difference. In this sense, national identity, like ethnic or communal identity, is contingent and relational: it is defined by the social or territorial boundaries drawn to distinguish the collective self and its implicit negation, the other.[9]

Identity as opposition—the complementary definition of "us" and "them"—has frequently been interpreted in light of the distinctively twentieth-century political modalities, fascism and nationalism. To Carl Schmitt, writing in the 1920s, the opposition of "friend and enemy" lay at the foundation of all political life. For Schmitt, the political referred to international relations in the period of the sovereign state, roughly since the eighteenth century. But because state and society penetrated each other so thoroughly, everything became grouped under the essential friend/enemy antithesis, which the state alone was able to define. Thus the political enemy was

> the other, the stranger; and it is sufficient for his nature that he [sic] is, in a specially intense way, existentially something different and alien, so that in the extreme case conflicts with him are possible. These cannot be decided by a previously determined general norm nor by the judgment of a disinterested and therefore neutral third party.[10]

But the sense of difference that defines the other does not depend on a continuous state of war. It is neither an exclusive product of political life in the twentieth century, nor is it necessarily founded on the experience of complete unrelatedness or foreignness. In many ways the sense of difference is strongest where some historical sense of cooperation and relatedness remains, as in the Catalan borderland of France and Spain.

Frontier regions are privileged sites for the articulation of national distinctions. Freud remarked how "it is precisely communities with adjoining territories, and related to each other in other ways as well, who

---

9. On the definition of ethnic groups in the ancient world as "differential and oppositional," see Benveniste, *Le vocabulaire des institutions indo-europeennes,* 1: 368; and Armstrong, *Nations before Nationalism,* 3–11. On communities, see A. P. Cohen, *The Symbolic Construction of Community* (London, 1985), esp. 115; on nations, see the remarks in the Introduction to Grillo, ed., *"Nation" and "State,"* 10–11; and Gellner, *Nations and Nationalism,* 7.

10. C. Schmitt, *The Concept of the Political,* trans. G. Schwab (New Brunswick, N.J., 1976), esp. 27–37.

are engaged in constant feuds and in ridiculing each other—like the Spaniards and Portuguese, for instance, the North Germans and South Germans, the English and the Scots, and so on." Freud called this phenomenon, so well illustrated in the Cerdanya, the "narcissicism of minor differences."[11] Yet the oppositional nature of identities and counter-identities is not dependent on the existence of a shared boundary. It is not only communities living in the borderland of national states which adopt, through the process of segmentary opposition and unity, national identities, for the "us/them" grouping can be extended beyond (or below) the state to elucidate certain images of conflict relations and identity among European rural communities before and during the formation of the modern national state.

On the one hand, the village community, both before and during the formation of the national territorial state, is often described as if its constituent households were either in a "state of war" or existed without coherence, "much as potatoes in a sack form a sack of potatoes," to borrow Marx's infamous caricaturization. In this view, if peasant families were not directly competing against each other, they nonetheless shared no conception of a common good beyond the material interests of the household. On the other hand, the village community inevitably formed a united front in opposition to outsiders, whether neighboring villages, privileged groups, judicial officers, or overly curious anthropologists. The identity of the community took shape in opposition to the other.[12]

Endless lists of popular and often scatological insults exchanged between adjoining neighborhoods and villages in Europe attest to the fact that self-definition depends on antithesis, identity on counter-identity.[13] In much the same way, the segmentary model of opposition and unity could be applied to the identity of districts or cantons, regions or provinces, and even nations. The words for "Spaniards" and for "Catalans," for example, were originally used by foreigners to define these

11. S. Freud, *Civilization and Its Discontents,* trans. and ed. J. Strachey (New York, 1960), 61.

12. K. Marx, *The Eighteenth Brumaire of Louis Bonaparte* (New York, 1963), 124; for an extreme description of "amoral familism" in a southern Italian village, see E. C. Banfield, *The Moral Basis of a Backward Society* (New York, 1958); for examples of the contextual fission and fusion of communal life, see J. Pitt-Rivers, *People of the Sierra* (Chicago, 1957); and L. Wylie, *Village in the Vaucluse* (Cambridge, Mass., 1974).

13. For Catalonia, Sales, "Naturals i aliénígenes," 700–701; and more generally, Caro Baroja, "El sociocentrismo," 264–292; for a partial bibliography of French *blasons populaire,* which include such insults, see A. Van Gennep, *Manuel de folklore français,* 4 vols. (Paris, 1938–1945), 4: 750–762.

peoples and only adopted subsequently by the peoples themselves. Whether designated by foreigners or expressed by a nation, the terms of identity depend on the existence of an other.[14] As the liberal theorist of the emerging nation-state, John Stuart Mill, wrote in the 1840s, "The evils of Spain flow as much from the absence of nationality among the Spaniards themselves as the presence of it in relations with foreigners."[15]

Mill was wrong to dismiss the oppositional definition of nationality as a defect of Spanish national character. For the contextual and differential affirmation of national identity was present even in France, the nineteenth century's paradigm of national and cultural unity. Yet it is true that the segmentary model of contrast and opposition fits the Spanish case extremely well: in the early modern period, Spanish unity was constructed from the conjoining of relatively autonomous kingdoms and provinces, whose unity was defined in relational and oppositional terms. In France, by contrast, these kingdoms and provinces, although preserving their customs and privileges, were nonetheless merged within an encompassing—and king-centered—nation. North of the Pyrenees, the concentric circle model seems to describe a historical experience.

Yet on both sides of the Pyrenees, the models of segments and circles are appropriate illustrations of the modalities of local and national identities. The two models are not incompatible, as the case of the Cerdanya illustrates. Expressions of loyalty and identity in the borderland were grounded in relational and oppositional terms. But individuals and communities also identified in various degrees with wider and encompassing communities. Central to the coexistence of both models is the importance of modifying this notion of "encompassment," especially in France. Michelet commented in 1833 on the extremely centralized character of French political and cultural life, claiming that

14. On the identity of cantons (*comarques*) in Catalonia, which developed oppositionally, see P. Vila, *La divisió territorial de Catalunya* (Barcelona, 1979) 47–49; on the dependency of early forms of "nationalism" on contrast and opposition, see H. Koht, "The Dawn of Nationalism in Europe," *American Historical Review* 52 (1947): 265–280; and, for the early modern period, Ranum, "Counter-Identities," passim. For the origins of the words for "Spaniard," see A. Castro, *The Spaniards: An Introduction to Their History,* trans. W. F. King and S. Margaretten (Berkeley, Los Angeles, London, 1971), 10 ff.; and for "Catalan," T. Bisson, "L'essor de la Catalogne: Identité, pouvoir, et idéologie dans une société du XIIe siècle," in *Annales* 39 (1984): 456. For a contemporary example of national counter-identities, see E. M. Lipiansky, "L'imagérie de l'identité: le couple France-Allemagne," in *Ethnopsychologie* 34 (1979): 273–282.

15. John Stuart Mill, *System of Logic* (London, 1847), bk. 6, chap. 10, sec. 5.

"the Parisian genius, as the highest and most complex form of France," necessarily entailed "the annihilation of all local spirit, all provincialism." The idea was echoed by Tocqueville, writing about the role of Paris under the Old Regime, and by others since.[16] This "myth of inevitable conformity," to use Cohen's term,

> suggests that the outward spread of cultural influences from the center will make communities on the periphery less like their former selves—indeed, will dissipate their distinctive cultures—and will turn them, instead, into small-scale versions of the center itself.[17]

By identifying with more encompassing communities, however, local communities lost neither their distinctiveness nor a sense of place. Peasants, privileged groups, and village communities adopted national identities without losing their local ones, and did so well before their respective national states undertook consistent policies of linguistic and cultural unification.

This book has explored two dimensions of state formation and nation building since the seventeenth century. It has considered the emergence of a national political boundary and conceptions of territorial sovereignty in France and Spain, and it has focused on the historical formation of national identities in the French–Spanish borderland. Although these two processes were intimately related, they unfolded differently in time.

The Treaty of the Pyrenees named the "Pyrenees Mountains" as the "limit of the two kingdoms," but the boundary line separating distinct national territories was only formally delimited by the Treaties of Bayonne. Between 1659 and 1868, the conception and realization of the boundary and the nature of sovereignty evolved through four distinct stages. The later seventeenth century (1660–1723), a period of perpetual warfare between France and Spain, was dominated by a military frontier that incorporated the totality of the Cerdanya valley, although in the final instances of warfare during those years both states came to recognize the territorial division of the two Cerdanyas at the heart of the military frontier. During the eighteenth century (1723–1789), a period of peace between the two allied crowns, the French and Spanish governments cooperated and competed over the administra-

---

16. Jules Michelet, *Tableau de France* (Paris, [1833] 1948), 89; Alexis de Tocqueville, *The Old Regime and the French Revolution*, trans. S. Gilbert (New York, 1955), 72–96.

17. Cohen, *The Symbolic Construction of Community*, 36.

tion of justice, ecclesiastical affairs, taxation, and contraband trade in the borderland, in the process defining their separate territories. In the second half of that century, the French state sought to delimit formally its territorial boundaries, although it failed in that task along the Pyrenean frontier. The French Revolution (1789–1799) and Spain's reaction was marked by the appearance of distinctively national territories in the borderland and, in France, the administrative consolidation of territorial sovereignty in the form of the *départements*. Finally, the nineteenth century (1799–1868) witnessed the growing politicization of national boundaries and the simultaneous demilitarization of the frontier, as the two states sought to repress the "territorial violations" that gave expression to their national territorial sovereignties.

In the Cerdanya itself, although the evolution of local boundaries and the appearance of national identities were processes that ran parallel to the definition of the political boundary of national territories, the mechanism and timing of their appearance differed significantly. Until the eighteenth century, local society in the Cerdanya largely ignored the division of the valley into two parts, just as the village communities, peasants, and rural notables in the French Cerdagne tended to resist the (albeit limited) process of "incorporation" imposed from above, expressing instead their inherited antipathy toward the French. But with the end of the military frontier, the Cerdans sought increasingly to manipulate and utilize the jurisdictional differences introduced by the division of the valley: in the process, over the course of the eighteenth century, the village communities of both Cerdanyas and the valley's largely Spanish ruling class came to assert their national identities as Frenchmen and Spaniards. To be sure, this stage in the development of national consciousness and territoriality was grounded in the pursuit of local interests and advantages. The Cerdans continued to demonstrate, in a variety of contexts, their resistance to the centralizing states. In using national identities, however, they came to nationalize their interests, as they strategically asked government officials to delimit the territorial boundary.

The French Revolution was a critical period in the formation and development of national identities in the Cerdanya, just as it was for the appearance of a national territorial state. It was during these years that the perspectives of state and society became conjoined, despite the fact that most of the French Cerdans opposed the intrusions of the revolutionary state by welcoming the Spanish troops in 1793, fleeing to Catalonia with the French reconquest of the valley, or remaining in the Cerdagne but resisting the demands of the revolutionary administration. Yet by the end of the 1790s, many of those who had fled came back to reclaim and to use their French nationality. That nationality was defined by the territory in which they lived and worked; conversely, their nationality took on, the principles of jus sanguinis aside, a territorial dimension. The nineteenth century witnessed a consolidation of national identities similar to that of local and national territories in the borderland. Even as the Cerdans continued to resist the imposition of the nation from above, they asserted their nationalities by excluding the "other"—either those individuals who abused the masks of nationality, or the village communities across the political boundary with whom they competed over increasingly scarce resources. The paradoxical appearance of national identities occurred within a twofold structure of continuity. First, expressions of French and Spanish identity took place within an intensified demographic continuity across the political boundary, itself occasioned by the political process of national territorial differentiation. And second, national identity found expression within the continuity of Catalan culture and ethnicity, and of shared bonds of resistance to the centralizing states.

I have argued in this book how the initially arbitrary political division of 1660 became a meaningful structure of national identities and how local society in the Cerdanya was instrumental in creating the political division of France and Spain. The formation of the national territorial boundary line and the expression of national identities in the Cerdanya were two-way processes. The states did not simply impose the boundary or the nation on local society. By defining their own social and territorial boundaries, village communities, peasants and nobles, made use of the national state and its boundaries. By repeatedly using their nations, by bringing the nation into the village, members of local society ended up as national citizens. Through their role in constructing the national boundary, the Cerdans created new identities for themselves as Frenchmen and Spaniards, never abandoning in either process their local interests or sense of place.

Photograph 1. The French village of La Tor de Carol, the abandoned hamlet of Sant Pere de Cedret, and the boundary stone of 1868, as seen from the Spanish side of the boundary.

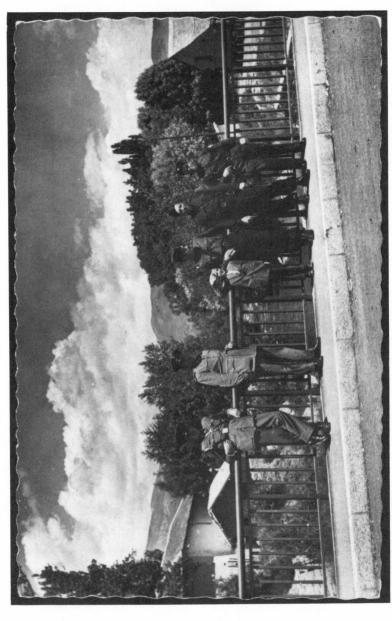

Photograph 2. French *gendarmes* and Spanish *guardias civiles* mark the boundary at the Puigcerdà–Bourg-Madame bridge over the Raour River in this postcard from the late 1950s.

Epilogue: States and Nations since 1868

The delimitation of the international boundary of Spain and France in 1868 coincided with a watershed period in the history of state building and nation formation on both sides of the Pyrenees. During the last third of the nineteenth century, the political and cultural histories of the Spanish and French nation-states led in strikingly different directions. France was consolidated from the center, just as Spain began to disintegrate at the periphery: the one crystallized into a single "Nation, One and Indivisible," while the other broke apart into autonomous nations.

In Spain, the "Glorious Revolution" of 1868 touched off six years of struggles among Monarchists, Republicans, and a nascent working class in the cities, climaxing in the shortlived Federal Republic of 1873, the Third Carlist War, and the restoration of Alfonso XII in 1874.[1] More than just another episode in the turbulent Spanish politics of the nineteenth century, the "Six Years" were marked by the appearance of a new political ideology. Early autonomist aspirations of Catalan nationalism found expression during the debates over the Federal Republic, as Catalan ethnicity and identity became both the vehicle and substance of new claims within a changing political culture in Spain. The nationalist movements for political autonomy in Catalonia and the

1. C. A. M. Hennessy, *The Federal Republic in Spain* (Oxford, 1962).

279

Basque provinces constituted Spain's "counterexperience" in the late nineteenth and twentieth centuries.[2]

In France, the establishment of the Third Republic followed France's defeat by Prussia and the fall of the Commune in 1871. Despite a succession of unstable coalition governments during the next three decades, the French state launched a coherent policy of nation building. The Republic furthered the policies of road and railway construction begun under the Second Empire and introduced new laws of mandatory military service and universal compulsory primary schooling, thus marking the technological and cultural assimilation of "peasants into Frenchmen."[3]

These different political and national histories of France and Spain had more than casual reverberations in the Cerdanya. By 1868, the two Cerdanyas were definitively tied to the historical trajectories of their respective polities: the French Cerdagne experienced an economic assimilation and cultural integration into France, whereas the Spanish Cerdaña partook of the structural underdevelopment of Spain and of the claims of Catalan nationalism. It is important to point out that the experiences of the French and Spanish sides of Catalonia ran counter to the more general opposition of France and Spain: Roussillon remained largely rural and underdeveloped compared with both France and with Catalonia, the most industrialized and wealthiest region of an otherwise underdeveloped Spain. This paradox helps to make sense of the divergence of economy, politics, and culture in the borderland during the last century. The dramatic differentiation of local society in the Cerdanya, however, was not simply a result of forces imposed on the valley from distant political centers; it was also the final stage in the process of self-differentiation that had begun after the division of the valley in 1659.

STATE BUILDING AND DIFFERENTIATION IN THE BORDERLAND

During the Third Carlist War, the inhabitants of Puigcerdà confirmed and deepened their liberal and constitutional loyalties, as be-

2. P. Vilar, *La Catalogne,* 1: 131–165; S. Payne, "Catalan and Basque Nationalism," *Journal of Contemporary History* 6 (1971): 15–51; and the sources on Catalan nationalism listed below, n. 21.

3. L. Girard, *La politique des travaux publics du Second Empire* (Paris, 1951); R. D. Challner, *The French Theory of the Nation in Arms, 1866–1939* (New York, 1965); and E. Weber, *Peasants into Frenchmen: The Modernization of Rural France, 1870–1914* (Stanford, Calif., 1976).

tween August 20 and September 3, 1874, they defended the walls of the town against the repeated assaults of Carlist troops. Their heroic defense won Puigcerdà the title of "Always Invincible," and 1875 produced a flood of memorials and congratulations from the towns of Catalonia and from dozens of members of the political and industrial elite of Barcelona.[4] The Barcelona bourgeoisie, whose celebratory telegrams and letters were collected in a commemorative volume, had a special stake in the town, for Puigcerdà had recently become their summer home, their mountain retreat, and their Pyrenean paradise.

"Outsiders," especially from Barcelona, had long been drawn to the beauty, climate, and attractions of the Cerdanya. An early seventeenth-century eulogy of the valley described its "baths and springs," where "foreigners, who have heard how good they are, come from far away."[5] But it was only in the late 1860s that Puigcerdà was "discovered" by the Barcelona elite and reconstructed as its summer home. A town of some 2,500 inhabitants in the mid-1870s, the population of Puigcerdà more than tripled during the summer months. In 1881, 5,000 "outsiders" arrived in the Cerdan capital between July 15 and August 21.[6] Many of the wealthier families had constructed summer villas, Victorian-style palaces in the district surrounding the town's reservoir, which was rechristened "the lake." The municipality instituted a number of festivals, sometimes reviving traditional ones such as the Feast of the Rosary, sometimes establishing new ones such as the Feast of the Lake. A theater and a casino were built to accommodate the summer residents, who often took day trips to the Roman baths and other sites in both the French and Spanish Cerdanyas.[7]

The easiest route from Barcelona to Puigcerdà in the late 1870s, however, required passing through France—by rail or coach to Perpignan, then up the Tet Valley along the paved road that had linked the French Cerdagne to the plain in 1853. It was only in 1914 that the Spanish government completed the public paved road from Ripoll to

---

4. *El sitio de Puigcerdà por los carlistas desde el dia 20 de agosto hasta el 3 de septiembre de 1874* (Barcelona, 1875); *Libro de honor de la heroica e invicta villa de Puigcerdà* (Barcelona, 1876).

5. BC MS 184, fol. 5: "Descripción de la tierra y condado de Cerdaña."

6. *La Voz del Pirineo* (Puigcerdà), no. 29 (July 17, 1881).

7. On local life and estival culture of the late nineteenth and early twentieth centuries, see *La Voz del Pirineo* (founded 1879), and subsequent newspapers (*Gazeta Cerdaña, Ceretania,* and *El Pirineo*); Jaume Bragulat Sirvent, *Vint-i-cinc anys de vida puigcerdanesa, 1901–1925* (Barcelona, 1969); and P. Sola i Gussinyer, *Cultura popular, educació i societat al nord-est català (1887–1959)* (Girona, 1983), 435–446.

Puigcerdà, across the Tossa Pass, allowing travelers to go directly from Barcelona to the Cerdanya without crossing French territory.[8]

Differences in road networks in the Catalan borderland long antedated these developments. Crossing the Perthus pass from the Empurdà plain into Roussillon in the late 1780s, the English traveler Arthur Young was astonished by the contrasts:

> Here we take leave of Spain and reenter France: the contrast is striking. . . . From the natural and miserable roads of Catalonia, you tread at once on a noble causeway, made with all the solidity and magnificence that distinguish the highways of France. Instead of beds of torrents you have well-built bridges; and from a country wild, desert, and poor, we found ourselves in the midst of cultivation and improvement. Every other circumstance spoke the same language. . . . The more one sees, the more I believe we shall be led to think, that there is but one all-powerful cause that instigates mankind, and that is GOVERNMENT! Others form exceptions, and give shades of difference and distinction, but this acts with permanent and universal force. The present instance is remarkable; for Roussillon is in fact a part of Spain; the inhabitants are Spaniards in language and customs, but they are under a French government.

Young failed to recognize the distinctiveness of Catalan language and customs, just as he undoubtedly conflated a political distinction of governments with a geographical contrast within the Mediterranean landscape.[9] Unfortunately, he also failed to visit the Cerdanya and the more mountainous part of the borderland, where the contrasts would have likely been much less striking. For although there is no doubt that the contrasting roadbuilding policies of Louis XV in France and the absence of improvements under Charles III in Spain left their mark on the two Cerdanyas, it was not until the arrival of paved roads in the mid-nineteenth century that the differentiation became marked.[10]

The construction of paved roads to the French Cerdagne a half-century before their arrival on the Spanish side of the boundary at once symbolized and created a more general set of differences that left the Spanish Cerdaña "underdeveloped" with respect to the French side. (It should be noted that the geographical obstacles to linking Barcelona or Lérida to the Cerdaña by paved road were much greater, and the infrastructure much more costly, than were the obstacles to joining Per-

---

8. Sermet, "Communications pyrénéennes," 119.

9. A. Young, *Travels through France in the Years 1787, 1788, and 1789*, 2 vols. (London, 1793), 1: 59–60; Vilar, *La Catalogne*, 1: 180, 2: 223–224.

10. On improvements in communications in the eighteenth-century French Cerdagne, especially after the creation of the Administration of Bridges and Roads (*Ponts et Chaussées*) in 1736, see the dossiers in ADPO C 1191.

pignan or Toulouse to the Cerdagne. But this geographical contrast became relevant as a function of culture and politics, since it was the political division that structured the earlier "integration" of the French Cerdagne, compared to the Spanish side of the valley.)[11] More generally, road networks, public services, and telegraphs together constituted a set of structural differences that came to distinguish the two Cerdanyas in the late nineteenth century. Contemporaries were struck by the disparities, among them Valenti Almirall, the principal voice of early Catalan nationalism:

> For shame, the Puigcerdanos, for all their liberalism, for all their thirst for progress and betterment, for all their sacrifices made to the patria, have to contemplate sadly, with their hands over their faces, how the French side surpasses them. . . . Still today, to go with comfort from Barcelona to Puigcerdà, one has to pay tribute to the French roads and railways. A town of 2500 inhabitants, Puigcerdà has no telegraph. To communicate with the world, they have to use the French station at Bourg-Madame, a hamlet of four houses. And worst of all, the Cerdans can do nothing about it. . . . If they were part [of France], the state of the Cerdaña would be very different.[12]

By the 1880s, the two Cerdanyas had indeed gone their different ways, expressing the fates of their respective states. Such was the case of border regions all along the Pyrenees, where the contrasting development of what Gomez-Ibañez calls the "infrastructures of change"—administration, communications, energy, and education—led to the earlier and more complete development of the French side of the boundary.[13] In the eastern Pyrenees, the economic divergences between the Roussillon and Empurdà Plains date from the midnineteenth century, when the former "discovered it was the most southern part of France" while the latter "remained the most northern part of Spain."[14]

Travelers of the late nineteenth century often exaggerated the differences between the two Cerdanyas. Victor Dujardin, a "man from the north" of France who visited the Cerdanya in the late 1880s, was astonished by the contrasts:

11. Sermet, "Communications pyrénéennes," passim; on the contrasting policies of France and Spain more generally, see R. Price, *The Modernization of Rural France: Communications Networks and Agricultural Market Structures in Nineteenth-Century France* (New York, 1983); and D. R. Ringrose, *Transportation and Economic Stagnation in Spain, 1750–1850* (New York, 1970).

12. Reprinted in *La Voz del Pirineo,* no. 26, June 26, 1881.

13. D. A. Gomez-Ibáñez, *The Western Pyrenees: Differential Evolution of the French and Spanish Borderland* (Oxford, 1975), 56–133.

14. P. Deffontaines, "Parallélle entre les économies de l'Ampurdan et du Roussillon: Le role d'une frontière," RGPSO 38 (1967): 244–258.

From the last French village to the first one of Spain, the contrast is striking: on this side, in France, usable paths, superb roads, a comfortable population, well-constructed, clean, and comfortable houses, good schools, magnificent fountains and drinking troughs; on the other side, in Spain, awful, muddy ruts of goat trails, fallen down and dirty farms, and misery everywhere.[15]

The assessment, which reproduced stereotypical images of France and Spain, falls short of describing the actual circumstances: if he had visited Puigcerdà, he could not have missed the construction boom in housing and the textile mills that surrounded the town. Yet historians and geographers lend support to such images of the differential evolution of the two Cerdanyas since the midnineteenth century. The French Cerdagne entered the modern world well before the Spanish Cerdaña experienced similar transformations. Writing in the mid-1920s, Pau Vila noted that the French Cerdagne had 140 kilometers of paved roads compared with 78 on the Spanish side; only 40 houses in France were not serviced by a road, while over 850 in Spain still used the "medieval paved paths."[16] More recently, Antoni Tulla has documented the shift since 1850 from "a precapitalist society with a certain autarky . . . to a predominance of capitalist relations" in the French Cerdagne, whereas no such movement was visible on the Spanish side. The shift, most marked in the first four decades of this century, involved the marginalization of mountain villages (and the pastoral way of life more generally), the decline of the system of impartible inheritance, the growth of employment in the "secondary" and "tertiary" economic sectors and the decline of agriculture, and the overall decline of residents born in French Cerdagne. Such processes began to be felt in the Spanish Cerdaña in the 1950s, but south of the boundary, the economy retained greater links to the inherited agro-pastoral way of life.[17]

Though differences have leveled off in the last three decades, important contrasts remain. Demographically, more than half the population of the French Cerdagne in 1968 were born outside the department of the Pyrénées-Orientales, while 75 percent of Spanish Cerdans were natives of the valley. Today there are few peasants still working the

15. *Souvenirs du Midi par un homme du Nord* (Perpignan, 1891), quoted in Brousse, *La Cerdagne française*, 68.

16. Vila, *La Cerdanya*, 193–194.

17. A. Tulla, "Les deux Cerdagnes: Example de transformations économiques asymétriques de part et d'autre de la frontière des Pyrénées," RGPSO 48 (1977): 409–424; on the relative archaism of the Spanish Cerdaña in the early 1960s, see A. Gil Coriélla, "Vida y derecho popular: Estampas de la Cerdaña española," *Revista de derecho notarial* 58 (1965): 199–224. In commenting on recent developments in the Cerdanya, I will be drawing on interviews and data collected during June and July 1985.

land in the French Cerdagne. The village of Dorres, for example, had twenty-nine estates still functioning in 1955, ten in 1970, and only one in 1985. The rural population has left, taking jobs in government service or in the urban centers of Toulouse, Perpignan, or Barcelona. Tourists and summer residents, employees from the ski resorts, retirees, interns of the governmental vacation and health colonies, and engineers at the experimental solar power installations have replaced peasants working the land. The Spanish Cerdaña, by contrast, remains more "traditional": the development of milk cooperatives, in particular, has allowed more peasants to work on their farms.[18] Yet similar shifts toward tourism and the construction of secondary residences have begun in earnest on the Spanish side of the border, expedited by the completion of the Cadi Tunnel in 1985. The tunnel, avoiding the Tossa Pass, reduces driving time from Barcelona to Puigcerdà to under two hours. The focus of "development" has shifted to the Spanish Cerdaña, although if the projected tunnel on the French side avoiding the Puigmorens Pass were to be built, it would make the Cerdanya as accessible from Toulouse, and building and tourist activity would shift back again to France.

Such revolutions in transport and communication mark important stages in the process of national integration and assimilation. But the construction of national unity—the making of France and Spain— involves much more than the technological achievements that allow people of different regions to communicate with one another, to participate in the same national markets, or to be oriented toward the same cultural centers, as most studies of nation building would have it.[19] The functionalist and positivist models of national assimilation assume that people must participate in a shared life before they can experience a shared identity—that peasants must literally take part in the life of the nation before they will consider themselves members of a national community. But nations, like other forms of community, are

18. Tulla, "Les deux Cerdagnes," 422; A. Tulla, "Procés de transformació agrària en àreas de muntanya. Les explotacions de producció lletera com a motor de canvi a les comarques de la Cerdanya, El Capcir, l'Alt Urgell, i el Principat d'Andorre" (Tesi doctoral, Universitat Autònoma de Barcelona, 1981); *La Cerdanya: Recursos economics i activitat productiva* (Barcelona, 1981), 51–54; H. Arpajou, "L'agriculture catalane: Statistique agricole," *Revue "Terra Nostra"* 18 (Prades, 1975); and *Les problèmes fonciers en Basse-Cerdugne,* Ministère de l'Agriculture, Direction départemental d'études économiques d'aménagement rural (Perpignan, 1978).

19. For example, Deutsch, *Nationalism and Social Communication;* G. Ardant, "Financial Policy and Economic Infrastructure of Modern States and Nations," in C. Tilly, ed., *The Formation of National States in Western Europe* (Princeton, 1975), 164–242; and Gellner, *Nations and Nationalism.*

not only symbolic constructions entailing a shared sense of member-ship.[20] Nations and other communities gain their identities from a sense of distinction, from the boundaries drawn between themselves and outsiders.

Well before the arrival of paved roads, public services, and the other "infrastructures" that differentiate the two Cerdanyas, the people of the Cerdanya had developed their own sense of national differences. The sense of difference antedated the experience of differentiation; it coexisted with the increasing continuity of social relations across the boundary and with the persistence of the Catalan language and a dis-tinctive Catalan ethnicity. The sense of difference emerged from the meaningful appropriation of the fact of their political dependency. Whether they resisted the authority of the state or sought its assistance, the Cerdans came to identify with their respective states, Spanish and French. Indeed, the more they resisted the state, and the more they developed an instrumental stance with respect to their nationality, the more they identified the nation as their own.

Living on the boundary, the Cerdans conceptualized the differences of French and Spanish territory and nationality long before these differ-ences became apparent to the two states. Through their local struggles and disputes, the landowners, peasants, and municipalities of the two Cerdanyas gave meaning to these distinctions, bringing the nation into the village and the village into the nation. The struggles of propertied elites during the eighteenth century to win tax exemptions for their properties in the French Cerdagne led them to identify themselves as Spaniards; communities, in their opposition to these exemptions, saw themselves as French. The disputes of village communities across the boundary over limited ecological resources led them to adopt new na-tional identities while never abandoning their local ones; and the com-munities' exclusion of "outsiders" from their midst amounted to an assertion of their own nationality. The Cerdans came to identify them-selves as French or Spanish, localizing a national difference and na-tionalizing local ones, long before such differences were imposed from above.

CATALAN NATIONALISM IN THE BORDERLAND

The structural differentiation of economy and landscape in the two Cerdanyas beginning in the late nineteenth century was matched by

20. See the studies cited above, Introduction, n. 18.

emerging differences in culture and politics on both sides of the boundary. In Spain, Catalan nationalism made its appearance: the "nation" became Catalonia, and it became politicized. But the claims of Catalan nationalism found little resonance in the Catalan counties north of the border, where the nation remained France.

Catalan nationalism is hardly a unitary phenomenon. The diverse claims of "federalism" and "autonomy" and the profusion of political parties since the late nineteenth century took shape within a historical oscillation from the political right to the left and back again.[21] But all programs within Catalan nationalism have been founded on an identity constructed from the distinctiveness of Catalan language and culture. The Catalan "character," defined in large part by the Catalan language, lies at the foundation of modern *catalanisme*. During the early nineteenth century, the Catalan language was refined, made grammatically more rigorous, and purged of some Castilian influence. The *renaixença* of Catalan language and literature had its early literary expressions in the patriotic poetry of Aribau, the works of Jacint Verdaguer, Joan Maragall, and others, but it was the recreation of the medieval "Floral Games" in Barcelona beginning in 1859 which gave the movement an institutional identity.[22]

Literary revivals of similarly submerged languages took place throughout Europe in the later nineteenth century, but few such "literary nationalisms" were converted into significant expressions of political nationalism.[23] In Roussillon, a literary renaissance of the Catalan language appeared belatedly in the 1880s and 1890s but had virtually no political significance. Nor did the renaissance of troubador poetry in Provençal and the literary movement of Felibrige which celebrated an Occitan or Provençal nation defined by its distinctive language. In striking contrast to northern Spain, where literary revivals led to the formation of major political parties, southern France saw few organized political claims; where such movements developed, they tended to be marginal, conservative, and mostly Catholic responses to the unifying policies of the republican regime.[24]

21. Recent interpretations of Catalan nationalism include J. M. Poblet, *Història bàsica del catalanisme* (Barcelona, 1975); and J. Termes, *Federalismo, anarcosindicalismo y catalanismo* (Barcelona, 1976). On contemporary Catalonia, see K. A. Woolard, "The 'Crisis in the Concept of Identity' in Contemporary Catalonia, 1976–82"; and S. M. DiGiacomo, "Images of Class and Ethnicity in Catalan Politics, 1977–1980," in G. W. McDonogh, ed., *Conflict in Catalonia: Images of an Urban Society* (Gainesville, Fla., 1986), 54–71 and 72–92.

22. L. A. Kauffman, "Politicizing Identity: The Jocs Florals of Barcelona and the Creation of Catalan Nationalism, 1859–1902" (Senior thesis, Princeton University, 1987).

23. Anderson, *Imagined Communities*, 66–79.

The frontiers of Catalan nationalism are implicit in the "Hymn of Cerdanya," a song dating from the late 1890s, which included a refrain that was to become a widely known aphorism about the valley. Written and sung in Catalan, the song was commissioned by Josep Maria Marti, mayor of Puigcerdà during the Carlist invasion of 1874 and subsequent organizer of the "Floral Games" held annually in Puigcerdà around the turn of the century.

> We are the heirs of the mountain
> we are the sons of the Pyrenees,
> sons of France or of Spain,
> we are all brothers at Font Romeu.
> Here is our cradle and our tombstone
> which is our natural patria;
> from the Carlit to the Tossa
> from the Cadi to the Puigmal.
> Half of France, half of Spain
> there is no land like the Cerdanya.[25]

Evoking a local pride of place, the song describes a set of "natural frontiers" of the patria which fail to delimit the political division of the valley between France and Spain.

Yet the inhabitants of the Cerdanya, like the cultural elites of Perpignan and the *catalanista* writers and politicians in Barcelona during the late nineteenth century, were aware of the differences that the political boundary had introduced and of the transformed Catalan culture on the French side. Valenti Almirall, the founder of the first daily newspaper in Catalan (the *Diari Catala,* in 1881), wrote that "The Catalans of Roussillon have lost more of their personality than those on this side of the Pyrenees, and their example cannot tempt us [to tie ourselves to France]." Most of the perceived "loss" centered around the question of language. Responding to an invitation to the Floral Games in 1868, Lluis de Bonnefoy wrote that, although "We are still Catalans by name in the Roussillon, we are losing day by day the language of the patria." Indeed, the Catalan language had been reduced to the debased status of a patois, a dialect. Although such disdain had been leveled against local tongues since before the Revolution, by the midnineteenth century

24. H. Guiter, "Catalan et français en Roussillon," *Ethnologie française* 3–4 (1973): 300–302; and V. Nguyen, "Aperçus sur la conscience d'Oc autour des années 1900 (vers 1890–vers 1910)," in C. Gras and G. Livet, eds., *Régions et régionalismes en France du XVIIIe siècle à nos jours* (Paris, 1977), 241–256.
25. Xandri, *La Cerdaña,* 298–300; see also Bragulat Sirvent, *Vint-i-cinc anys,* 61–63.

Catalan was used as a literary language only in burlesque and vulgar poetry.[26]

In the absence of a literary and political renaissance, Catalan culture thus gained the status of folklore north of the Pyrenees: language or costume could be used in festive or literary contexts as a cultural sign of difference, but not as everyday markers of national identity. Once loosened from the fabric of daily life, things Catalan became potentially available for signification in other, completely unrelated projects, in everything from museums to businesses. Only politics in French Catalonia remained exempt from this process of folklorization. The contrast between the folkloric status of things Catalan in France and the political and national identity of Catalonia across the boundary is striking today. On the Spanish side, Catalan has become the official language of an autonomous region and the language of everyday life. On the other side, Catalan identity is either for sale (a Catalan insurance agent next to a Catalan Inn or a Catalan Driving School), or makes its appearance during festive celebrations of Catalan origins, held during the high tourist season.[27]

Yet since the 1970s, "Northern Catalonia" has witnessed the development of political claims of Catalan nationalism—a movement drawing at once on the emerging strengths of other French "minority nationalisms," especially since 1968, and on the revival of Catalan nationalism across the boundary toward the close of the Franco era. But the participants and supporters of demands for autonomy and separatism north of the Pyrenees are few in number.[28] In the 1986 Legislative Assembly elections, for example, three catalanista parties together received only 2 percent of the vote in the French Cerdagne. As one catalanista from La Tor de Carol explained, "We are an insignificant minority." And even this man claimed that his "catalanism" was "purely sentimental. I would not go out carrying the Catalan flag, but sentimentally, I am closer to the Catalans than to the French government."

Few natives of the French Cerdagne today would deny their ties to France. At best, their position echos that of the Perpignan archivist Bernard Alart in 1868 who, accepting an invitation to the Floral Games

26. V. Almirall, *Lo catalanisme* (Barcelona, 1979 [1886]), 94; Bonnefoy, in BC Jocs Florals, Consistori (Comunicacions, 1868); Guiter, "Catalan et français," 300.

27. On the "folklorization" of Catalan culture in Roussillon, see Planes, *El petit llibre de Catalunya-Nord*, 100–102.

28. Planes, *Petit llibre*, 94–96.

in Barcelona, responded that "[e]ven if separated from other parts of the great kingdom of Aragon, and joined forever by ties of great affection to its natural nation, France, the Roussillon cannot ignore the grandeur of its past."[29] Today, the French Cerdans do not deny their historical ties to Catalonia, yet they proudly call themselves "French Catalans"—even if, like most Frenchmen, they remain suspicious of the government in Paris.

Yet it is practically unthinkable for natives of the Spanish Cerdaña to identify themselves as "Spanish Catalans." Catalonia is the "nation," the Catalan language its defining characteristic. Spain is the "state" or, worse yet, "the empire." The refrain of around 1900—"Half of France, half of Spain"—has been turned on its head. Today, many Spanish Cerdans will say "Neither France nor Spain, the Cerdanya is Catalan." The new aphorism may have had its origins after the formation of the Second Republic in 1931, or when Puigcerdà in 1936 became an anarcho-syndicalist stronghold under the dictatorial powers of Antonio Martin, "the cripple from Malaga." But the saying is not simply an anarchical rejection of the political authority of an abstract state: it is a statement about the identity of a nation divided between two states. The so-called mutilation of Catalonia, its division between France and Spain by the Treaty of the Pyrenees, is widely decried among historians, journalists, and advocates of Catalan nationalist claims today on both sides of the boundary. That the final accord of the treaty was never approved by Philip IV is reiterated as a claim of its nonbinding character. Yet all recognize that the incorporation of the Roussillon and part of the Cerdanya into France resulted in profound and irreversible changes in the meaning of the nation.

THE USE AND ABUSE OF THE NATION

Two kinds of difference distinguished the two Cerdanyas beginning in the second half of the nineteenth century. On the one hand, the economic integration of the French Cerdagne left the Spanish side by comparison underdeveloped in terms of road networks, communications, and public services. On the other hand, the appearance and growth of Catalan nationalism remained restricted, in large measure, to the Spanish side of the boundary. How were these transformations related?

29. BC Joczs Florals, Consistori (Communicacions, 1868).

In the early 1930s, Joaquim Cases Carbó, catalanista from Barcelona, asked a landowner in the upper Vallespir Valley, in Conflent, about why Catalan nationalism had not spread to the Roussillon. The peasant replied—in local dialect—with a whim of regret:

> You all, you can be catalanistes! Your government in Madrid treats you very badly. We cannot be [catalanistes] since our government in Paris treats us very well. We ask for a road, they build it right away. We want a telegraph, they put one in. We ask for a school, they give us one. We can not be catalanistes, but you all, you can be catalanistes.

"He said it with remorse, as if he were jealous," commented Cases Carbó, "the ancestral spirit filling his heart."[30] Unintentionally, perhaps, he revealed how identity was a function of local interests: the French Catalans were bound to the French state by its ability to satisfy their needs. The French state had created the ties of loyalty and identity instrumentally, by fulfilling the material needs of its citizens. For the people of the borderland, the greater advantage lay in being French than in defining their national identity as Catalans, which would only be meaningful in opposition to the Spanish state.

The strategic or instrumental use of national identity had long been practiced by the citizens of the French–Spanish borderland: pursuing their local interests, people sought out the nationhood that best fulfilled them. In the process, they nationalized their local economic or political interests and their local identities. But although Spanish and French citizenships had each had their uses under the Old Regime, French nationality became more useful after the French Revolution. The movement of Spaniards and Catalans seeking political refuge, asylum, and permanent residence in France began in the early nineteenth century and culminated in the massive resettlements during the Civil War of 1936–1939 and the consolidation of Franco's regime in Spain. And the differentiation of economy and public services in the later nineteenth century suggest a crystallization of a national and territorial distinction where the advantages, more often than not, lay in France.

The twentieth-century welfare state in France offers its citizens a range of services and benefits much greater than its Spanish counterpart, and Spaniards continue to immigrate in large numbers to the French Cerdagne. Although many are from the villages at the far end of the Spanish Cerdaña and from Catalonia, increasingly they are Gali-

---

30. *Catalunya francesa* (Barcelona, 1934), 30.

cians from the northwest and Andalusians from the south of Spain. As they settle in the French Cerdagne, many seek to become naturalized French citizens, for there are clear benefits in doing so. The French welfare state provides much more than a safety net—it means insurance, education, social services. Today, more than ever, the instrumental notion of nationhood seems to dominate: the Cerdans today are conscientious and unabashed manipulators of identity in the service of their interests, masters of the techniques of shaping identities.

Yet there are clear and well-defined limits to this manipulation of identity today, boundaries that took shape in the late eighteenth and nineteenth centuries. For one, although individuals within certain boundaries may opt to assume French nationality, the communities of the Cerdanya are not free to do so. The events of 1985 are particularly revealing in this regard. Faced with the possibility that the Madrid government might cut the train service between Puigcerdà and Ripoll, the municipality of Puigcerdà threatened in declarations to the Spanish and Catalan press that they would seek the renegotiation of the Treaty of the Pyrenees and ask to become French. It was a blatant and well-chosen ploy, and—along with 5,000 postcards sent to Prime Minister Felipe Gonzalez demanding the keeping of the rail line—it worked. The incident brought a range of responses in the two Cerdanyas. Many thought it was "a good joke," but some expressed doubts about the (in)sincerity of the request. A whiff of France surrounds Puigcerdà and has for the last two centuries. The town has always had its "collaborators"—from the revolt of the Catalans in 1640 to the last French military occupation in 1823. Puigcerdà today relies on French business and tourists; most shopkeepers speak French and all accept French currency (at favorable rates of exchange). And yet there is nothing to suggest that the municipal council's statement was anything but a "good joke." As more than one French Cerdan remarked when interviewed, "Puigcerdà no more wants to become French than we want to become Spanish."

During the French Revolution, the French government had refused to allow a plebiscite in the Cerdagne concerning their nationality; but from the midnineteenth century onward, French officials constantly urged the adoption of a plebiscite for the enclave of Llívia, convinced as they were that Llívia would want to become French. In 1867, General Callier believed that all France had to do to acquire the enclave was to demonstrate to the inhabitants of Llívia the disadvantages of remaining Spanish. He thereby ordered the construction of the road from Mont-Louis to Bourg-Madame to bypass the enclave entirely, arguing that "political interests oblige us to leave Llívia in the neglect which it suffers, so that its prolonged isolation makes it feel each day the burdens of its foreign nationality."[31] In the early 1890s the question of France's incorporation of Llívia was again raised, this time in diplomatic circles, and in 1912, in a still unresolved debate, Emmanuel Brousse, representative to the General Council of the Pyrénées-Orientales, described the situation of the enclave.

> Spain knows them only to crush them with taxes. It undertakes no amelioration of their lives: neither roads nor schools nor fountains nor paths nor lighting nor telegraph nor telephone nor post-office—as there are in all the French communes which encircle the enclave. . . . One thousand of the 1,200 inhabitants would be delighted to become French; of the remaining two hundred, 150 want the annexation, but they don't dare say so because of the landowners who make their large fortunes in smuggling.[32]

Though Llívia, like Puigcerdà, had always had its collaborators, there is little to suggest that the municipality itself would ever have sought to change nationalities. In the past, individuals born in Llívia and the Spanish Cerdaña have sought to gain French citizenship, and some still do today. But despite this flow of people across the boundary, the communities continue to emphasize their national identities and differences.

At the same time, the people of the two Cerdanyas do not deny their common identity or history and, indeed, seek to affirm their unity across the boundary in ritual terms. In 1981, the Institute for Cerdan Studies in Puigcerdà, in conjunction with other cultural associations of the valley, organized what would be the first *Diada de Cerdanya*—a daylong festival celebrating the unity of the valley to alternate annually between a village of the Spanish and French Cerdanyas. The first Diada was not an overwhelming success, but by 1985 the annual celebration drew local and international support. Members of the Catalan parliament and elected representatives from France were present; in Bourg-Madame/la Gingueta there were tables laid out with local specialties of the region, and instrumental groups played traditional Catalan songs. A West German, Tilbert Stegmann, delivered an address in Catalan, to the astonishment and delight of many, in which he spoke of the ways in which the Treaty of the Pyrenees had "mutilated" the valley and

31. FAMRE CDP 10, fols. 338–342 (Callier to foreign minister, August 16, 1867).
32. *L'Indépendent* (Perpignan), February 12, 1912; on opposition to the incorporation in the 1890s, see, for example, the articles in the *Voz del Pirineo*, December 1892.

Catalonia more generally. In the evening, a public ball drew several hundred people.

There was no national revolt of the Cerdans in September 1985, as had occurred during the Feast of the Rosary in July 1825. To the contrary: the festival celebrated a self-conscious recreation of the unity of the two Cerdanyas, with the Catalan identity of the valley as the common term of that unity (see illustration 10). That in the 1980s the Cerdans ritually marked their unity is suggestive of how much the political boundary had become an organizing structure of their lives, even as talk of its artificiality is much in evidence.

Today the Cerdans discuss the boundary in ways which only indirectly illuminate the history of its formation. When asked about the meaning of the political boundary, their first and instinctive reaction is to deny its existence. In both Cerdanyas, older peasants point out the unity of social relations that have continued since the Treaty of the Pyrenees. They note how Puigcerdà is *la vila* (the town) that brings the two Cerdanyas together. They remark on the marriages of young people who had met in Puigcerdà's most famous discothèque, the Gatzara. They tell their own family histories, emphasizing how their parents or grandparents came from France or Spain—in other words, from villages on opposing sides of the boundary. Most often, they summarize their social relations by the phrase: "We know each other." The more catalanista among them, especially on the Spanish side, underscore the Catalan identity and language which unite the two sides of the valley; others simply point to the ties of family and neighborhood which make them part of a larger unity.

The Treaty of the Pyrenees has an important place in this discourse about the boundary. On the one hand, the Treaty of the Pyrenees has an exceptional status as a founding event, a charter of social relations in the borderland. The Cerdans believe that the Treaty of 1660 guaranteed them their rights to cultivate lands across the boundary, to move freely between France and Spain, and to exercise their rights of pasture in the other country. All the provisions assuring the continuity of local society across the boundary as explicitly stipulated in the delimitation Treaties of Bayonne in 1866 and 1868 are attributed to the Treaty of the Pyrenees. The Treaties of Bayonne have left no trace in the memories of the Cerdans: the boundary itself—the border stones—are attributed to the earlier accord. The Treaties of Bayonne are, for them, a nonevent; once instituted by the Pyrenees Treaty, the boundary has virtually no history.

Illustration 10. *Diada de Cerdanya* (1985), the poster announcing the annual "Day of Cerdanya" festival, which alternates between French and Spanish villages. The two sides of the valley are shown to be united here by the red and yellow Catalan national flag (*els quatre barres*).

On the other hand, the Cerdans assert that the boundary has only recently become a reality that divides them. They point to the experiences of warfare—the two World Wars and the Spanish Civil War— as the moments during which the boundary lost its permeability. In his memoirs spanning the First World War, Jaume Bragulat Sirvent recalled:

> The frontier, at the beginning of the century, was something a little less than illusory, and continued to be so until the First World War, that of 1914. Everyone crossed where he wanted and when he wanted. The Treaty of the Pyrenees had no meaning in our Cerdanya, neither practically nor politically.[33]

Today, the Cerdans are more likely to point to the closing of the boundary between 1939 and 1947, from Franco's triumph over the Spanish Republicans until the reestablishment of diplomatic ties with France. Between November 1942 and June 1944, German soldiers and their dogs patrolled the boundary line in the Cerdanya, all but closing it down.[34] Neither Spain nor France rationalized the closed boundary by evoking the need for a sanitary cordon, as had occurred in 1821. The different political histories of France and Spain in the twentieth century assumed the distinctiveness of French and Spanish territory and simultaneously reinforced the territorial separation of Spain and France. As the Cerdans recall these events, they note how roads across the boundary were closed (and have not since reopened), how news from families across the boundary became more limited, and how relations with the Cerdans on the other side "became cold." There are even suggestions of national animosities—such as the hostility of French Cerdans toward the tens of thousands of Spanish and Catalan Republican refugees who fled with Franco's victory in 1939 and camped in the stables and fields of the French Cerdagne before being interned.

By focusing on the more recent past, the Cerdans today efface a history of the formation and consolidation of the territorial boundary. Although some recognize that perhaps, in the past, communities across the boundary struggled for the control of water, pastures, and territory, most even deny this history of disputes. They cannot imagine a seriously contested boundary, nor can they imagine peasants going out at

33. *Vint-i-cinc anys,* 26.
34. On the experience of the occupation in the Cerdagne, see R. Renault-Roulier, *La ligne de démarcation* (Paris, 1969), 15: 9–110; (Paris, 1970), 19: 109–152; and (Paris, 1972), 21: 243–310. On border crossings and resistance in the Pyrenees during the war, see E. Eychenne, *Montagnes de la peur et de l'espérance* (Toulouse, 1980), and his *Les Pyrénées de la Liberté* (Paris, 1983).

night and moving stones. "The stones are sacred," says one, "only crazy tourists would do that." The Cerdans recognize the authority of the state in having placed the boundary stones, and in maintaining them. When Spanish and French officials and soldiers arrive in the valley for their annual inspection of the boundary line, they are accepted and occasionally feasted by the local inhabitants.

These soldiers and officials come to the Cerdanya as representatives of the International Commission of the Pyrenees. Created in 1875, the commission has met every two years—except during the wars—as an international administrative and jurisdictional body designed to maintain the "good neighborly" relations of Spain and France. In addition to maintaining the boundary markers, the commission has performed two other functions. As an international court hearing expert testimony from both nations, it has settled the occasional dispute among frontier inhabitants—from conflicts over rights of passage, to exchanges of communally owned property on the Carlit Mountain, to the construction of a French dam at the Etang de Lanoux, which changed the amount of water reaching the Carol River in Spain. And as an international administrative commission, it has negotiated agreements between the two states such as the joining of national roads and railways in the Cerdanya and elsewhere.[35]

Today, the Cerdans accept the existence and jurisdiction of the International Commission, although it is a distant entity largely irrelevant to their daily concerns. Their acceptance of it, and acceptance of the distinctions of territory and nationhood in the two Cerdanyas, is the reason why the Pyrenean boundary has been considered a "dead" frontier in the twentieth century. In gaining that status over the course of three centuries, the national boundary came to structure the practice of daily life. Certain features that made their appearance in the eighteenth and nineteenth centuries continue to affect local relations in the borderland. Today Spanish proprietors own and cultivate the larger estates of the French Cerdagne; individuals and families organize the practical aspects of their lives around administrative and economic centers on opposing sides of the valley.[36] And if in festive moments the Cerdans express their unity, border disputes still surface which underscore the

35. On the origins, functions, and jurisdiction of the International Commission, see Fernandez de Casadevanti Romani, La frontera hispano–francesa, esp. 319–338.

36. Almost 30 percent of the inhabitants of the French Cerdagne were born in the Spanish Cerdaña, and 20 percent of the estates in the French Cerdagne—most of the larger and better-quality ones—belong to Spaniards: see Tulla, "Transformació agrària," 272–274; and Les problèmes fonciers en Basse-Cerdagne, 20–21.

difference a state makes. In 1971, it was the "war of the stop signs," a dispute over the right-of-way where the French highway crossed the "neutral road" from Llívia to Puigcerdà. In 1986, it was the canal of Puigcerdà, when mayors from Enveig, La Tor de Carol, Porta, and Porté blocked the canal and stopped the water from reaching their "enemy brothers" in Puigcerdà. According to the mayor of Enveig, "cutting the canal is the only weapon we have, and we will use it. We are only asking for the reestablishment of a right." The "right" in question involved claims to open a road, closed since 1936, which links the Carol Valley more directly to Puigcerdà. And according to a Perpignan newspaper, the incident was designed to assure the application of the provisions . . . of the Treaty of the Pyrenees.[37]

The development of national identities and the formation of a territorial boundary line in the Cerdanya were historically variable and hardly unilinear processes. There were moments in the history of the Cerdanya when inhabitants of the two Cerdanyas felt themselves to be distinct, just as there were times when they felt a common identity against the imposition of political authority from above. Today, the Cerdans maintain a certain continuity of social relations across the boundary, developing a shared myth about the "artificiality" of the division, in which the Cerdans deny the role of the state in differentiating them. Yet the national boundary remains accepted and uncontested, and the Cerdans deny their own role in the making of France and Spain.

37. *L'Indépendent,* July 8, 1986.

Appendix A
Texts of the Division of the Cerdanya, 1659–1660 and 1868

EXCERPTS FROM THE TREATY OF THE PYRENEES, 1659–1660

1. ARTICLE 42 OF THE TREATY OF THE PYRENEES, NOVEMBER 7, 1659

And as concerns the countries and places on the Spanish side, taken by the Arms of France during this war: As it had been formerly agreed in the negotiations begun in Madrid in the year 1656, upon which this present Treaty is founded, that the Pyrenees Mountains, which anciently divided the Gauls from the Spains, shall henceforth be the division of the two said kingdoms.[1] It has been concluded and agreed that the Most Christian King [of France] shall remain in possession and shall effectively enjoy the whole County and Viguery of Roussillon, the said County and Viguery of Conflent, [and the] countries, towns, castles, boroughs, villages, and places which make up the said Counties and Vigueries of Roussillon and Conflent; and to his Catholic Majesty [of Spain] shall remain the County and Viguery of Cerdanya, and the entire Principality of Catalonia, with the vigueries, places, towns, castles, boroughs, hamlets, and countries that make up the said County of Cerdanya and the whole Principality of Catalonia. Provided that if there be found any place of the County and Viguery of Conflent only, and not of Roussillon, on the side of the said Pyrenees Mountains toward Spain, it shall remain of his Catholic Majesty; and likewise, if any place of the said County and Viguery of Cerdanya only, and not of Catalonia, is found to be on the side of the said Pyrenees Mountains toward France, it shall remain of the Most Christian King. And that the said

---

1. The Spanish text reads: " . . . which commonly have always been held to be the division of the Spains and the Gauls."

division might be concluded, commissioners shall presently be appointed on both sides, and shall together, in good faith, declare which are the Pyrenees Mountains, which according to the content of this article, should hereafter divide the two kingdoms, and they shall mark the limits they ought to have. And the said commissioners shall meet on the premises one month at the latest following the signature of the present Treaty, and in the course of one more month shall have met together and declared by common consent that which has been stated. It is understood that if then they cannot agree among themselves, they shall make known the basis of their opinions to the two plenipotentiaries of the two kings, who, taking note of the difficulties and differences encountered, shall come to agreement between themselves on this point, so that arms will not be taken up about this matter.

(SOURCES: French: FBN MS 4309, fols. 2–3; Spanish: ACA Cancilleria Real, vol. 5586.)

2. EXPLICATION OF ARTICLE 42 OF THE TREATY OF THE PYRENEES,
   SIGNED BY CARDINAL MAZARIN AND DON LUIS DE HARO ON MAY 31, 1660

And the Catholic King will retain the entire Principality of Catalonia and all the County and Viguery of Cerdanya in whatever part are situated the cities, places, towns, hamlets, and sites which make up the said Principality of Catalonia and the said County of Cerdanya with the exception of the Valley of Carol, in which are found the Castle of Carol and the Tor Cerdanya, and within a continuation of the territory which joins the said Valley of Carol and the Capcir, of the Viguery of Conflent, thirty-three villages which together will remain of His Most Christian Majesty, and which should be composed of those in the said Valley of Carol and those found to be in the said passage joining Carol and Capcir; and if there are not enough villages in the said valley and the said passage, the number of thirty-three will be supplemented by other villages of the said County of Cerdanya which will be found among the most contiguous.

(SOURCE: French: FBN MS 4309, fols. 132–135.)

3. LLÍVIA ACCORD CEDING THIRTY-THREE VILLAGES TO FRANCE,
   SIGNED ON NOVEMBER 12, 1660

We, Don Miguel de Salvà i Vallgornera, Knight of the Order of Santiago, member of His Majesty's Supreme Council of the Crown of Aragon, and Hyacinthe Serroni, Bishop of Orange, Counciller to his Most Christian Majesty in His Council of State; commissioners sent by the Catholic and Most Christian Majesties to execute the last article made and agreed upon by the plenipotentiaries of Spain and France on the Isle of the Pheasants, on May 31 of this year 1660.

Having come to the Cerdanya and held many conferences, after having communicated and exchanged copies of our respective commissions, and having considered the reasons from both sides, after seeing and making note of all the villages and jurisdictions,[2] we resolved and concluded that the thirty-three vil-

2. French: *villages et jurisdictions*; Spanish: *villajes y terminos.*

lages which shall remain of his Most Christian Majesty in the Cerdanya in virtue of the article of the above-mentioned treaty, will be the following: Carol, including all of its valley, which shall be counted with all the villages in it as two; Ur and Flori, [which shall be counted] as one; Vilanova and Les Escaldes as one; Dorres, Angostrina, Targasona, Palmanill, Egat, Odello, Via, Bolquera, Vilar d'Ovensa, Estavar, Bajanda, Sallagosa, Ro, Vedrinyans, La Perxa, Rohet, Llo, Eyna, Sant Pere dels Forcats, Santa Llocaya and Llus as one, Err, Planes, Càldegues and Onzès as one, Nahuja, Osseja, Palau, and Hix.

And since the territory of Hix extends to the other side of the river which comes from Ur, called the Raour River, we the commissioners have declared and declare, that as concerns the other villages, the division of Spain and France will be understood as the division of their territories and jurisdictions; nonetheless, as concerns the village of Hix only, the division of Spain and France will be taken as the said river following its natural course, and will follow its course until it enters and meets the territory of Aja, which will remain of Spain; so that half of the said river and of the bridge called Llívia will remain of Spain, that is to say, that bank which faces Puigcerdà; and the other side shall remain of France, that is to say, that bank which faces Llívia and the Perxa Pass; this division does not separate the said territory from the said village of Hix as to that which concerns the seigneurie, property, fruits, pastures, or any other such relevant thing, so that this separation is understood as dividing Spain from France, and not the seigneurie or private property of the said territory, which shall always remain united to the said village of Hix.

As to that which concerns Llívia and its jurisdiction, we the commissioners declare that it shall forever remain united to His Catholic Majesty on the condition that at no time can his Majesty fortify Llívia nor any other place or post of its jurisdiction and territory, and the said commissioner of Spain is obliged specifically to ascertain in the ratification of this accord and convention that Llívia cannot be fortified nor any other place or post of its jurisdiction or territory, in which case only does the French commissioner consent that Llívia and its territory remain of his Catholic Majesty. And to go from one village or another of those which remain of France, or to go from Llívia to Puigcerdà, or Puigcerdà to Llívia, passing through the territories of several villages of France, we the commissioners declare that whatever kind of merchandise or provisions which shall pass through the said territories will not pay any taxes to officials of the fairs, customs officers, tax farmers, nor any other tax collectors of the two kingdoms. We further declare that the said passages and roads necessarily taken to go from one village to another of those remaining part of France, or to go from Llívia to Puigcerdà or Puigcerdà to Llívia, shall remain free for the respective subjects of the two kings, without there being the possibility that officers of the two said kings restrict these passages for whatever reason. Nonetheless, this freedom of passage cannot serve for crimes or infractions which might be committed on the said roads or passages, since the arrest and punishment of the said crimes shall belong to the officers on whose territory the violation was committed.

(SOURCES: French: FBN MS 4240, fols. 347–350; Spanish: ACA CA leg. 231, no. 233.)

EXCERPTS FROM THE FINAL ACT OF DEMARCATION, JULY 11, 1868

Act of Demarcating the Boundary [amojonamiento de la frontera] of the Province of Gerona and the Department of the Pyrénées-Orientales.

Following the prescriptions of article 17 of the Delimitation treaty signed at Bayonne on May 26, 1866, the plenipotentiaries of Spain and of France... have proceeded, with the assistance of delegates of the interested Spanish and French municipalities, with the definitive determination and marking of the dividing line between the Spanish province of Gerona and the French department of the Pyrénées-Orientales.

FIRST SECTION: DEMARCATION OF THE BOUNDARY LINE [AMOJONAMIENTO DE LA LINE FRONTERIZA] FROM THE VALLE DE ANDORRA TO THE MEDITERRANEAN

The boundary markers [señales de limites] consist of border stones or of crosses, except those markers which encircle the fort of Bellegarde. The stones or pillars are prismatically shaped, 80 centimeters high and with a square base of 50 centimeters on each side. The crosses are 20 centimeters with four equal arms engraved in firm rock within a rectangle 40 centimeters high by 35 wide. A cardinal number, which is written here at the beginning of each paragraph, and which designates the site and the kind of signal corresponding to it, is engraved on each of the stones, starting with the number 427, which follows immediately after the last one named in the Act of Marking the Boundary signed on February 27, 1863, as the first annex to the Treaty of limits of April 14, 1862, which includes [the boundary between] western Navarre and the Valle de Andorra.

Border Stone 427:

From the Pico [Peak] den Valira, situating at the crest of the Pyrenees between France and Andorre, [the boundary line] follows an abutment south, and there meets a pass well known by the name of Coll den Gait or Portella Blanca de Andorra. Here marker 427 has been placed on the northern edge of the path, at a point common to Spain, France, and Andorra.

The borderline follows from marker 427 on the ridge of the same abutment, climbing the peak called Toseta de la Esquella by the Spanish and Camp Couloumer by the French. From this ridge, which forms a tabletop, two other ridges break off: one goes south, and encloses Spain; the other goes east, and is called by the Spaniards, Sierra de la Esquella, on the crest of which runs the boundary line, passing by the Coll [Pass] and Pico de Bresoll, arriving at the Portella den Gourts or de Maranges.

Border Stone 428: At the said Portella.

The international line continues following the same crest to the Pico de Puig Pedros, where it leaves the ridge to go directly to the source of the Bovedó [stream].

Border Stone 429:

Below the last one, at 210 meters, on the left bank [of the canal of San Pedro de Cedret]. At this point, the international line leaves the canal of San Pedro, turning south at an angle of 147 degrees.

. . . Border Stone 450:

In the same direction, 43 meters, at the place which the Spaniards call Coll de Sansobell and the French call Coll de la Madalene.

Border Stone 451:

In the same direction, and following a wall close to the meadow of Casa Mitjana, at a southeast angle from the meadow, the boundary stone [is placed] at 217 meters.

Border Stone 452:

Going straight at an angle of 171 degrees, arriving at an embankment forming a place called las Costas de San Pedro or Deves de Roco, and on the slope, a marker [is] placed at 451 meters.

Boundary Stone 453:

Following the same line for 123 meters, arriving at the canal of the stream called Rech de Llinás or de la Salancas, at which point of intersection a stone is placed on the left edge of the stream.
The international division descends along the Rech de Llinás.

. . . Border Stone 483:

Pillar at 10 meters from the left bank of the Segre River, at 51 meters 60 centimeters from number 482 on the French side, at the end of the mud wall [described in number 482].

Border Stone 484:

Following the direction of the same wall, which forms an angle of 162 degrees with marker 482 of the French line; and 235 meters away is planted a stone in a bend in the wall.
Here the boundary forms an angle of slightly more than 90 degrees from the wall, which after 25 meters changes to another angle of 90 degrees.

Border Stone 485:

Pillar at the extreme of this site which marks the frontier at 110 meters measured in a straight line from the previous one. This line and that of markers 483 and 484 opens to an angle of 170 degrees.

Border Stone 486:

Forming an angle of 156 degrees at 305 meters, a stone at the western edge of the road from Bourg-Madame to Aja. The line follows the winding crest of the ridge known as the Riba de la Coma del Mas Blanch up to stone 489.

. . .

SECOND SECTION: DEMARCATION OF THE BOUNDARY OF THE
ENCLAVE OF LLÍVIA [*AMOJONAMIENTO DEL TERMINO ENCLAVADO
DE LLÍVIA*]

Crosses have been placed and stones carved with their corresponding cardinal numbers to mark the limits of the jurisdictional perimeter of Llívia. The crosses are the same as those described in the first section of this act; but the pylons are only 60 centimeters high, and the unequal sides of the base are 30 and 35 centimeters respectively. These markers have a double "L" engraved on the bordering side of Llívia, and on the other side, the initial of the name of the French village corresponding.

Border Stone 1:

> The first stone is situated on the northeast side of the road from Puigcerdà to Llívia, at the site called Pontarro de Xirosa, next to the old stone which had been the boundary of Llívia, Ur, and Càldegues.

Following what was done in the boundary marking from the Valle de Andorra to the Mediterranean, the angles are taken with respect to the direction from which the line arrived, and the distances from the last marker, unless otherwise described.

The first line of the perimeter forms an angle of 45 degrees with the road cited above, and ends at the Mojon den Puñet, which is number 3. As a general rule, it goes in a straight line from one signal of limits to the next one, unless otherwise described.

Border Stone 2:

> Stone in the same direction, at 480 meters in the Paso dels Bous, next to the wall which meets the road from Llívia to Onzès from the west.

Border Stone 3:

> At 302 meters, at the site of the old Mojón den Puñet, a new one is established 20 meters from the right bank of the Segre River.

. . . Border Stone 44:

> At an angle of 174 degrees, a boundary stone at 400 meters on the southern side of the road which goes from Ur to Llívia.

Border Stone 45:

> In the same direction, the last marker is at 325 meters, in the Tosal de Piedra Lagre or Peyre Llargue, which is an overhang of a small stream.
>
> A line 585 meters long joins boundary stone 45 to the Pontarro de Xidosa [number 1], and closes the perimeter, forming with the Piedra Lagra a salient angle of 142 degrees and one of 126 degrees from the Pontarro de Xidosa.

(SOURCE: *Acuerdos fronterizos con Francia y Portugal,* Ministerio de Hacienda [Madrid, 1969].)

(Note: Spellings of toponyms, in Castilian and Catalan, have been preserved.)

Appendix B
Population, Marriage, and Property between the Seventeenth and Nineteenth Centuries

POPULATION OF THE CERDANYA, 1642–1877

In 1754, the viguier of the French Cerdagne, François Sicart, complained in a memoir to the intendant about the "false information given by the bailiffs, syndics, and principal inhabitants" concerning their properties and incomes and, as a result, their tax assessments. The idea that proprietors naturally tend to "exaggerate their needs and dissimulate their revenues," and that Sicart had to react against such tendencies, is a sobering reminder about the value of statistical data.[1] This was true especially under the Old Regime, and especially regarding those tabulations prepared by government and military officials with their own political projects at stake. Tables B.1 and B.2* are drawn from a variety of such sources—declarations to the French armies when they occupied the Cerdanya; surveys by administrative reformers of the enlightened state; different tax rolls and official censuses of the nineteenth-century bureaucratic state. Although not as fantastical as many of the possible data bases, they nonetheless provide only an approximate measure—a global view—of population growth in the two Cerdanyas.

THE GEOGRAPHY OF MARRIAGE

The differential survival in the French and Spanish Cerdanyas of parish registers reveals the contrasting political experiences of the two sides of the valley since the late eighteenth century. "With the [Napoleonic] invasion," wrote

---

* Tables are at end of appendix B.

1. Quoted in Brutails, "Notes sur l'économie rurale," 229–230; see also the viguier's remarks in ADPO C 984: "Mémoire en réponse sur la différence qui s'est trouvée entre les états fournis en 1741 pour la Dixième et ceux fournis pour l'abonnement des deux Vingtièmes en 1758."

Pau Cot Verdaguer in his manuscript notes of 1928, the Spanish Cerdaña "lost the best of its riches. Everything in the village archives was robbed or burned, as was everything else of whatever value in the parish archives. Everything was burned or destroyed."[2] The Spanish Civil Wars of the 1820s and 1830s likewise involved the destruction of parish archives, as did the anticlericalism of 1936. Whereas the French Revolution made short shrift of the Old Regime, the crisis of the Old Regime in Spain was a prolonged, disruptive struggle—as evidenced by the fate of parish records. With the absence of such records for the Spanish Cerdaña, those of the French side provide only a partial understanding of the way in which the political division of the valley shaped household recruitment in the borderland.

The parish records of five French Cerdan communities—Santa Llocaya, Càldegues and Onzès, Osseja, Carol, and Odello—were sampled for periods of ten or more years in the late seventeenth, late eighteenth, and midnineteenth centuries. Table B.3 measures the percentage of "endogamous marriages"— those in which both spouses were born in the community—and the percentage of native spouses. Table B.4 describes the origins and percentages of "non-native spouses," marriage partners of both genders born outside the village during each of the sampled periods.[3]

"It is the custom that the marriage is celebrated in the village of the wife, and not that of the husband," wrote a peasant from Dorres to the Departmental Administration in 1800, concerning his own marriage to Marguerite Soler of Llívia.[4] The data presented in tables B.3 and B.4 thus concern sites where marriages were celebrated, not the villages where the newlyweds went to live—although by the nineteenth century these were predominantly on the French side of the boundary (see above, chap. 6 sec. 3).

The five chosen communities differed in size, wealth, resources, distance from the political boundary, and interest to the state. Santa Llocaya, on the valley floor, was a wealthy community with a large percentage of privileged landowners with obvious ties to the other notable families in the town of Puigcerdà, across the boundary. The hamlets of Càldegues and Onzès adjoined Santa Llocaya, but they were less wealthy and had fewer ties to the elite families of Puigcerdà and Llívia. Osseja and Carol were both "valley–communities" with large numbers of packeteers and merchants involved in the large-scale contraband trade that moved through the valley; but they were also the site of cus-

2. BC MS 1535, fols. 28: "Notes folkloriques pirinenques" (1928).

3. Monographs that inspired the use of these criteria include: R. J. Johnston and P. J. Perry, "Déviation directionelle dans les aires de contact: Deux examples de relations matrimoniales dans la France rurale du XIXe siècle," *Etudes rurales* 46 (1972): 23–33; P. E. Ogden, "Expression spatiale des contacts humains et changement de la société: L'example de l'Ardeche, 1860–1970," *Revue de géographie de Lyon* 49 (1974): 191–209; and M. Jollivet, "L'utilisation des lieux de naissance pour l'analyse de l'espace social d'un village," *Revue française de sociologie* 6 (1965): 74–95. I have not adopted the concentric-circle model used—implicitly or explicitly—in most studies of marriage recruitment; in addition to its conceptual inadequacies, it tends to be useful only when measuring much larger distances from a parish (generally circles of ten, twenty, and fifty kilometers); see, for example, Balfet, et al, "De la maison aux lointains," 27–75.

4. ADPO L 690; the customary practice of matrifocality was confirmed in interviews with older peasants during the summer of 1985.

toms posts and military garrisons whose personnel sometimes married local women. Odello was a poorer, mountain village on the slopes of Carlit Mountain, far from the political boundary.

More detailed descriptions and discussions of methodology and conclusions concerning the geography of marriage can be found in Sahlins, "Between France and Spain," 440–454, and 881–885.

PROPERTY HIERARCHIES IN THREE FRENCH CERDAN VILLAGES

Tables B.5 and B.6 represent property hierarchies in the three French villages of Angostrina, Santa Llocaya, and Estavar and Bajanda at three moments: 1694, 1775, and 1829–1830. Once again, the absence of documentation on the Spanish side precludes the possibility of comparison. And as in the case of population statistics, the fact that the data was collected by military and government officials for the purposes of levying duties or taxes, and that these officials had different projects in each period, makes the data less than reliable. In addition, the nature of "ownership" and "property" differed, not only in each period but also among social classes in the same period. Still, historians have used such sources with some success, at least in giving an overall sense of property hierarchies and their general movement.[5]

Table B.5 lists percentages of proprietors falling within the standardized categories of "very small," "small," "medium," and "large" proprietors—the Cerdanya is notable for an absence of "very large" proprietors.[6] Table B.6 represents this data graphically, adding a contrast between the approximate percentage of proprietors in each category (top half) and the percentage of the property (as a proportion of all property in the village) which falls into each category (bottom half).

As in the study of marriage recruitment, note should be taken of the differences in relative size and wealth of the villages in question. Santa Llocaya, with its many large landowners, contrasts strikingly with Angostrina, with its overwhelming percentage of very small proprietors. Estavar and Bajanda represented something of a mean in 1694, but by 1829 they resembled Angostrina much more. Although the statistical data may be flawed, the overall movement it suggests can be confirmed: as the population grew, and a certain amount of additional land was put under cultivation, smaller properties fragmented while larger ones were consolidated, thus increasing the gap between wealthy landowners and small proprietors. The insufficiency of the smaller plots meant that, increasingly, small proprietors became rural proletarians dependent on seasonal migration, domestic industry, and smuggling. By the nineteenth century, the poorer peasantry was migrating permanently from the valley.

5. For a more complete discussion of the sources of their problems, see Gavignaud, *Propriétaires–viticulteurs*, 1: 64–65 and nn. 12–13 (on the 1775 tax rolls); and 233–237, nn. 45–49 (on the Napoleonic cadastre). See also Sahlins, "Between France and Spain," 478–482.

6. The amount of land used to determine these categories varies from region to region; I have followed the amounts used by Gavignaud, *Propriétaires–viticulteurs*, 1: 237–238.

TABLE B.1 POPULATION OF THE SPANISH CERDAÑA, 1642–1877

	1642	1694	1718	1787	1830	1857	1877
Alp	127	348	263	427	536	648	551
Baronia de Bellver	—	756	815	1,813	1,452	2,012	1,678
Bolvir	129	174	240	166	190	385	371
Talltorta	26	45	24	56	44	—	—
Das & Tartera	179	229	190	299	310	348	340
Mossol	20	32	39	45	38	—	—
Sanabastre	54	65	62	61	75	—	—
Ellar	—	114	75	122	98	254	147
Olopte & Cortas	—	84	180	202	223	—	—
Isobol	62	62	61	60	74	417	281
All	66	89	49	117	85	—	—
Ger	73	269	215	582	470	631	634
Saga	—	8	15	35	26	—	—
Grus & Riu	175	222	161	184	480	239	161
Guils	72	170	112	235	223	394	393
Saneja	—	59	73	91	98	—	—
Lles	—	—	200	229	246	—	—
Coborriu	—	—	53	28	24	—	—
Llívia	—	682	680	777	910	968	1,070
Meranges	204	235	154	377	385	391	304
Montella	—	—	345	932	705	1,212	994
Musser	—	—	130	49	83	—	—
Aranser	—	—	129	136	108	520	457
Prats & Sampsor	105	142	90	191	195	268	227
Prullans & Ardovol	—	—	250	375	466	654	548
Puigcerdà	—	1,349	1,130	1,754	2,030	2,342	2,396
Ventajola	—	26	17	34	29	—	—
Rigolisa	—	86	115	121	116	—	—
Queixans	118	137	133	130	222	274	246
Les Pereres	—	45	63	42	32	—	—
Talltendre & Orden	200	118	120	139	303	390	174
Urtx & Vilar	141	190	98	239	116	544	427
Estoll	36	65	69	131	102	—	—
Vilallobent	—	58	59	113	98	271	249
Aja	—	113	67	131	143	—	—
Villec & Estana	—	—	120	121	116	406	199
Beixac	—	—	50	36	63	—	—
TOTALS	1,787	5,972	6,646	10,580	10,914	13,568	11,847

TABLE B.1 POPULATION OF THE
SPANISH CERDAÑA, 1642–1877 (CONTINUED)

SOURCES:

1642: "Manifest de blat i perçones," 1642, AHP.

1694: "Estat du nombre de Familles . . . de Biens fonds que chacun possede . . . et le nombre des Forains, possedans biens dans le Terroir dudit lieu," census based on the declarations of batlles and consuls of the villages controlled by the French armies, ADPO C 1418.

1718: "Noticia del Principado de Cathaluña," 1718, published in J. Iglésies, *Estadistiques de població de Catalunya del primer vicenni del segle XVIII*, 3 vols. (Barcelona, 1974).

1787: "Censo del Conde Floridablanca," published in *El cens del comte Floridablanca, 1787: Part de Catalunya*, ed. J. Iglésies (Barcelona, 1969). For a careful scrutiny of the 1718 and 1787 censuses, see Vilar, *La Catalogne*, 2: 16–185.

1830: "Subdelegación de Policia de Puigcerdà," 1830 AHP Poblacio.

1857: *Censo de la población de España* (Madrid, 1858).

1877: *Censo de la población de España* (Madrid, 1878).

TABLE B.2   POPULATION OF THE FRENCH CERDAGNE, 1642–1877

	1642	1694	1730	1787	1792–1793 year II	1798–1799 year VIII	1806	1836	1856	1866	1877
Angostrina	155	197	201	388	416	219	394	506	454	448	518
Bolquera	—	173	173	350	270	278	288	338	403	463	456
Càldegues & Onzès	69	57	73	79	94	69	92	163	191	187	170
Dorres	158	156	168	293	253	175	310	334	341	300	320
Egat	—	60	48	92	89	65	78	89	115	109	125
Enveig	124	120	104	365	254	258	265	420	432	424	392
Err	255	322	320	524	600	592	672	736	728	738	689
Estavar	101	—	191	204	249	192	230	343	322	309	335
Bajanda	42	—	—	48	69	64	77	—	—	—	—
Eyna	—	195	95	167	225	216	238	269	306	303	253
Hix/Bourg-Madame	36	49	81	97	133	115	128	220	263	289	280
Llo & Rohet	—	274	418	387	351	267	336	417	420	422	410
Nahuja	—	91	47	159	130	48	81	164	160	172	171
Odello	—	177	176	166	265	241	279	501	505	461	476
Via	—	82	95	190	170	—	—	—	—	—	—
Osseja	—	418	283	904	503	814	969	880	1,014	1,053	1,067
Valsebollera	—	—	—	—	—	—	—	329	335	334	302
Palau	—	112	145	269	234	180	260	269	287	310	256
Sallagosa	—	197	363	369	333	311	360	631	607	608	540
Ro	—	47	36	68	170	69	—	—	—	—	—
Vedrinyans	—	47	46	79	77	76	—	—	—	—	—
Santa Llocaya	—	119	91	100	132	55	74	108	118	128	102

TABLE B.2   POPULATION OF THE FRENCH CERDAGNE, 1642–1877 (CONTINUED)

	1642	1694	1730	1787	1792–1793 year II	1798–1799 year VIII	1806	1836	1856	1866	1877
Targasona	—	101	103	93	117	82	97	141	146	158	137
Ur & Flori	104	37	163	333	243	186	221	280	299	311	279
Vilanova	67	66	60	141	134	59	152	200	190	157	158
Carol Valley	410	299	348	1,122	1,073	1,295	1,318	1,635	695	734	631
Porta	—	—	123	—	—	—	—	—	978	563	404
Porté	—	—	103	—	—	—	—	—	—	420	371
TOTAL	1,521	3,396	4,054	6,987	6,584	5,926	6,319	8,973	9,309	9,401	8,844

SOURCES:

1642:   "Manifest de blat i perçones," 1642, AHP.

1694:   "Estat du nombre de Familles . . . de Biens fonds que chacun possede . . . et le nombre des Forains, possedans biens dans le Terroir dudit lieu," ADPO 2050–2097 ("Communautés de la Cerdagne française").

1730:   "Etat des noms des hants . . . du nombre des enfants males et femelles, de ce que chaque chef de famille possede en journeaux de terre labourable, en pred, et des quantitéz de bestiaux de toute espèce, en l'année 1730," ADPO C 2045.

1787:   "Procès-verbal de l'Assemblée du district des Vigueries du Conflent et de Cerdagne, tenue à Prade, le 17 novembre 1787," ADPO C 2108.

1792–1793 (year II):   "Tableaux des communes de la population du district de Prades, Perpignan, et Ceret," ADPO L 469.

1798–1799 (year VIII):   "Dénombrement de la population du département des Pyrénées-Orientales," ADPO L 469.

1806, 1836, 1856, 1866, and 1877: Population censuses published in M. Battle and R. Gual, "Fogatges" Catalans. Revue "Terra Nostra," 11 (Prades, 1973).

TABLE B.3   ENDOGAMOUS MARRIAGES AND NATIVE SPOUSES, SEVENTEENTH TO NINETEENTH CENTURY

Village (parish)	Number Marriages	Number Endogamous Marriages (percent)	Number Native Spouses (percent)
Odello			
1681–1695	31	10 (31.2)	37 (59.7)
1777–1789	31	21 (67.7)	50 (63.3)
1853–1862	37	12 (32.4)	47 (63.5)
La Tor de Carol			
1676–1690	37	17 (45.9)	52 (70.3)
1777–1788	46	22 (47.8)	64 (69.6)
1853–1864	50	13 (26.0)	67 (67.0)
Osseja			
1566–1602	34	8 (23.5)	41 (60.3)
1674–1685	39	21 (53.8)	50 (64.1)
1774–1783	40	24 (60.0)	63 (78.8)
1855–1865	53	15 (28.3)	58 (54.7)
Càldegues and Onzès			
1664–1689	13	2 (15.4)	10 (38.5)
1770–1785	15	0 (0)	15 (50.0)
1853–1871	35	1 (2.9)	24 (34.3)
Santa Llocaya			
1661–1686	18	3 (16.7)	18 (50.0)
1774–1789	19	4 (21.1)	22 (57.9)
1853–1867	24	0 (0)	24 (50.0)

TABLE B.4   GEOGRAPHIC ORIGINS OF NON-NATIVE SPOUSES,
SEVENTEENTH TO NINETEENTH CENTURY

	Percentage of Spouses Born in:				
	French Cerdagne	Spanish Cerdaña	Roussillon	France	Catalonia
Odello					
1681–1695	58	5	37	0	0
1777–1789	50	17	33	0	0
1853–1862	63	22	15	0	0
La Tor de Carol					
1676–1690	64	23	13	0	0
1777–1788	46	39	11	0	4
1853–1864	40	40	9	0	11
Osseja					
1566–1602	26	41	0	33	0
1674–1685	28	17	38	6	11
1774–1783	53	18	6	0	23
1855–1865	35	27	22	8	8
Càldegues and Onzès					
1664–1689	56	44	0	0	0
1770–1785	65	32	0	3	0
1853–1871	45	44	5	0	5
Santa Llocaya					
1661–1686	67	17	11	0	5
1774–1789	37	31	13	6	13
1853–1867	63	30	3	0	3

SOURCES:

Odello: Parish of Sant Marti; marriage partners from the adjoining hamlet of Via are counted as natives of Odello, although Via was an independent parish (Santa Colomba).

La Tor de Carol: Parish of Sant Esteve, including hamlets of Iravals, Quers, and Corbassil; "native spouses" also includes spouses born in the adjoining villages and hamlets of the Carol Valley, which formed a coherent "valley–community"; Carol (parish of Sant Marçal), Porta (Sant Joan Bautista) and Porté (Navitat de Nostra Senyora).

Osseja: Parish of Sant Pere, including the hamlets of Valsebollera, Le Puig, Mascarell, and Concellabre, which together formed the dependent parish of Sant Feliu.

Càldegues and Onzès: Parish of Sant Roman.

Santa Llocaya: Parish of Santa Llocaya.

TABLE B.5   DISTRIBUTION OF LANDOWNERS IN THREE FRENCH VILLAGES, 1694–1830

Percentage of Landowners

Village	Very small (less than 10 jornals)	Small (10–28)	Medium (28–56)	Large (greater than 56 jornals)
Angostrina				
1694	89.7	5.6	3.7	0
1775	71.5	23.2	5.4	0
1830	76.2	20.5	4.5	0
Santa Llocaya				
1694	42.2	7.6	23.1	26.9
1775	43.4	0	13.0	43.3
1829	19.0	12.9	6.5	35.5
Estavar and Bajanda				
1694	46.7	30.0	20.0	3.3
1775	65.1	11.0	2.8	5.5
1829	82.0	9.5	5.2	3.7

SOURCES for tables B.5 and B.6:

1694: "Estat du nombre de Familles . . . de Biens fonds que chacun possede . . . et le nombre des Forains, possedans biens dans le Terroir dudit lieu," ADPO 2050 (Angostrina); ADPO C 2063 (Santa Llocaya); and ADPO 2060 (Estavar and Bajanda).

1775: "Etat Général de tous les Biens Fonds . . . appartenant aux Habitants non-exempts, qu'aux exempts et privilégiés," ADPO C 2050 (Angostrina); ADPO C 2063 (Santa Llocaya); and ADPO 2060 (Estavar and Bajanda).

1829–1830: "Cadastre. Etat des sections des communes," ADPO 3 P.

TABLE B.6   DISTRIBUTION OF LANDOWNERS AND PROPERTIES IN THREE FRENCH VILLAGES, 1694–1830

Angostrina

1694
50 houses
54 proprietors
216 jornals/77 ha

% proprietors

% properties

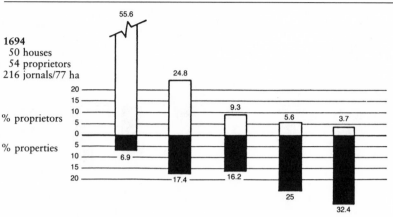

1775
79   houses
56   proprietors
437.25 jornals/155.7 ha

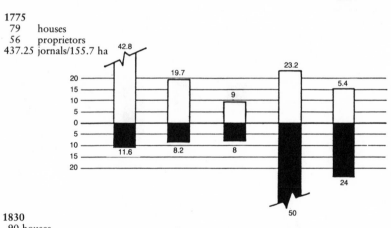

1830
90 houses
112 proprietors
847 jornals/302.5 ha

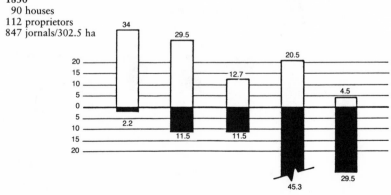

TABLE B.6 DISTRIBUTION OF LANDOWNERS AND PROPERTIES IN THREE FRENCH VILLAGES, 1694–1830 (continued)

Estavar and Bajanda

1694
34 houses
30 proprietors
546 jornals/167.5 ha

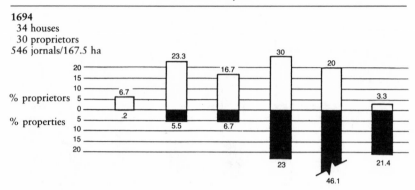

1775
53 houses
73 proprietors
668 jornals/204.9 ha

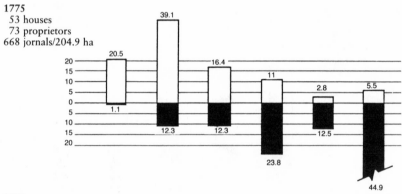

1830
46 houses
166 proprietors
1,156 jornals/354.6 ha

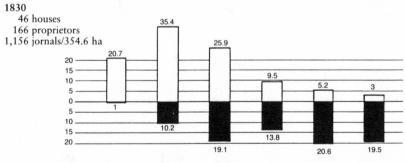

TABLE B.6 DISTRIBUTION OF LANDOWNERS AND PROPERTIES IN THREE FRENCH VILLAGES, 1694–1830 (continued)

Santa Llocaya

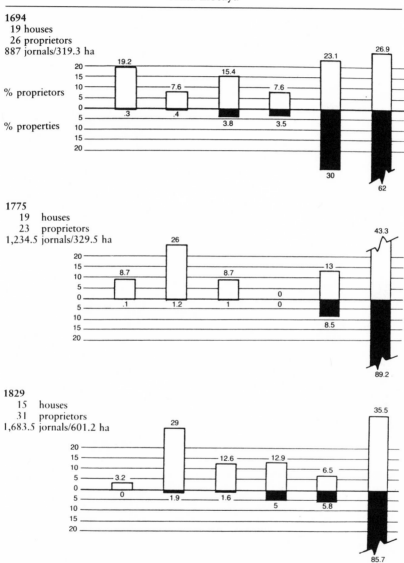

Bibliography

LIST OF ABBREVIATIONS AND ARCHIVAL
SOURCES CONSULTED

FRANCE

FAN   *France, Archives Nationales* (National Archive, Paris)

AF   Archives du pouvoir exécutif de 1789 à 1815:

        (II) 134 (1034–1037); 255 (2166); 256 (2160, 2174); 257 (2176–2180); 258 (2181); 264
        (III) 61 (244, 245); 62 (250); 63 (251)

D   Révolution: I: 34–39; III: 198–230

F(1)   Administration générale: E(72)

F(2)   Administration départementale: 2001.

        (I) 116, 106 (34), 447
        (II) Pyrénées-Orientales 1

F(3)   Administration communale: 11, 1211

F(7)   Police générale: 3608, 4130–4132, 4152, 5497–5507, 5812, 6138 (3–7), 6641, 8480, 9692

F(8)   Police sanitaire: 74 (1)

F(11)   Subsistences: 396, 637–639

F(12)   Commerce et industrie: 1263, 1889, 1965, 6370

F(14)   Travaux publiques: 169, 6377

F(17)   Instruction publique: 139, 1361, 1365, 2657, 9271, 10459

F(20)   Statistiques: 243, 435

G   Administrations financières: (7) 506–509

K   Monuments historiques: (VIII) 1221, 1332, 1623; KK (VI) 1247

M   Mélanges: (IV) 1020

Mi   Microfilmed records: 12 Mi 17, 202, 204 (K 1406, 1623–1625: copies of AGS SE "Negociaciones de Francia")

Q   Titres domaniaux: (1) 968–969; (2) 134

BB   Ministère de la justice: (18) 61, 671, 971, 1022, 1370(A)

FMG AAT *France, Ministère de la Guerre, Archives de l'Armée de la Terre* (War Ministry, Land Army Archive, Château de Vincennes)

MR   Mémoires et reconnaissances: 1083–1084, 1221–1224, 1325–1333, 1339, 1341, 1349, 2338–2339

A   Archives anciennes: (1) 219, 611, 1014, 1017, 1106–1109, 1287–1289, 1602, 1887–1888, 1891–1892, 1897, 2053–2054, 2257, 2405–2406, 3075, 3719

B   Révolution: (4) 101, 103

C   Premier Empire: 280

E   Monarchie libéral: (4) 23, 39, 43, 51

FAIG   *France, Archives de l'Inspection du Génie* (Archive of the Superintendence of Engineers, Château de Vincennes):

         Article 4, secs. 3, 6
         Article 12, sec. 1
         Article 8, sec. 1

FAMRE   *France, Archives du Ministère des Relations Extérieures* (Archive of the Ministry of Foreign Relations, Paris)

MD   Mémoires et documents

         FF: Fonds France: nos. 321–325
             PFR: Petit Fonds Roussillon: 1744–1748
         FD: Fonds Divers (Espagne): 32, 50, 53–54, 57–63, 153, 183
         FDS: France et divers, supplément: 402, 2160, 2193

CP   Correspondence Politique, Espagne: vols. 20, 32, 37–39, 44, 47–48, 762, 763, 769, 801, 833, 868, supp. 5

PA   Papiers d'agents: Desage, vols. 28–30

ADP   Affaires diverses politiques, France: 1–3, 6–10, 12, 14, 18

Limites: volumes 7, 432–446, 459, 461, 463
          CDP (Commission de Délimitation des Pyrénées): vols 1–11
          Fonds Callier (Papiers du Général Callier): 3, 4, 7, 12, 13, 16

ADPO   *Archives Départementales des Pyrénées-Orientales* (Archive of the
          Pyrénées-Orientales Department, Perpignan)

B   Cours et juridictions avant 1790
          2B (Actes du Conseil Souverain): 29, 41

Bnc   B Non-classée (Juridictions secondaires, Viguerie de la Cerdagne
          française): 281, 305, 327, 342, 426–470

C   Administration provinciales et contrôle des actes avant 1790

          Intendance du Roussillon et pays de Foix: 167–168, 170–171,
          257, 275–294, 326, 330–332, 406, 433, 437, 457, 602–604, 669,
          684, 688, 743–744, 793–794, 797, 800, 804, 843–846, 851,
          863–864, 866–868, 883, 984–985, 1007, 1011–1017, 1024,
          1027, 1034, 1039–1041, 1045, 1051, 1059–1062, 1064–1065,
          1067–1069, 1071, 1090, 1101, 1188–1191, 1201, 1273, 1282,
          1289, 1335, 1354–1356, 1359, 1361, 1411, 1413, 1418, 1485–
          1486; 1500–1501, 1513, 2042–2064, 2076–2097

2C   Enregistrement, Bureau de Saillagouse: 991

L   Révolution, Administrations et tribuneaux (1790–1800): 106–
          115, 129, 131, 138, 184, 293, 299, 358–359, 388–389, 393, 403,
          409, 421–422, 429, 438–440, 464, 572, 577, 582, 591, 617–618,
          625, 641, 688, 690, 724, 748, 761, 777, 883–884, 927, 1014,
          1022, 1024, 1070, 1083, 1115, 1151–1152, 1163, 1312, 1314,
          1327, 1358, 1361, 1411, 1411 bis, 1420

M   Administration générale et économie, 1800–1940: 3M(1) 20/A-C,
          43, 71, 73–74, 137–138, 145/1–2

Mnc   M non-classée (unclassed): 831/1–3, 832/1–4, 833, 836–837,
          851/1–2, 872/1–4, 878/2, 897/1, 1461, 1722/2–3, 1824/1, 1825/
          1, 1828/1–2, 1837, 1839–1840, 1842, 1844/1–3, 1850, 1854/1–
          4, 1858/1–3, 1859/1–3, 1861, 1874/1, 1882, 1883/1–2, 1900,
          1907, 1909, 1911, 1918/1, 1921, 1924/1, 1925/1, 1932/3–5,
          1943–1944, 1951/3, 1964/1–2, 1978/5, 2100/1–3, 2233, 3054/1,
          3072/1, 3466/1–2, 3472, 3479/2–4, 3779, 3797/2, 3816, 3817,
          3818/1–7

N   Administration et comptabilité départementales, 1800–1940: 1 N
          11

O   Administration et comptabilité communales, 1800–1940: Angoustrine, Bourg-Madame, Caldegas-Onzes, Dorres, Egat, Enveitg, Err, Estavar, Llo, Osseja, Porta, Saint Léocadie, Ur, Valcebollère, La Tour de Carol

P   Finances, Cadastres, Douanes, Postes, 1800–1940:

      3P (Cadastres): maps, sections, matrices of selected French Cerdan communities
      5P (Douanes): 33 W 6

Q   Domaines: 5, 38–39, 117, 120–124, 138–141, 148–150, 153, 163, 283, 293–294, 297, 299, 360, 368–370, 383–385, 434, 437–444, 464, 519, 734, 740

Q Supp: 17–18, 24, 34, 45

S   Travaux publics: XIV S 102–105, 108–113

T   Enseignement: 1T 181

U   Justice: 2688(B), 2762–2764, 3117

V   Cultes (dépuis 1800): 2V 6–12, 46, 50–51

W   Versements, 1940— : 114W 669, 697, 995

J   Documents entrés par voie extraordinaire:

      2J: 81/1–3, 91
      30–35J: Parish registers, Targassona, Angostrin, Enveig, Càldegues, Santa Llocaya, Odello, Osseja
      7J (Fonds Salsas): 1–30, 51–60, 68–74, 89–91, 101–102, 105

AC   Archives communales (Communal Archives) deposited in the ADPO: Bourg-Madame (Hix), Caldegas-Onzes, Angoustrine, Enveigt, Dorres, La Tour de Carol, Osseja, Palau, Nahuja, Saillagouse, Estavar-Bajanda, Llo

Biblio   Bibliothèque

MS   Manuscripts: 32

SPAIN

ACA   *Archivo de la Corona de Aragón* (Archive of the Crown of Aragon, Barcelona)

CA   Consejo de Aragón (Council of Aragon): legajos 208–211, 221–222, 227–235, 254–256, 314–342, 409–412, 468, 530–531, 539–542

RA   Real Audiència:

      volumenes (registros de consultas, cartas acordadas): 3, 7–8, 10, 16, 23, 120, 123, 128–131, 134, 136–137, 140–149, 165–169,

187, 222–223, 391–392, 467, 469, 471, 565–566, 573–574, 805–810, 821, 823, 1158, 1160, 1165

legajos (Papeles de su Excelencia): 6, 8, 9, 30, 34, 81, 86, 87, 89, 91, 98, 99, 111, 118, 123, 125, 151, 162, 174–175, 237, 302

RP Real Patrimonio:

Bailia General de Cataluña:
1a C(18)
2a Ab (16); Bg, HH (1–11)
7a C(12)

Maestro Racional: 256–257

Procesos de Bailia Moderno

GI Guerra de la Independencia: cajas II, V–VIII, X, XI, XIV, XV, XVI, XIX, XX, XXI–XXIII, XXVI, XXX, XXXV, XXXXI, XXXXVI, XXXXVII, LIII, LIV, LVI, LX

DI Diversos:

Commandancia de Ingeneros: 2413–2416, 2479, 2609, 2613

SAHN *Spain, Archivo Histórico Nacional* (National Archive, Madrid)

Estado:

libros 105, 672–675d, 993d
legajos 65A, 215, 383–316, 1603, 2845, 2934, 3369, 3373, 4734, 4814, 4828, 6228, 6439, 6804–6806, 6816, 8363, 8696

Hacienda: legajos 247, 2341

SAMRE *Spain, Archivo del Ministerio de Relaciones Exteriores* (Archive of the Ministry of Foreign Relations, Madrid)

C Correspondencia: legajos 2013–2019, 2449–2450, 2746, 2873–2874

TN Tratados, Negociaciones: 221–223 (uninventoried)

SBCM Spain, Biblioteca Central Militar (Central Military Library, Madrid)

Documentos

Aparici: Fifty-eight volumes of sixteenth to eighteenth century documents recopied in the midnineteenth century from the AGS by Colonel D. José Aparici y Garcia

RSM *Real Sociedad Económica Matritense de Amigos del País* (Madrid Royal Economic Society of the Friends of the Country, Madrid)

AGS   *Archivo General de Simancas* (General Archives of Simancas)

SE   Secretaria de Estado: leg. 4682 (see also FAN Mi)

SG   Secretaria de Guerra: leg. 1849; supplemento 226

DGR   Dirección General de Rentas: 2536–2541

SSH   Secretaria y Superintenencia de Hacienda: legs. 844–845, 916, 1017

LOCAL ARCHIVES

AMLL   *Arxiu Municipal de Llívia* (Municipal Archive of Llívia): The AMLL is currently being classified; in addition to the (tentative) series numbers cited, the following were also consulted:

    MA   Manuel d'Accords
            Catastro

AHP   *Arxiu Històric de Puigcerdà* (Puigcerdà Historical Archive): Much of the AHP remains unclassified and uninventoried; when possible, I have tried to name the bound volumes or collections from which citations were drawn. These include:

    MA   Manuels d'Accords (Municipal Registers, seventeenth to nineteenth century)
            Protocols
            Correspondencia, including two bound volumes "A" and "B"
            "Morbo de Marzella"
            Internacional
            "Libro de Acequia"
            Revolucio françesa
            Catastro

ADSU   *Arxiu Diocesà de la Seu d'Urgell* (Diocesan Archive of the Seu d'Urgell)
        leg. "Andorra, Cerdanyas, y Valle d'Aran," 1742–1808 (bound volume)
        Processos de visitas

ACSU   *Arxiu Capitular de la Seu d'Urgell* (Chapter Archive of the Seu d'Urgell)
        "Pleitos sobre la Cerdaña francesa"

EP   Etude Ponsaillé (Notarial Records of Ponsaillé, Sallagosa)

MANUSCRIPT COLLECTIONS

SBN   *Spain, Biblioteca Nacional* (National Library, Madrid):
        MSS: 615, 2384, 2413, 2406, 2398–2406, 6332

FBN   *France, Bibliothèque National* (National Library, Paris):
MSS (Fonds français): 1801, 4193–4195, 4240, 4309, 8021, 11801, 22419

BRP   *Biblioteca del Real Palacio* (Library of the Royal Palace, Madrid): MS 2436

BC   *Biblioteca de Catalunya* (Library of Catalonia, Barcelona):
MSS 184, 186, 222, 313, 347, 778–789, 1535, 2371
FB (Fullets Bonsoms): 937, 943, 1026, 1640, 2397, 2477, 2479, 2491, 4024, 4091, 6835, 6838, 6840, 6898, 7324, 10096, 10838

AHMB   *Arxiu Històric Municipal de Barcelona* (Municipal Archive, Barcelona):
MS A 18, B 45, B 184

BA   *Bibliothèque de l'Arsénal* (Arsenal Library, Paris): MS 6935

BMP   *Bibliothèque Municipal de Perpignan* (Perpignan Municipal Library):
MS 203

PRIVATE ARCHIVES

Delcor (Palau)
Garreta (La Tor de Carol)
Carbonell (Gorguja)
Montellà (Santa Llocaya)

MAP COLLECTIONS

British Museum
Archives Nationales de France
Servicio Geográfica del Ejercito, Cartoteca Histórica
Biblioteca Central Militar
Biblioteca de Catalunya
Archivo Histórico Nacional
Archivo General de Simancas

OTHER SOURCES CITED

ABBREVIATIONS

BSASLPO: *Bulletin de la société agricole, scientifique, et littéraire du département des Pyrénées-Orientales*

CERCA: *Centre d'études et recherches catalanes*

RGPSO: *Revue géographique des Pyrénées et du Sud-Ouest*

RHAR: *Revue historique et archéologique du Roussillon*

CERDANYA, ROUSSILLON, CATALONIA

Alart, B. "Correspondance inédite de l'archiviste Alart et du Général Callier au sujet de la délimitation de la frontière dans les Pyrénées-Orientales, 1864–1867." *Ruscino* 3 (1913): 195–231; 4 (1914): 5–44, 129–176; 5 (1914): 397–472; 6 (1915): 125–144; 7 (1916): 83–129.

———. *Notices historiques sur les communes du Roussillon*, 2 vols. Perpignan, 1878.

———. *Inventaire-sommaire des archives départementales, antérieures à 1790*. Vol. 2 (série C). Paris, 1877.

Alcoberro, A. "Entre segadors i vigatans: L'ocupació francesa de 1694–1698." *L'Avenç* 109 (1987): 40–46.

Almirall, V. *Lo catalanisme*. Barcelona, 1979 (1886).

Amelang, J. *Honored Citizens of Barcelona: Patrician Culture and Class Relations, 1490–1714*. Princeton, 1986.

Aragon, H. *La campagne de 1719 des armées du Roussillon et d'Espagne: Lettres inédites du Maréchal de Berwick et du Maréchal le Blanc, Ministère de la Guerre*. Perpignan, 1923.

———. "Les Ducs de Noailles." *Ruscino* 10 (1919): 49–68; 11 (1919): 137–145; 12 (1920): 5–24.

Armangau, J. F. "Contribution à l'étude de la garde nationale dans les Pyrénées-Orientales pendant la période révolutionnaire, 1789–1799." Mémoire de maîtrise d'histoire contemporaine, Université de Montpellier III, 1971.

Ardit, M., A. Balcells, and N. Sales. *Història dels països catalans, de 1714 a 1975*. Barcelona, 1980.

Armengol, J., M. Batlle, and R. Gual. *Materials per una bibliografia d'Andorra*. Perpignan, 1978.

Assier-Andrieu, L. *Le peuple et la loi: Anthropologie historique des droits paysans en Catalogne français*. Paris, 1987.

———. *Coutumes et rapports sociaux: Etude anthropologique des communautés paysannes du Capcir*. Paris, 1981.

Badosa i Coll, E. "La indústria rural a Catalunya a finals del s. XVIII." In *Actas del primer congrés d'història moderna de Catalunya*. 2 vols. Barcelona, 1984. I: 345–359.

Balcells, E. "Vicisitudes historicas de las comarcas descritas (Alto Urgell, Alto

Bergada, Cerdana y Andorra)." In *Actas del 7e congreso internacional de estudios pirenaicos*. Jaca, 1974: 117–133.

Balouet, J. "L'élévage en Roussillon au XVIIIe siècle." *CERCA* 4 (1959): 165–184.

Batlle, M., and R. Gual. *"Fogatges" Catalans. Revue "Terra Nostra"* 11 (1973).

Becat, J. *Atlas de Catalunya Nord*. 2 vols. Prades, 1977.

Bernardo, D., and B. Rieu. "Conflit linguistique et revendications culturelles en Catalogne-Nord." *Les temps modernes* 324–326 (1973): 302–332.

Bisson, T. "L'essor de la Catalogne: Identité, pouvoir, et idéologie dans une société du XIIe siècle." *Annales E.S.C.* 39 (1984): 454–479.

Blanchon, J. P. *Vie de l'ancienne vallée d'Osseja*. Toulouse, 1973.

———. "En Cerdagne: Les écoles de la Belle Epoque." *Font de Segre* 7 (1981): 7–125.

———. "Hix. Les Guinguettes d'Hix. Bourg-Madame." *Conflent* 45 (1968): 111–125.

———. "La Cerdagne devant la rivalté franco–espagnole au XVIe siècle." *Conflent* 29 (1965): 7–24.

Bouille, M. "La décadence de la langue catalane en Roussillon au XVIIIe siècle." *Tramontana* 434–435 (1960): 89–95.

Bosc, A. *Sumari, índex o epítome dels . . . títols d'honor de Catalunya, Rosselló, i Cerdanya*. Perpignan, 1629.

Bosom Isern, S. *Homes i oficis de Puigcerdà al segle XIV*. Puigcerdà, 1982.

Bragulat Sirvent, J. *Vint-i-cinc anys de vida puigcerdanesa 1901–1925*. Barcelona, 1969.

Broc, N. "Géographes et naturalistes dans les Pyrénées catalanes sous l'Ancien Régime." In *Trois Siècles de Cartographie des Pyrénées*. Lourdes, 1978: 55–72.

Brousse, E. *La Cerdagne française*. Perpignan, 1926.

Brunet, M. *Le Roussillon: Une société contre l'Etat, 1780–1820*. Toulouse, 1986.

Brutails, J. A. *Etude sur la condition des populations rurales au moyen age*. Perpignan, 1891.

———. "Notes sur l'économie rurale du Roussillon à la fin de l'Ancien Régime." *BSASLPO* 30 (1889): 225–451.

Cahner, M. E. "La guerra gran a través de la poesia de l'època, 1793–1795." *Germinabit* 58 (1959): 1–5.

Calmette, J. *La question des Pyrénées et la marche d'Espagne au moyen age*. Paris, 1925.

Capeille, J. *Dictionnaire de biographies roussillonaises*. Perpignan, 1914.

———. "Le couvent de Notre Dame de Belloch." *RHAR* 5 (1904): 257–267.

Carrera i Pujal, J. *Historia política de Cataluña en el siglo XIX*, 7 vols. Barcelona, 1958.

———. *Historica econòmica de Catalunya*, 4 vols. Barcelona, 1946.

Carrere, J. B. *Voyage pittoresque de la France . . . Le province de Roussillon*. Paris, 1788.

Caseneuve, P. de. *La Catalogne française, où il est traité des droits du Roi sur la Catalogne*. Toulouse, 1644.

Cases Carbó, J. *Catalunya françesa*. Barcelona, 1934.

La Cerdanya: Recursos econòmics i activitat productiva. Caixa d'Estalvis de Catalunya. Barcelona, 1981.

Chevalier de la Grave. *Essai historique et militaire sur la province du Roussillon*. London, 1787.

Cholvy, G. "Réligion et politique en Languedoc méditerranéen et Roussillon à l'époque contemporaine." In *Droite et Gauche de 1789 à nos jours*. Actes du colloque de Montpellier, 9–10 juin 1973. Montpellier, 1975: 33–74.

Clerc, G. "Recherches sur le Conseil Souverain de Roussillon, 1660–1790: Organisation et compétence." Thèse de droit, Université de Toulouse, 1974.

Colomer Preses, I. M. *Els mapes antics de les terres catalanes*. Granollers, 1967.

———. *Els cent primers mapes del Principat de Catalunya, segles XVI–XIX*. Barcelona, 1966.

Conard, P. *Napoléon et la Catalogne*. Paris, 1910.

Corbera, Esteva de. *Catalunya illustrada*. Naples, 1678.

Costa i Costa, J. "Aproximació lingüística al català de la Cerdanya." In *Primer congrés internacional d'història de Puigcerdà*. Puigcerdà, 1983: 207–213.

Cotxet, B. *Noticia històrica de la imatge de Nostra Senyora d'Err . . . y una curta relació de la seguedat de 1847*. Perpignan, 1855.

Deffontaines, P. "Parallele entre les economies de l'Ampurdan et du Roussillon: Le role d'une frontiere." RGPSO 38 (1967): 244–258.

Defourneaux, M. "La contrabande roussillonnaise et les accords commerciaux franco–espagnols après le Pacte de Famille (1761–1786)." *Actes du 94e congrès national des sociétés savantes*. Pau, 1969 (Paris, 1971): 1, 147–163.

Delcor, M. *Les vierges romanes de Cerdagne et de Conflent*. Barcelona, 1970.

Descallar, X. de. "Episodes de la domination française dans les comtés de Roussillon et de Cerdagne sous le regne de Louis XI." RHAR 3 (1902): 359–371.

Deverell, F. H. *Border Lands of Spain and France: With an Account of a Visit to the Republic of Andorra*. London, 1856.

DiGiacomo. S. M. "Images of Class and Ethnicity in Catalan Politics, 1977–1980." In G. W. McDonogh, ed., *Conflict in Catalonia: Images of an Urban Society*. Gainesville, Fla., 1986: 72–92.

Elliott, J. H. *The Revolt of the Catalans: A Study in the Decline of Spain*. Cambridge, Eng., 1963.

Fabrègue-Pallarès, P. de la. "L'affaire de l'échange des Pays-Bas Catholiques et l'offre de retrocession du Roussillon à l'Espagne, 1668–77." BSALSPO 68 (1953): 47–63, 115–133.

Faucher, B. "Une annexion éphémère, ou le Val d'Aran de 1812 à 1814." *Actes du 2e congrès international d'études pyrénéennes, Luchon, 1954*, 7 vols. Toulouse, 1962. 7: 39–49.

Ferro, V. *El dret públic català: Les institucions a Catalunya fins al decret de Nova Planta*. Vic, 1987.

Fontana, J., *El crisis del Antiguo régimen, 1808–1830*. Barcelona, 1973.

Fontana, J. et al. *La invasió napoleònica*. Barcelona, 1981.

Font-Rius, J. M. *Estudis sobre els drets i instituciós locals en la Catalunya medieval*. Barcelona, 1985.

———. *Cartas de población y franquicia de Cataluña*, 2 vols. Madrid and Barcelona, 1969.

Fornier, E. "Les relations frontaliers en Cerdagne et Haut-Ariège." 2 vols. Diplôme d'études supérieures, Section géographie, Université de Toulouse, 1966.

Frenay, E. L'école primaire dans les Pyrénées-Orientales, 1833–1940. Perpignan, 1983.

————, ed. Cahiers de Doléances du Roussillon. Perpignan, 1979.

Galceran Vigué, S. "Els privilegis reials de la Vila de Puigcerdà." In Primer congrés internacional d'història de Puigcerdà. Puigcerdà, 1983: 124–133.

————. La indústria i el comerç a Cerdanya. Barcelona, 1978.

————, ed. Dietari de la fidelíssima vila de Puigcerdà. Barcelona, 1977.

————. L'antic sindicat de Cerdanya: Estudi socio-econòmic. Girona, 1973.

————. La ocupación francesa de Cerdaña desde 1462 al 1493 y Nuestra Señora de Gracia. Ripoll, 1971.

García Cárcel, R. Historia de Cataluña, siglos XVI–XVII, 2 vols. Barcelona, 1985.

Gavignaud, G. Propriétaires–viticulteurs en Roussillon, 2 vols. Paris, 1983.

————. "L'organisation économique traditionnelle communautaire dans les hauts pays catalans." In Conflent, Vallespir, et Montagnes Catalanes. Actes du 51e congrès de la Fédération historique du Languedoc méditerranéen et du Roussillon. Montpellier, 1980: 201–215.

————. "La frontière pyrénéenne et la partie française de la Catalogne depuis 1659." Actes du 101e congrès national des société savantes, Lille, 1976. Paris, 1978: 155–170.

Gil Coriella, A. "Vide y derecho popular: Estampas de la Cerdaña española." Revista de derecho notarial 58 (1965): 199–224.

Gouges, M. "Le monastère de Belloch." Tramontana 465–466 (1963): 55–58.

————. "L'affaire Picas, ou l'alignement de la Rahur." Tramontane 453–454 (1962): 58–60.

Grabolosa, R. Carlistes i liberals: Història d'unes querres. Barcelona, 1974.

Guiler, J. M. Unitat històrica del Pirineu. Barcelona, 1964.

Guiter, H. "Sobre el Tractat de Corbeil." Revista catalana 38 (1978): 13–15.

————. "Catalan et français en Roussillon." Ethnologie française 3–4 (1973): 291–304.

————. Atlas linguistique des Pyrénées-Orientales. Paris, 1966.

Henry, D. M. G. Histoire du Roussillon, 2 vols. Perpignan, 1935.

Hoffman, J. F. La peste à Barcelone. Paris, 1964.

Iglésis, J., ed. Estadistiques de població de Catalunya del primer vicenni del segle XVIII, 3 vols. Barcelona, 1974.

————, ed. El cens del comte Floridablanca, 1787: Part de Catalunya. Barcelona, 1969.

Isern, M. "L'évolution de l'immigration catalane en Roussillon à partir du XIXe siècle." Mémoire de maîtrise en espagnol, Université Paul Valéry, Montpellier, 1972.

Kauffman, L. A. "Politicizing Identity: The Jocs Florals of Barcelona and the Creation of Catalan Nationalism, 1859–1902." Senior thesis, Princeton University, 1987.

Libro de honor de la heroica e invicta villa de Puigcerdà. Barcelona, 1876.

Lalinde Abadia, J. La jurisdicción real inferior en Cataluña. Barcelona, 1966.

Lazerme, P. *Noblesa catalana,* 3 vols. Perpignan, 1975.

Llati, M. "La contrabande en Roussillonais au début du XIXe siècle." Diplôme d'études supérieures, Université de Montpellier, 1955.

Lluch, E. *El pensament econòmic a Catalunya, 1760–1840.* Barcelona, 1973.

Mach, C. "Production agricole et population du Roussillon au XVIIIe siècle." Mémoire de maîtrise, La Sorbonne, 1968.

Marca, P. de. *Marca hispanica sive limes hispanicus,* trans. into Catalan by J. Icart. Barcelona, 1965 (1688).

Marcet, A. "La Cerdagne après le Traité des Pyrénées." *Annales du Midi* 93 (1981): 141–155.

——. "Une révolte antifiscale et nationale: Les Angelets du Vallespir, 1663–72." *Actes du 102e congrès national des sociétés savantes, Limoges, 1977.* Paris, 1978: 35–48.

——. "Les conspirations de 1674 en Roussillon: Villefranche et Perpignan." *Annales du Midi* 86 (1974): 275–296.

Marti Sanjaume, J. *Las virgenes de Cerdaña.* Lérida, 1927.

Martzluff, M. "Aspects socio-politiques du Roussillon dans le contexte de l'expédition d'Espagne en 1823." Mémoire de maîtrise, Université de Montpellier, 1977.

McPhee, P. "The Seed-Time of the Republic: Society and Politics in the Pyrénées-Orientales." Ph.D. dissertation, University of Melbourne, 1977.

Mercader i Riba, J. *Felip V i Catalunya.* Barcelona, 1968.

——. *Catalunya i l'imperi napoleònic.* Barcelona, 1960.

——. "Puigcerdà, capital del departament del Segre." *Pirineos* 9–10 (1948): 413–457.

Meugniot, M. "La conscription dans les Pyrénées-Orientales sous l'Empire." Mémoire de maîtrise, Université de Montpellier, 1955.

Miquel de Riu, G. "Extrait des souvenirs de M. François Sicart d'Alougny." *BSASLPO* 33 (1897): 167–178.

Moreu Rei, E. *Els immigrants francesos a Barcelona.* Barcelona, 1959.

Nadal, J., and E. Giralt. *La population catalane de 1553 à 1717: L'immigration française.* Paris, 1960.

Ossorio y Gallardo, A. *Historia del pensamiento político catalan durante la guerra de España con la República francesa, 1793–1795.* Madrid, 1931.

Paillissé, M.-A. "Mont-Louis, place forte et nouvelle." Mémoire de maîtrise, Université de Montpellier, n.d.

Pastor, J. de. "La Cerdagne française (Etude de géographie humaine)." *Bulletin de la société languedocienne de géographie* 3 (1933): 119–144, 158–181; 4 (1933): 28–44, 53–82, 117–139, 184–196.

Payne, S. "Catalan and Basque Nationalism," *Journal of Contemporary History* 6 (1971): 15–51.

Pladevall, A., and A. Simon. *Guerra i vida pagesa a la Catalunya del segle XVII.* Barcelona, 1986.

Planes, Ll. *El petit llibre de Catalunya Nord.* Perpignan, 1974.

Poeydavant, M. "Mémoire sur la province de Roussillon et le pays de Foix," ed. M. E. Desplanque. *BSALPO* 37 (1889): 283–409; 51 (1910): 1–188.

Poblet, J. M. *Història bàsica del catalanisme.* Barcelona, 1975.

Pons, J.-S. *La littérature catalane en Roussillon au XVIIe et au XVIIIe siècles: L'esprit provincial. Les mystiques. Les goigs et le théâtre réligieux,* 2 vols. Toulouse, 1929.

Les problèmes fonciers en Basse-Cerdagne. Ministère de l'Agriculture, Direction départemental d'études économiques d'aménagement rural. Perpignan, 1978.

Puig i Oliver, L. M. de. *Girona françesa, 1812–4: L'Annexió de Catalunya a França i el domini napoleònic a Girona.* Barcelona, 1976.

Puyol Safont, A. *Los hilos illustres de Cerdaña.* Barcelona, 1926.

Ramonatxo, H. *Un cerdan témoin de l'histoire: François Garreta.* Luzech, 1978.

Rosset, P. "Culture et élévage dans la viguerie de Cerdagne à la fin de l'Ancien Régime." *Revue d'histoire moderne et contemporaine* 31 (1984): 131–142.

Rouse, E. *Histoire de Notre Dame de Font-Romeu.* Lille, 1890.

Sacquer, J. "La frontière et la contrabande avec le Catalogne dans l'histoire et l'économie du département des Pyrénées-Orientales de 1814 à 1850." Diplôme d'études supérieures, Histoire économique, Université de Toulouse, 1967.

Sagnes, J., ed. *Le pays Catalan.* Pau, 1983.

Sahlins, P. "The Nation in the Village: State-Building and Communal Struggles in the Catalan Borderland during the Eighteenth and Nineteenth Centuries." *Journal of Modern History* 60 (1988): 234–263.

————. "Between France and Spain: Boundaries of Territory and Identity in a Pyrenean Valley, 1659–1868." Ph.D. dissertation, Princeton University, 1986.

————. "An Aspect of Property Relations in the Borderland: Nationality, Residence, and the 'Capitation' Tax in the Eighteenth-Century French Cerdanya." In *Actas del primer congrés d'història moderna de Catalunya,* 2 vols. Barcelona, 1984. 1: 411–418.

Sales, N. "Naturals i aliénígenes: Un cop d'ull a algunes naturalitzacions dels segles XV a XVIII." In *Studie in honorem Prof. M. de Riquer.* Barcelona, 1987: 675–705.

————. "Classes ascendents i classes descendents a la Catalunya françesa d'antic régimen: La noblesa rossellonesa, arruïnada i disminuïda?" In R. Garrabou, ed. *Terra, treball, i propietat: Classes agraries i règim senyorial als Països Catalans.* Barcelona, 1986: 24–41.

————. "'Bandoliers espaignols' i guerres de religió françesa." *L'Avenç* 82 (1985): 46–55.

————. *Senyors bandolers, miquelets i botiflors: Estudis sobre la Catalunya dels segles XVI al XVIII.* Barcelona, 1984.

————. *Sobre esclavos, reclutas y mercaderes de quintos.* Barcelona, 1974.

Salsas, A. *La Cerdagne espagnole.* Perpignan, 1890.

Sanabre, J. *La resistència del Rosselló a incorporar-se a França.* Barcelona, 1970.

————. *El tractat dels Pirineus i la mutilació de Catalunya.* Barcelona, 1961.

————. *La acción de Francia en Cataluña en la pugna por la hegemonía de Europa, 1640–1659.* Barcelona, 1956.

Sanllehy i Sabi, M. Angels. "L'afillament a les communitats araneses (segles XVII–XIX)." *L'Avenç* 115 (1988): 32–37.

Sarrete, J. *La confrérie du Rosaire en Cerdagne.* Perpignan, 1920.

———. "Les paysans de Cerdagne sous l'Ancien Régime." *Guide-annuaire du Roussillonnais 1910:* 149–174. Perpignan, n.d.

———. "La Cerdagne pendant la révolution Catalane et jusqu'au Traité des Pyrénées." *Guide-annuaire du Roussillonnais* 1910: 175–213.

———. "La condition des non-privilégiés sous l'Ancien Régime à Osseja." *RHAR* 6 (1905): 84–94, 108–115.

———. "La paroisse de Hix." *BSASLPO* 46 (1905): 313–347.

———. "La Cerdagne de 1642 à 1652." *RHAR* 4 (1903): 183–190.

El sitio de Puigcerdà por los carlistas desde el dia 20 de agosto hasta el 3 de septiembre de 1874. Barcelona, 1875.

Soldevila, F. *Història de Catalunya,* 3 vols. Barcelona, 1962.

Sola i Gussinyer, P. *Cultura popular, educació, i societat al nord-est català (1887–1959).* Girona, 1983.

Sole Sabaris, L. "Del paisaje de Cerdaña." *Ilerda* 19 (1955).

Stivil, G. *Le régime administratif du Roussillon, depuis le Traité des Pyrénées jusqu'à la Révolution.* Paris, 1927.

Termes, J. *Federalismo, anarcosindicalismo, y catalanismo.* Barcelona, 1976.

Teule, E. de. *Etat des juridictions inférieures du comté de Roussillon avant 1790.* Paris, 1887.

Thiers, A. *Les Pyrénées et le Midi de la France pendant les mois de novembre et décembre 1822.* Paris, 1823.

Torras, J. "Peasant Counterrevolution." *Journal of Peasant Studies* 5 (1977): 66–88.

———. *Liberalismo y rebeldia campesina 1820–23.* Barcelona, 1976.

Torras i Ribe, J. M. *Els municipis catalans de l'Antic Règim, 1453–1808.* Barcelona, 1983.

Torreilles, P. "Le role politique de Marca et de Serroni durant les guerres de Catalogne, 1644–60." *Revue des questions historiques* 69 (1901): 59–98.

———. "L'oeuvre de Vauban en Roussillon." *BSALSPO* 42 (1901): 181–288.

———. "La délimitation de la frontière en 1660." *RHAR* 1 (1900): 21–32.

———. "L'organisation administrative du Roussillon en 1660." *RHAR* 1 (1900): 263–273.

———. "Troubles et guerre en Roussillon." *RHAR* 1 (1900): 321–328.

———. *La diffusion de français à Perpignan, 1660–1700.* Perpignan, 1910.

———. *Histoire du clergé dans les Pyrénées-Orientales pendant la Révolution française.* Perpignan, 1890.

Torreilles, P., and E. Desplanques. "L'enseignement élémentaire en Roussillon dépuis ses origines jusqu'au commencement du XIXe siècle." *BSASLPO* 50 (1895): 145–398.

Torres i Sans, X. "Les bandositats de Nyerros i Cadells a la Reial Audiéncia de Catalunya (1590–1630): Policia o alto govierno?" *Pedralbes* 5 (1985): 145–171.

———. "Els bandols de 'Nyerros' i 'Cadells' a la Catalunya Moderna." *L'Avenç* 49 (1982): 345–350.

Tulla, A. "Transformació agrària en areas de muntanya: Les explotacions de producció lletera com a motor de canvi a les comarques de la Cerdanya, El Capcir, l'Alt Urgell, i el Principat d'Andorre." Tesi doctoral, Universitat Autònoma de Barcelona, 1981.

―――. "Les deux Cerdagnes: Example de transformations économiques asymétriques de part et d'autre de la frontière des Pyrénées." *RGPSO* 48 (1977): 409–424.

Vassal Reig, C. *La prise de Perpignan.* Paris, 1939.

―――. *Richelieu et la Catalogne.* Paris, 1935.

Ventura Subirats, J. *Els heretges catalans.* Barcelona, 1963.

Verdaguer, P. *El Català al Rosselló: Gallicismes, Occitanismes, Rossellonismes.* Barcelona, 1974.

Vidal, P. *Histoire de la Révolution dans le département des Pyrénées-Orientales,* 3 vols. Perpignan, 1885–1886.

Vidal Pla, J. *Guerra dels segadors i crisis social: Els exiliats filipistes (1640–1652).* Barcelona, 1984.

Vigo, E. *La política catalana del gran Comité de Salut Publica.* Barcelona, 1956.

Vila, P. *La Cerdanya.* Barcelona, 1985 (1926).

―――. *La divisió territorial de Catalunya.* Barcelona, 1979.

Vilar, P. *La Catalogne dans l'Espagne moderne,* 3 vols. Paris, 1962.

Woolard, K. A. "The 'Crisis in the Concept of Identity' in Contemporary Catalonia, 1976–82." In G. W. McDonogh, ed., *Conflict in Catalonia: Images of an Urban Society.* Gainesville, Fla., 1986: 54–71.

Xandri, J. *La Cerdaña.* Madrid, 1917.

Zamora, F. de. *Diario de los viajes hechos en Cataluña,* ed. R. Boixareu. Barcelona, 1973.

OTHER WORKS CITED

Acuerdos fronterizos con Francia y Portugal. Madrid: Ministerio de Hacienda, 1969.

Agulhon, M. *Marianne au Combat.* Paris, 1979.

―――. *The Republic in the Village: The People of the Var from the French Revolution to the Second Republic,* trans. J. Lloyd. Cambridge, Mass., 1982 (1970).

Akerman, J. A. "Cartography and the Emergence of Territorial States." In *Proceedings of the Tenth Annual Meeting of the Western Society for French History* (Lawrence, Kans., 1984): 84–93.

Allison, J. *Thiers and the French Monarchy.* New York, 1968 (1926).

Alliès, P. *L'invention du territoire.* Grenoble, 1980.

Anderson, B. *Imagined Communities: Reflections on the Origin and Spread of Nationalism.* London, 1983.

André, L. *Louis XIV et l'Europe.* Paris, 1950.

Anes, G. *El Antiguo régimen: Los Borbones.* Madrid, 1981.

Ardant, G. "Financial Policy and Economic Infrastructure in Modern States and Nations." In C. Tilly, ed., *The Formation of National States in Western Europe.* Princeton, N.J., 1975: 164–242.

Ardascheff, P. *Les intendants de province sous Louis XVI*, trans. L. Jousserandot. Paris, 1909.

Armstrong, J. *Nations before Nationalism*. Chapel Hill, N.C., 1982.

Artola, M. *La hacienda del Antiguo régimen*. Madrid, 1982.

Assier-Andrieu, L. "La communauté villageoise: Objet historique, enjeu théorique." *Ethnologie française* 17 (1987): 351–360.

Aymes, J. R. *La guerre d'indépendence en Espagne 1808–1814*. Paris, 1973.

Balfet, H., C. Bromberger, and G. Pavis-Giordani. "De la maison aux lointains: Pour une étude des cercles de référence et d'appartenance social en méditerranée nord-occidentale." In *Pratiques et représentations de l'espace dans les communautés méditerranéennes*. Paris, 1976: 27–75.

Banfield, E. C. *The Moral Basis of a Backward Society*. New York, 1958.

Barth, F., ed. *Ethnic Groups and Boundaries: The Social Organization of Cultural Difference*. Boston, 1967.

Battifol, L. "Richelieu et la question d'Alsace." *Revue historique* 138 (1921): 161–200.

Baudrillart, H. *Populations agricoles de la France*, 3 vols. (Paris, 1893).

Beaune, C. *Naissance de la nation France*. Paris, 1986.

Behar, R. *Santa María del Monte: The Presence of the Past in a Spanish Village*. Princeton, N.J., 1986.

Bell, D. "National Character Revisited." In E. Norbeck, ed., *The Study of Personality*. New York, 1968.

Bendix, R. *Nation-Building and Citizenship*. New York, 1964.

Benveniste, E. *Le vocabulaire des institutions indo-europeennes*, 2 vols. Paris, 1969.

Berenson, E. *Populist Religion and Left-Wing Politics in France, 1830–1852*. Princeton, N.J., 1985.

Berlet, C. *Les tendances unitaires et provincialistes en France à la fin du XVIIIe siècle*. Nancy, 1913.

Berthaut, C. *Les ingénieurs géographes militaires, 1624–1831*, 2 vols. Paris, 1902.

Birot, L. *Etude comparée de la vie rurale pyrénéenne dans le pays de Pallars et de Couserans*. Paris, 1937.

Bloch, M. *French Rural History: An Essay on Its Basic Characteristics*, trans. J. Sondenheimer. Berkeley and Los Angeles, 1966.

Blum, J. "The European Village Community: Origins and Functions." *Agricultural History* 45 (1971): 157–178.

———. "The Internal Structure and Polity of the European Village Community from the Fifteenth to the Nineteenth Century." *Journal of Modern History* 43 (1971): 541–576.

Boislisle, A. de, ed. *Correspondence des contrôleurs généraux des finances avec les intendants de province*, 3 vols. Paris, 1874–1897.

Bonenfant, P. "A propos des limites médiévales." *Hommage à Lucien Febvre: Eventail de l'histoire vivant*, 2 vols. Paris, 1953: 1, 73–79.

Bonnain, R. "Droit écrit, coutume pyrénéenne, et pratiques successorales dans les Baronnies de 1769 à 1836." In *Los Pirineos: Estudios de antropologia social e historia*. Coloquio Hispano-Francés, Casa de Velazques. Madrid, 1986: 232–240.

Bosher, J. F. *The Single Duty Project: A Study of the Movement for a French Customs Union in the Eighteenth Century.* London, 1964.

Boulainvilliers, Comte de. *Etat de France: Extrait des mémoires dressés par les intendants du royaume et par ordre du roi,* 5 vols. London, 1737.

Boutier, J. A. Dewerpe, and D. Nordman. *Un tour de France royal: Le voyage de Charles IX, 1564–66.* Paris, 1984.

Braudel, F. *L'identité de la France,* vol. 1. Paris, 1986.

————. *The Mediterranean and the Mediterranean World in the Age of Philip II,* trans. S. Reynolds, 2 vols. New York, 1966.

Brette, A. *Les limites et divisions territoriales de la France en 1789.* Paris, 1907.

Brives, A. *Pyrénées sans frontière: La vallée de Barèges et l'Espagne du XVIIIe siècle à nos jours.* Argelès-Gazost, 1984.

Brun, A. *L'introduction de la langue française en Béarn et en Roussillon.* Paris, 1923.

Brunhes, J., and C. Vallaux. *La géographie de l'histoire.* Paris, 1921.

Brunot, F. *Histoire de la langue française des origines à 1900,* 12 vols. Paris, 1905–1953.

Buisseret, D. "The Cartographic Definition of France's Eastern Boundary in the Early Seventeenth Century." *Imago Mundi* 36 (1984): 72–80.

Buot de l'Epine, A. *Le comité contentieux des départements.* Paris, 1972.

Bussy-Rabutin, Roger de. *Mémoires de Messire Roger de Rabutin, Comte de Bussy, lieutenant général des armées du Roy,* 2 vols. Paris, 1696.

Carlos Garcia, *La oposición y conjunción de los dos grandes luminares de la tierra, o la antipatía de franceses y españoles.* Edmunton, 1979 (Madrid, 1617).

Caro Baroja, J. *Razas, pueblos, y linajes.* Madrid, 1957.

Carr, R. *Spain, 1808–1975.* Oxford, 1982.

Caseneuve, P. de. *La Catalogne française, où il est traité des droits du roy sur la Catalogne.* Toulouse, 1644.

Castel, J. *España y el tratado de Münster, 1644–48.* Madrid, 1956.

Castro, A. *The Spaniards: An Introduction to Their History,* trans. W. F. King and S. Margaretten. Berkeley, Los Angeles, London, 1971.

Catudal, H. M. *The Exclave Problem of Western Europe.* University, Ala., 1979.

Cavailles, H. "Une fédération pyrénéenne sous l'Ancien Régime." *Revue historique* 55 (1910): 1–32, 241–274. Reprinted in J. F. Nail et al., *Lies et passeries dans les Pyrénées.* Tarbes, 1986: 1–67.

Cénac Montcaut, J. E. M. *Histoire des peuples et des états pyrénéens,* 5 vols. Paris, 1860.

Censo de la población de España, 1857. Madrid, 1858.

Censo de la población de España, 1877. Madrid, 1878.

Certeau, M. de, D. Julia, and J. Revel. *Une politique de la langue: La Révolution française et les patois.* Paris, 1975.

Challner, R. D. *The French Theory of the Nation in Arms, 1866–1939.* New York, 1965.

Chevalier, M. *La vie humaine dans les Pyrénées ariegeoises.* Paris, 1956.

Chotard, H. *Louis XIV, Louvois, Vauban, et les fortifications du nord de la France.* Paris, 1890.

Church, F. "France." In O. Ranum, ed., *National Consciousness, History, and Political Culture in Early Modern Europe*. Baltimore, 1975: 43–66.

———. *Richelieu and Reason of State*. Princeton, N.J., 1972.

Clark, G. *The Seventeenth Century*. New York, 1961.

Cohen, A. P. *The Symbolic Construction of Community*. London, 1985.

———, ed. *Symbolizing Boundaries: Identity and Diversity in British Cultures*. Manchester, 1986.

Colletet, F. *Journal contenant la relation véritable et fidèle du voyage du Roy*. Paris, 1659.

———. *Entrevue et conférence de son Eminence le Cardinal Mazarin et Dom Luis d'Aro, le 13 août 1659*. Paris, 1659.

Contamine, P. "Mourir pour la Patrie, X–XXe siècle." In P. Nora, ed., *Les lieux de mémoire*, 2 vols. Paris, 1986: 2 (pt. 3): 11–43.

Contrasty, J. *Le clergé français exilé en Espagne, 1792–1802*. Toulouse, 1910.

Comellas García-Llera, J. *El Trienio Constitucional*. Madrid, 1963.

Cordero Torres, J. M. *Fronteras hispanicas: Geografía e historia*. Madrid, 1960.

Costa Martinez, J. *Colectivismo agrario en Espana*. Madrid, 1915.

Coverdale, J. *The Basque Phase of Spain's Carlist War*. Princeton, N.J., 1984.

Dainville, F. "Cartes et contestations au XVe siècle." *Imago Mundi* 24 (1970): 99–121.

———. *Le langage des géographes: Termes, signes, couleurs des cartes anciennes*. Paris, 1964.

———. *La géographie des humanistes*. Paris, 1940.

Darnton, R. *The Great Cat Massacre and Other Episodes in French Cultural History*. New York, 1985.

Davis, N. Z. *Society and Culture in Early Modern France*. Stanford, Calif., 1975.

Daveau, S. *Les régions frontalières de la montagne jurassienne: Etude de géographie humaine*. Lyon, 1959.

Defourneaux, M. *Inquisicion y censura de libros en Espana*. Madrid, 1963.

De la Blottière and Roussel. *Légende de tous les cols, ports, et passages dans les Pyrénées, 1716–1719*, ed. J. Escarra. Paris, 1915.

Descheemaeker, J. "La frontière pyrénéenne de l'Ocean à l'Aragon." Thèse de droit, Université de Paris, 1945.

Désorges, R. "Le Briançonnais, une région frontière de montagne." Thèse de droit, Université de Grenoble, 1956.

Desplat, C. "Le parlement de Navarre et la définition de la frontière franco–navarraise à l'extrême fin du XVIIIe siècle." In J. F. Nail et al., *Lies et passeries dans les Pyrénées*. Tarbes, 1986: 109–120.

Deutsch, K. *Nationalism and Social Communication: An Inquiry into the Foundations of Nationality*. Cambridge, Mass., 1953.

Deutsch, K., and W. Foltz, eds. *Nation-Building*. New York, 1963.

Dion, R. *Les frontières de la France*. Paris, 1947.

Dockès, P. *L'espace dans la pensée économique du XVIe au XVIIIe siècles*. Paris, 1969.

Doucet, R. *Les institutions de la France au XVIe siècle*, 2 vols. Paris, 1948.

Duplessis-Bésançon, B. *Mémoires,* ed. H. de Beaucaire. Paris, 1892.

Dupont-Ferrier, G., "L'incertitude des limites territoriales en France du XIIIe au XVIe siècle." *Académie des Inscriptions et Belles Lettres: Comtes Rendus,* 1942: 62–77.

————. "Le sens des mots 'patria' et 'patrie' en France au moyen age et jusqu'au début du XVIIe siècle." *Revue historique* 188 (1940): 81–104.

————. "Sur l'emploi du mot 'province,' notamment dans le langage administratif de l'ancienne France." *Revue historique* 160 (1929): 241–267.

Dupuy, J. *Traité touchant les droits du Roy.* Paris, 1655.

Elliott, J. *Richelieu and Olivares.* Cambridge, Eng., 1984.

————. *Imperial Spain, 1469–1716.* New York, 1963.

Espoz y Mina, *Memorias.* Biblioteca de autores españoles, vols. 146–147. Madrid, 1962.

Evans-Pritchard, E. E. *The Nuer.* Oxford, 1940.

Eychenne, E. *Les Pyrénées de la Liberté.* Paris, 1983.

————. *Montagnes de peur et de l'espérance.* Toulouse, 1980.

Febvre, L. "*Frontière*: The Word and the Concept." In P. Burke, ed., *A New Kind of History: From the Writings of Febvre.* London, 1973: 208–218.

————. "Langue et nationalité en France au XVIIIe siècle." *Revue de synthèse historique* 42 (1926): 19–40.

————. "Politique royale ou civilisation française: Remarques sur un problème d'histoire linguistique." *Revue de synthèse historique* 38 (1924): 37–53.

Fernandez, J. "Enclosures: Boundary Maintenance and Its Representations in Asturian Mountain Villages [Spain]." In E. Ohnuki Tierney, ed., *Symbols in Time.* Stanford, Calif., in press.

Fernandez de Casadevanti Romani, C. *La frontera hispano–francesa y las relaciones de vecindad (Especial referencia al sector fronterizo dels pais vasco).* San Sebastian, 1986.

Fervel, J. N. *Histoire des campagnes de la Révolution,* 3 vols. Paris, 1823.

Fontana, J. *La crisis del Antiguo régimen, 1808–1833.* Barcelona, 1979.

Ford, R. *Gatherings from Spain.* London, 1970 (1846).

Fortes, M., and E. E. Evans-Pritchard. *African Political Systems.* London, 1940.

Foucher, M. *L'invention des frontières.* Paris, 1987.

Fraelich, A. "The Manuscript Maps of the Pyrenees by Roussel and La Blottière." *Imago Mundi* 15 (1960): 94–104.

Freud, S. *Civilization and Its Discontents,* trans. and ed. by J. Strachey. New York, 1960.

Galaty, J. G. "Models and Metaphors: On the Semiotic Explanation of Segmentary Systems." In L. Holy and M. Stichlik, eds., *The Structure of Folk Models.* New York, 1981: 63–92.

Gallois, L. *Les variations de la frontière française du Nord et du Nord-Est depuis 1789.* Paris, 1918.

Gaguère, F. *Pierre de Marca, 1594–1662: Sa vie, ses oeuvres, son gallicanisme.* Lille, 1932.

Gellner, E. *Nations and Nationalism.* Ithaca, N.Y., 1983.

Geoffrey de Grandmaison. *L'expédition française d'Espagne en 1823.* Paris, 1928.

Gilbert, R. "La condición de los extranjeros en el antiguo derecho español." In *L'étranger*. Receuils de la société Jean Bodin pour l'histoire comparative des institutions, vol. 10. Brussels, 1958.

Girard, A. *Le commerce français à Séville et Cadix au temps des Hapsbourg: Contribution à l'étude du commerce étranger au XVIe et XVII siècles*. Paris, 1932.

Girard d'Albissin, N. *Genèse de la frontiere franco–belge: Les variations de limites septentrionales de la France de 1659 à 1789*. Paris, 1970.

————. "Propos sur la frontière." *Revue historique du droit français et étranger* 47 (1969): 390–407.

Girard, L. *La politique des travaux publics du Second Empire*. Paris, 1951.

Godechot, J. "Nation, patrie, nationalisme et patriotisme en France au XVIIIe siècle." In *Patriotisme et nationalisme en Europe à l'époque de la Révolution française et de Napoléon*. XIIIe Congrès international des sciences historiques. Paris, 1973: 7–27.

————. *Les institutions de la France sous la Révolution et l'Empire*. Paris, 1951.

————. *La Grande Nation: L'expansion révolutionnaire de la France dans le monde de 1789 à 1799*, 2 vols. Paris, 1956.

Gomez-Ibañez, D. A. *The Western Pyrenees: Differential Evolution of the French–Spanish Borderland*. Oxford, 1975.

Goron, L. "Les migrations saisonnières dans les départements pyrénéens au début du XIXe siècle." *RGPSO* 4 (1933): 230–272.

Gottman, J. *The Significance of Territory*. Richmond, 1973.

Greer, D. *The Incidence of the Emigration during the French Revolution*. Cambridge, Mass., 1951.

Grillo, R. D., ed. *"Nation" and "State" in Europe: Anthropological Perspectives*. London, 1980.

Gross, R. "Registering and Ranking of Tension Areas," in *Confini e Regioni: il potenziale di sviluppo e di pace delle periferie* Trieste, 1973: 317–328.

Gualdo Priorato, G. *Histoire du traité de Paix conclue à Saint Jean de Luz entre les deux couronne en 1659*. Cologne, 1675.

Guenée, B. "Les limites." In M. François, ed., *La France et les français* (Paris, 1972): 50–60.

————. "Des limites féodales aux frontières politiques." In P. Nora, ed., *Les lieux de mémoire*, 2 vols. Paris, 1986: 2 (pt. 2): 11–33.

————. "Etat et nation en France au Moyen Age." *Revue historique* 237 (1967): 17–30.

Guez de Balzac. *Le Prince*. Paris, 1631.

Gullickson, G. *Spinners and Weavers of Auffay: Rural Industry and the Sexual Division of Labor in a French Village, 1750–1850*. Cambridge, Eng., 1986.

Gutierrez, A. *La France et les français dans la littérature espagnole: Un aspect de xénophobie en Espagne, 1598–1665*. Saint-Etienne, 1977.

Gutton, J. P. *La sociabilité villageoise dans l'Ancienne France*. Paris, 1979.

Hargreaves-Mawdsley, W. N. *Spain under the Bourbons, 1700–1833: A Collection of Documents*. London, 1973.

Hennesy, C. A. M. *The Federal Republic in Spain*. Oxford, 1962.

Herbin, R., and A. Pebereau. *Le cadastre français.* Paris, 1953.

Herr, R. *The Eighteenth Century Revolution in Spain.* Princeton, N.J., 1958.

Hertz, F. *Nationality in History and Politics: A Psychology and Sociology of National Sentiment and Nationalism.* London, 1944.

Higonnet, P. "The Politics of Linguistic Terrorism and Grammatical Hegemony during the French Revolution." *Social History* 5 (1980): 41–69.

Hinojosa, E. de. *El elemento germánico en el derecho español.* Madrid, 1915.

Hirschman, A. *Exit, Voice, Loyalty: Responses to Decline in Firms, Organizations, and States.* Cambridge, Mass., 1970.

Hood, J. N. "Revival and Mutuation of Old Rivalries in Revolutionary France." *Past and Present* 82 (1979): 82–115.

Hufton, O. *The Poor in Eighteenth-Century France.* Oxford, 1974.

Hugo, A. *Histoire de la campagne d'Espagne en 1823,* 2 vols. Paris, 1824–1825.

Hume, D. *Moral and Political Philosophy,* ed. H. D. Aiken. Oxford, 1963.

Hunt, L. *Politics, Culture, and Class in the French Revolution.* Berkeley, Los Angeles, London, 1984.

Hyslop, B. F. *French Nationalism in 1789, According to the General Cahiers.* New York, 1934.

Images de la montagne: Catalogue et essais. Paris, 1984.

Johnston, R. J., and P. J. Perry. "Déviation directionelle dans les aires de contact: Deux examples de relations matrimoniales dans la France rurale du XIXe siècle." *Etudes Rurales* 46 (1972): 23–33.

Jollivet, M. "L'utilisation des lieux de naissance pour l'analyse de l'espace social d'un village." *Revue française de sociologie* 6 (1965): 74–95.

Jones, S. B. *Boundary Making: A Handbook for Statesmen, Treaty Editors, and Boundary Commissioners.* Washington, D.C., 1945.

Jover Zamora, J. M. "Monarquia y nación en la España del s. XVII." *Cuadernos de historia de España* 13 (1950): 101–150.

Kamen, H. *Spain in the Later Seventeenth Century.* London and New York, 1980.

———. "The 'Decline of Spain': An Historical Myth?" *Past and Present* 81 (1978): 24–50.

———. *The War of Succession in Spain.* Bloomington, Ind., 1969.

Kantorowicz, E. *The King's Two Bodies: A Study in Medieval Political Theology.* Princeton, N.J., 1957.

Koch, C. *Table des traités entre la France et les puissances étrangères dépuis la paix de Westphalie jusqu'à nos jours,* 2 vols. Basel, 1802.

Koenigsberger, H. "Spain." In O. Ranum, ed., *National Consciousness, History, and Political Culture in Early Modern Europe.* Baltimore, 1975: 144–172.

Kohn, H. *Prelude to Nation-States: The French and German Experience, 1789–1815.* Princeton, N.J., 1967.

Koht, H. "The Dawn of Nationalism in Europe." *American Historical Review* 52 (1947): 265–280.

Konvitz, J. *Cartography in France: Science, Engineering, and Statecraft, 1660–1848.* Chicago, 1987.

Lameire, I. *Théorie et pratique de la conquête dans l'ancien droit*, vol. 2: *Les occupations militaires en Espagne pendant les guerres de l'ancien droit*. Paris, 1905.

La Mothe le Vayer. *Discours de la contrariété d'humeurs qui se trouve entre certains nations, et singulièrement entre la française et l'espagnole*. Paris, 1636.

Lapeyre, H. "La cartographie des Pyrénées avant Sanson." *Annales du Midi* 67 (1955): 261–268.

Lapradelle, P. de. *La frontière: Etude de droit international*. Paris, 1928.

Larrieu, M. "Les réfugiés espagnols à Toulouse lors des guerres carlistes." Diplôme d'études supérieures, Université de Toulouse–Le Mirail, 1971.

Lecuir, J. "A la découverte de la France dans les abrégés d'histoire et de géographie des collèges jésuites du XVIIe siècle." In *La découverte de la France au XVIIe siècle*. 9e Colloque de Marseille organisé par le Centre Méridional de Rencontres sur le XVIIe siècle, 25–28 janvier 1979. Paris, 1980: 298–317.

Lefebvre, G. *Napoleon: From 18 Brumaire to Tilsit, 1799–1807*, trans. J. E. Anderson. New York, 1969.

————. *Napoleon: From Tilsit to Waterloo, 1807–1815*, trans. J. E. Anderson, New York, 1969.

Lemarignier, J. *Recherches sur l'hommage en marche et les frontières féodales*. Lille, 1945.

Lentacker, F. *La frontière franco–belge: Etude géographique des effets d'une frontière sur la vie des relations*. Lille, 1974.

Le Nail, J. F., et al. *Lies et passéries dans les Pyrénées*. Tarbes, 1986.

Le Roy Ladurie, E. *Montaillou: The Promised Land of Error*, trans. B. Bray. New York, 1979.

Levi, C. *Christ Stopped at Eboli*, trans. F. Frenaye. New York, 1947.

Lipiansky, E. M. "L'imagérie de l'identité: Le couple France–Allemagne." *Ethnopsychologie* 34 (1979): 273–282.

Livet, G. "Louis XIV et les provinces conquises." *XVIIe siècle* 22 (1952): 481–507.

Lognon, A. "Les limites de la France et l'étendue de la domination anglaise à l'époque de Jeanne d'Arc." *Revue des questions historiques* 18 (1875): 444–546.

Lynch, J. *Spain under the Habsburgs*, 2 vols. Oxford, 1981.

Macartney, C. A. *National States and National Minorities*. Oxford, 1934.

Marca, P. de. *Histoire de Béarn*. Paris, 1640.

Mage, G. *La division de la France en départements*. Toulouse, 1924.

Marechal, C. *Spain, 1834–1844*. London, 1977.

Margadant, T. W. "French Rural Society in the Nineteenth Century: A Review Essay." *Agricultural History* 53 (1979): 644–651.

Marion, M. *Dictionnaire des institutions de la France aux XVIIe et XVIIIe siècles*. New York, 1968 (1923).

Marqués del Saltillon. "Don Antonio Pimentel de Prado y la Paz de los Pirineos." *Hispania* 7 (1947): 24–124.

Marquis de Roux. *Louis XIV et les provinces conquises*. Paris, 1938.

Martens, G. F. *Receuil des principaux traités d'alliance, de paix, de trève . . . depuis 1761 jusqu'à present.* 6 vols. Göttingen, 1791–1806.

Martin, M. M. *The Making of France: The Origins and Development of the Idea of National Unity,* trans. B. and R. North. London, 1951.

Marx, K. *Pre-Capitalist Economic Formations,* ed. E. J. Hobsbawm. New York, 1964.

—————. *The Eighteenth Brumaire of Louis Bonaparte.* New York, 1963 (1852).

Matilla Tascon, A. *Catalogo de la colección de ordenes generales de rentas,* 2 vols. Madrid, 1962.

Mazarin, Jules, Cardinal de. *Lettres du Cardinal Mazarin,* ed. M. A. Cheruel, 9 vols. Paris, 1879–1906.

—————. *Lettres du Cardinal Mazarin où l'on voit le secret de la négociation de la Paix des Pyrénées.* Amsterdam, 1693.

McManners, J. *The French Revolution and the Church.* New York, 1969.

Michelet, J. *Tableau de France.* Paris, 1833 (1948).

Mill, J. S. *System of Logic.* London, 1847.

Mitard, S. *La crise financière en France à la fin du XVIIe siècle: La première capitation (1685–1688).* Rennes, 1934.

Monfrin, J. "Les parlers en France." In M. François, ed., *La France et les français.* Paris, 1972: 745–775.

Morel-Fatio, A., ed. *Receuil des instructions données aux ambassadeurs et ministres de France,* vol. 11, no. 1: *Espagne, 1649–1700.* Paris, 1894.

Mousnier, R. *The Institutions of France under the Absolute Monarchy, 1598–1789,* trans. B. Pearce. Chicago, 1979.

Muñoz Perez, J. "Mapa aduanero del XVIII español." *Estudios geográficos* 16 (1955): 747–797.

Nguyen, V. "Aperçus sur la conscience d'Oc autour des années 1900 (vers 1890–vers 1910)." In C. Gras and G. Livet, eds., *Régions et régionalismes en France du XVIIIe siècle à nos jours.* Paris, 1977: 241–256.

Noël, J. F. "Les problèmes de frontières entre la France et l'Empire dans la seconde moitié du XVIIIe siècle." *Revue historique* 235 (1966): 333–346.

Nordman, D. "Frontiera e confini in Francia: Evoluzione dei termini e dei concetti." In C. Ossola, C. Raffestin, and M. Ricciardi, eds., *La frontiera da stato a nazione: Il caso Piemonte.* Rome, 1987: 39–55.

—————. "Des limites d'Etat aux frontières nationales." In P. Nora, ed., *Les lieux de mémoire,* 2 vols. Paris, 1986: 2 (pt. 2): 35–61.

—————. "Buache de la Neuville et la 'frontière' des Pyrénées." In *Images de la Montagne: Catalogue et essais.* Paris, 1984: 105–110.

—————. "L'idée de frontière fluviale en France au XVIIIe siècle." In *Frontières et contacts de civilisations.* Colloque universitaire franco–suisse. Neuchatel, 1979.

Ogden, P. E. "Expression spatiale des contacts humains et changement de la société: L'exemple de l'Ardeche, 1860–1970." *Revue de géographie de Lyon* 49 (1974): 191–209.

Ourliac, P. "La condition des étrangers dans la région Toulousain au moyen age." In *L'étranger.* Receuils de la société Jean Bodin pour l'histoire comparative des institutions, vol. 10. Brussels, 1958: 101–108.

Ozouf-Marignier, "De l'universalisme constituant aux interets locaux: Le debat sur la formation des departements en France (1789–1790)." *Annales* 41 (1986): 1193–1213.

Palacio Atard, V. *El tercer pacto de familia.* Madrid, 1921.

Palmer, R. *The Age of Democratic Revolution: A Political History of Europe and America, 1760–1800,* 2 vols. Princeton, 1964.

Pascal, B. *Pensées,* ed. L. Brunschvig. Paris, 1937.

Papy, M. "Mutilation d'un rite: La junte de Roncal et Barétous et la crise de nationalisme français dans les années 1890." In J. F. Nail, *Lies et passeries dans les Pyrénées.* Tarbes, 1986: 197–223.

Pastoureau, M. *Les atlas français, XVIe–XVIIe siècles: Répertoire bibliographique et étude.* Paris, 1984.

Peabody, D. *National Characteristics.* Cambridge, Eng., 1985.

Pernot, J. F. "Les chevauchées des ingénieurs militaires en France au XVIIe siècle, ou la maturité de voyages d'études politico-administratifs." In *La découverte de la France au XVIIe siècle.* 9e Colloque de Marseille organisé par le Centre Méridional de Rencontres sur le XVIIe siècle, 25–28 janvier 1979. Paris, 1980: 327–345.

Peyre, H. *La royauté et les langues provinciales.* Paris, 1933.

Pitt-Rivers, J. *People of the Sierra.* Chicago, 1957.

Plandé, R. "La formation politique de la frontière des Pyrénées." *RGPSO* IX (1938): 221–242.

Poitrineau, A. *Remues d'hommes: Les migrations montagnardes en France, XVIIe–XVIIIe siècles.* Paris, 1983.

Pop, I. *Voisinage et bon voisinage en droit international.* Paris, 1972.

Poumarède, J. "Les syndicats de vallée dans les Pyrénées françaises." In *Les communtautés rurales.* Receuils de la société Jean Bodin pour l'histoire comparative des institutions, vol. 43. Paris, 1984: 385–409.

Pounds, N. "France and 'les limites naturelles' from the Seventeenth to the Twentieth Centuries." *Annals of the Association of American Geographers* 44 (1954): 51–62.

———. "The Origin of the Idea of Natural Frontiers in France." *Annals of the Association of American Geographers* 41 (1951): 146–157.

Poussou, J. P. "Les movements migratoires en France et à partir de la France de la fin du XVe au début du XIXe siècle: Approches pour une synthèse." *Annales de démographie historique* (1970): 12–78.

Price, R. *The Modernization of Rural France: Communications Networks and Agricultural Market Structures in Nineteenth Century France.* New York, 1983.

Ranum, O. "Counter-Identities of Western European Nations in the Early-Modern Period: Definitions and Points of Departure." In P. Boerner, ed., *Concepts of National Identity: An Interdisciplinary Dialogue.* Baden-Baden, 1986: 63–78.

Ratzel, F. *Politische Geographie.* Paris and Munich, 1903.

Regla Campistol, J. "El tratado de los Pirineos de 1659," *Hispania* 11 (1951): 101–166.

———. *Francia, la Corona de Aragon, y la frontera pirenaica: La lucha por el valle de Aran, siglos XII–XV,* 2 vols. Madrid, 1951.

———. "La cuestión de los Pirineos a comienzos de la Edad Moderna." *Estudios de historia moderna* 1 (1951): 1–32.

Renault-Roulier, R. *La ligne de démarcation,* vols. 15, 19, and 21. Paris, 1969, 1970, and 1972.

Richelieu, Armand du Plessis, Cardinal Duc de. *Testament Politique du Cardinal de Richelieu,* ed. L. André. Paris, 1947.

———. *Lettres, instructions diplomatiques et papiers d'état du Cardinal de Richelieu,* ed. D. L. M. Avénel, 8 vols. Paris, 1853–1877.

Ringrose, D. R. *Transportation and Economic Stagnation in Spain, 1750–1850.* New York, 1970.

Root, H. L. *Peasants and King in Burgundy: Agrarian Foundations of French Absolutism.* Berkeley, Los Angeles, London, 1987.

Rousseau, C. *Les frontières de la France.* Paris, 1954.

Rowen, H. *The King's State: Proprietary Dynasticism in Early Modern France.* New Brunswick, 1980.

Rubio, J. *La emigración española a Francia.* Barcelona, 1984.

Sahlins, M. "The Segmentary Lineage: An Organization of Predatory Expansion." In R. Cohen and J. Middleton, eds. *Comparative Political Systems.* Austin, Tex., 1957: 89–115.

Sahlins, P. "The 'War of the Demoiselles' in Ariège, France (1829–1867)." Honors thesis, Harvard College, 1980.

———. "Natural Frontiers Revisited." In press.

Sales, N. "Ramblers, traginers, i mules (s. XVIII–XIX)." *Recerques* 13 (1984): 64–81.

Sanchez Mantero, R. *Los cien mil hijos de San Luis y las relaciones franco–españolas.* Seville, 1981.

———. *Liberales en exilio: La emigración política en Francia en la crisis del Antiquo régimen.* Madrid, 1975.

Schaffer, R. J. *The Economic Societies in the Spanish World, 1763–1821.* Syracuse, 1958.

Schmitt, C. *The Concept of the Political,* trans. G. Schwab. New Brunswick, 1976.

Scott, J. *Weapons of the Weak: Everyday Forms of Peasant Resistance.* New Haven, 1985.

Sermet, J. *La frontière hispano–française des Pyrénées et les conditions de sa délimitation.* Lourdes, 1983.

———. "Les communications pyrénéennes et transpyrénéennes." *Actes du 2e congrès international d'études pyrénéennes, Luchon-Pau, 1954.* Toulouse, 1962: 59–193.

Seton-Watson, H. *Nations and States: An Inquiry into the Origins of Nations and the Politics of Nationalism.* Boulder, Colo., 1977.

Sheppard, T. *Lourmarin in the Eighteenth Century.* Baltimore, 1971.

Smith, A. E. *The Ethnic Origins of Nation.* Chapel Hill, N.C., 1982.

Soboul, A. "The French Rural Community in the Eighteenth and Nineteenth Centuries." *Past and Present* 10 (1956): 78–95.

Solon, P. "French Cartography and Bourbon Ambition." *Proceedings of the*

Tenth Annual Meeting of the Western Society for French History. Lawrence, Kans., 1984: 94–102.

Sopheau, P. "Les variations de la frontière française des Alpes depuis le XVIe siècle." *Annales de géographie* 3 (1893–1894): 183–200.

Sorre, M. *Les Pyrenees mediterraneennes* Paris, 1913.

Soulet, J. F. *La vie dans les Pyrénées sous l'Ancien Régime du XVIe au XVIIIe siècle*. Paris, 1974.

Stahl, P. H. "Frontières politiques et civilizations paysannes traditionelles." In *Confini e regioni: Il potenziale di sviluppo e di pace delle periferie*. Trieste, 1973: 459–465.

Tapié, V. L. "Comment les français voyaient la patrie." *XVIIe siècle* 25–26 (1955): 37–58.

Tardy, R. *Le pays de Gex, terre frontalière*. Lyons, 1958.

Tate, R. *Ensayos sobre la historiografia peninsular del siglo XV*. Madrid, 1970.

Territoire et territorialité. Territoires no. 1. Ecole Normale Supérieure, Laboratoire de Sciences Sociales. Paris, 1983.

Tilly, C. "Did the Cake of Custom Break?" In J. Merriman, ed., *Consciousness and Class Experience in Nineteenth-Century Europe*. New York, 1979: 17–41.

―――. *The Vendée*. Cambridge, Mass., 1974.

Timbal, P. C. "De la communauté médiévale à la commune moderne en France." In *Les communautés rurales*. Receuils de la société Jean Bodin pour l'histoire comparative des institutions. Paris, 1984. 43: 337–348.

Timbal, P. C. "Les lettres de marques dans la France medievale." L'etranger 2: 110–138.

Tocqueville, A. de. *The Ancien Regime and the French Revolution*, trans. S. Gilbert. New York, 1955.

Toulgouat, P. *Voisinage et solidarités dans l'Europe du Moyen Age*. Paris, 1981.

Trudeau, D. "L'ordonnance de Villers-Cotterêts et la langue française: Histoire ou interprétation." *Bibliothèque d'humanisme et Renaissance* 45 (1983): 461–472.

Truyol y Serra, A. "Las fronteras y las marcas." *Revista española de derecho internacional* 10 (1957): 105–123.

Valfrey, J. *Hughes de Lionne: Ses ambassades en Espagne et en Allemagne, La Paix des Pyrénées*. Paris, 1881.

Vanel, M. "Evolution historique de la notion de français d'origine du XVIe siècle au Code Civil." Thèse de droit, Université de Paris, 1945.

Van Gennep, A. *The Rites of Passage*, trans. M. B. Vizedom and G. L. Caffee. Chicago, 1960.

―――. *Manuel de folklore français*, 4 vols. Paris, 1938–1945.

―――. *Traité comparatif des nationalités*, vol. 1: *Les éléments extérieurs de la nationalité*. Paris, 1922.

Varagnac, A. *Civilization traditionnelle et genres de vie*. Paris, 1948.

Vattel, E. de. *The Law of Nations, or Principles of Natural Law*, trans. C. G. Fenwick. Washington, D.C., 1916 (1758).

Vilar, P. "Quelques aspects de l'occupation et de la resistance en Espagne en

1794 et au temps de Napoleon." *Occupants-Occupes, 1792–1815.* Brussels, 1969.

Visscher, C. de. *Problèmes de confins en droit international public.* Paris, 1969.

———. *Theory and Reality in Public International Law,* trans. P. E. Corbett. Princeton, N.J., 1957.

Voltaire, F. M. Arouet, dit. *Le siècle de Louis XIV.* Paris, 1854 (1752).

Wallman, S. "The Boundaries of 'Race': Processes of Ethnicity in England," *Man* n.s. 13 (1978): 200–217.

Weber, E. "L'hexagone." In P. Nora, ed., *Les lieux de mémoire,* 2 vols. Paris, 1986: 2 (pt. 2): 97–116.

———. "Comment la Politique Vint aux Paysans: A Second Look at Peasant Politicization." *American Historical Review* 87 (1982): 357–389.

———. "The Second Republic, Politics, and the Peasant." *French Historical Studies* 11 (1980): 521–550.

———. *Peasants into Frenchmen: The Modernization of Rural France, 1879–1914.* Stanford, Calif., 1976.

Weber, H. "Richelieu et le Rhin." *Revue historique* 239 (1968): 265–280.

Wolff, C. *The Law of Nations,* trans. J. H. Drake. Washington, D.C., 1934 (1749).

Wylie, L. *Village in the Vaucluse.* Cambridge, Mass., 1965.

Young, A. *Travels through France in the years 1787, 1788, and 1789,* 2 vols. London, 1793.

Zeller, G. "Saluces, Pignerol, et Strasbourg: La politique des frontières au temps de la préponderance espagnole." *Revue historique* 193 (1942–1943): 97–110.

———. "Histoire d'une idée fausse." *Revue de synthèse* 11–12 (1933): 305–333.

———. "La monarchie d'Ancien Régime et les frontières naturelles." *Revue d'histoire moderne* 8 (1933): 305–333.

———. *L'organization défensive des frontières du Nord et de l'Est au XVIIe siècle.* Paris, 1928.

Zink, A. "Pays et paysans gascons sous l'Ancien Régime," 9 vols. Thèse de doctorat, Université de Paris, 1985.

———. *Azereix: La vie d'une communauté rurale à la fin du XVIIIe siècle.* Paris, 1969.

Index

Designer:	U.C. Press Staff
Compositor:	Prestige Typography
Text:	10/13 Sabon
Display:	Sabon
Printer:	Braun-Brumfield, Inc.
Binder:	Braun-Brumfield, Inc.